ISNM 97:
International Series of Numerical Mathematics
Internationale Schriftenreihe zur Numerischen Mathematik
Série Internationale d'Analyse Numérique
Vol. 97

Edited by
K.-H. Hoffmann, Augsburg; H. D. Mittelmann, Tempe;
J. Todd, Pasadena

Birkhäuser Verlag
Basel · Boston · Berlin

Bifurcation and Chaos: Analysis, Algorithms, Applications

Edited by

R. Seydel
F. W. Schneider
T. Küpper
H. Troger

1991

Birkhäuser Verlag
Basel · Boston · Berlin

Editors

R. Seydel
Abteilung für Numerik
Universität Ulm
Oberer Eselsberg
D–7900 Ulm

F. W. Schneider
Institut für Physikalische Chemie
Universität Würzburg
Marcusstr. 9/11
D–8700 Würzburg

T. Küpper
Mathematisches Institut
Universität Köln
Weyertal 88
D–5000 Köln

H. Troger
Institut für Mechanik
Technische Universität Wien
Wiedner Hauptstr. 8–10
A–1040 Wien

Deutsche Bibliothek Cataloguing-in-Publication Data

Bifurcation and chaos: analysis, algorithms, applications / ed.
by R. Seydel ... – Basel ; Boston ; Berlin : Birkhäuser, 1991
  (International series of numerical mathematics ; Vol. 97)
  ISBN 978-3-0348-7006-1   ISBN 978-3-0348-7004-7 (eBook)
  DOI 10.1007/978-3-0348-7004-7
NE: Seydel, Rüdiger [Hrsg.]; GT

© 1991 Birkhäuser Verlag, Basel
Softcover reprint of the hardcover 1st edition 199

Printed from the author's camera-ready manuscript on acid-free paper in Germany
ISBN 978-3-0348-7006-1

Contents

VIII

Introduction

This volume contains the proceedings of a conference held in Würzburg, August 20-24, 1990. The theme of the conference was *Bifurcation and Chaos: Analysis, Algorithms, Applications.* More than 100 scientists from 21 countries presented 80 contributions. Many of the results of the conference are described in the 49 refereed papers that follow. The conference was sponsored by the *Deutsche Forschungsgemeinschaft,* and by the *Deutscher Akademischer Austauschdienst.* We gratefully acknowledge the support from these agencies.

The science of *nonlinear phenomena* is evolving rapidly. Over the last 10 years, the emphasis has been gradually shifting. How trends vary may be seen by comparing these proceedings with previous ones, in particular with the conference held in Dortmund 1986 (proceedings published in ISNM 79). Concerning the range of phenomena, chaos has joined the *bifurcation* scenarios. As expected, the acceptance of *chaos* is less emotional among professionals, than it has been in some popular publications. *Analytical* methods appear to have reached a state in which basic results of singularities, symmetry groups, or normal forms are everyday experience rather than exciting news. Similarly, numerical *algorithms* for frequent situations are now well established. Implemented in several packages, such algorithms have become standard means for attacking nonlinear problems. The sophistication that analytical and numerical methods have reached supports the vigorous trend to more and more *applications.* Pioneering equations as those named after Duffing, Van der Pol, or Lorenz, are no longer exclusively the state of art. Many new demanding examples from all scientific areas are examined for their nonlinear phenomena. The conference (and these proceedings) has not only discussed mathematical tools, but has become a forum for fascinating applications.

Most papers are not devoted to one topic only. Frequently, papers discuss some aspect of analytical methods or of numerical algorithms, and show an interesting application. This diversity in many papers suggests not to group them into chapters that address one general theme. We prefer an alphabetic ordering. Being aware that it is not possible to do justice to every paper in this introduction, we specify general topics, and attempt to assign some of the results discussed in these proceedings. In what follows, we always mention the first author only, referring to the list of contents for the full list of authors and for the title.

Chaos is touched upon frequently, see the papers by Bajaj, Benedettini, El Naschie, Kaashoek, Krischer, Kunert, Lamarque, and Stelter. The tools of *manifolds* are developed and applied by Kirchgraber, and Rodríguez. *Symmetry* plays a dominating role in the contributions of Allgower, Dangelmayr, Mei, and Werner. Recent interest in *Takens-Bogdanov bifurcation* is reflected in the papers by Gamero, Kunkel, Rodríguez, and Werner. Theoretical bifurcation is investigated by Bazley, and Heinz. Vanderbauwhede proves the appearence of homoclinic orbits in conservative and reversible systems. A frequent topic is the characterization of dynamical systems; see Wedig for *invariant measures*, and Lamarque and Parlitz for cell-mapping approaches for *oscillators*. Further analytical and experimental work on oscillators can be found in the papers of Scheffczyk, and Szemplińska. Kirby, and Shaw capture dynamic behavior by only few modes or eigenfunctions. *Controllability* is addressed by Luce. On the edge between analytical and numerical methods lie procedures of *symbolic computation*, see Gamero, Kleczka, and Ponce-Núñez. *General numerical procedures* are discussed by Bayliss, Benedettini, Markus, and Roose. *Specific algorithmic topics* treated are the detection of Hopf bifurcation (Garratt), generalized turning points (Kunkel), coupled turning points (Pönisch), calculation of homoclinic orbits (Kuznetsov), and the calculation of derivatives without differencing (Griewank). Numerical aspects of exploiting *parallellism* are discussed by Markus, and Roose.

Among the applications, *chemical oscillations* play a prominent role for Amrehn, Eiswirth and Krischer. Bayliss, Boissonade and Busse address *pattern formation*. Applications include *convection* problems (Busse, Dangelmayr, and Kropp), *climate modeling* (Hetzer), *economy* (Kaashoek), *robot control* (Haller), *rolling motion of ships* (Falzarano), *motion of a moored pontoon* (Kleczka), *galvanostatic oscillations* (Krischer), *excitable systems* (Sevcikova), *dry friction* (Stelter), *rotating shafts* (Wauer), *an elastic model with continuous spectrum* (Domokos), *rings under hydrostatic pressure* (Labisch), *combustion* (Bayliss), *Turing structures* (Boissonade), and a *spinning satellite* (Guran).

In view of the diversity of the scientific areas from which the contributors stem, one might expect strong centrifugal trends. This has not been the case. In the contrary, the language of mathematics and its tools have shown to be generally accepted, and attracting. This ability to use the same language forms the basis for a further growing of the power, and of the influence that nonlinear phenomena will experience in the years to come.

December 1990   R. Seydel, F. W. Schneider, T. Küpper, H. Troger

International Series of Numerical Mathematics, Vol. 97, © 1991 Birkhäuser Verlag Basel

A Complete Bifurcation Scenario for the 2-d Nonlinear Laplacian with Neumann Boundary Conditions on the Unit Square

Eugene L. Allgower*
Colorado State University, Fort Collins, USA
Klaus Böhmer**
Philipps-Universität Marburg, 3550 Marburg/Lahn, FRG
Mei Zhen[†]
Philipps-Universität Marburg, 3550 Marburg/Lahn, FRG and Xi'an Jiaotong University, Xi'an, PRC

Abstract

In this paper we describe the complete scenario of solutions bifurcating from the trivial solution and their stability for a model problem in bifurcation analysis—the two-dimensional Laplacian with Neumann boundary conditions on the unit square. We show that at a corank ρ bifurcation point, there are exactly $(3^\rho - 1)/2$ different solution branches bifurcating from the trivial solution curve. Our results are easily extended to other boundary conditions, e.g. Dirichlet or Dirichlet and Neumann along different sides, etc. which can be embedded in the periodic boundary conditions.

AMS(MOS) Subject Classifications: 47H15, 58F14, 65J15

Key words: Nonlinear Laplacian with Neumann and Dirichelt boundary conditions, symmetry, equivariance, corank ρ bifurcation points, modified Lyapunov-Schmidt method, stability of solution branches.

1. Introduction

We study the model equation

$$\begin{cases} F(u,\lambda) := \Delta u + \lambda f(u) = 0 & \text{in } \Omega := [0,1]^2 \\ \text{with } \dfrac{\partial u}{\partial n}(x,y) = 0 & \text{on } \partial\Omega \end{cases} \tag{1.1}$$

where $\frac{\partial}{\partial n}$ denotes the normal derivative and f is assumed to be a smooth odd function and to be normalized as $f'(0) = 1$, and $f'''(0) \neq 0$. Due to $f(0) = 0$, we see $(u(x,y) \equiv 0, \lambda)$ is a (trivial) solution curve for (1.1). Similar problems with Dirichlet boundary conditions, mostly at simple or corank 2 bifurcation points, were discussed in Allgower/Chien/Georg [4], Berger [6], Budden/Norbury [10, 11], Lions [22], etc. The general corank ρ bifurcation for Dirichlet problems is discussed in Allgower/Böhmer/Mei [1]. The recent development

---

* Work supported by DAAD.
** Work supported by DFG and Colorado State University.
[†] Work supported by DFG.

in the equivariance bifurcation theory motivates an intensive study of bifurcations for
(1.1) with various boundary conditions at a general corank ρ bifurcation point, see e.g.
Allgower/Böhmer/Mei [1-3], Gomes [17], Healey/Kielhöfer [20], Mei [24]. In this paper we
want to determine all solutions bifurcating from the trivial solution curve $\{(0,\lambda),\lambda \in \mathbf{R}\}$.

First of all, the Frechet derivative of F at $(0,\lambda)$

$$F'(0,\lambda) = (D_u F(0,\lambda), D_\lambda F(0,\lambda)) = (\Delta + \lambda I, 0) \qquad \text{for all } \lambda \in \mathbf{R}$$

with Neumann boundary conditions, implies that $(u,\lambda) = (0,1)$ is always in the null space
of $F'(0,\lambda)$. The bifurcation points of (1.1) on the trivial solution curve have the form
$(0,\lambda_0)$, where $\lambda_0 \in \{(m^2+n^2)\pi^2 \mid m, n \in \mathbf{N} \cup \{0\}\}$ is an eigenvalue of $-\Delta$ on Ω with
Neumann boundary conditions, i.e. there is a nonvanishing v such that, for arbitrary
$\mu \in \mathbf{R}$,

$$\begin{cases} F'(0,\lambda_0)(v,\mu) = (\Delta + \lambda_0 I)v = 0 & \text{in } \Omega := [0,1]^2 \\[2mm] \text{with} \quad \dfrac{\partial v}{\partial n}(x,y) = 0 & \text{on } \partial\Omega. \end{cases} \qquad (1.2)$$

At a generic eigenvalue $\lambda_0\,(m \neq n)$ we have, with $N(A)$ indicating the null space of a linear
operator A,

$$\dim N(F_u(0,\lambda_0)) = 2, \qquad \dim N(F'(0,\lambda_0)) = 3, \qquad (1.3)$$

hence a bifurcation point of corank 2. Number theoretical results indicate that the multiplicity of an eigenvalue λ_0 can be arbitrarily high, e.g. $\rho = 4$ for $m_1^2 + n_1^2 = 4^2 + 7^2 =
1^2 + 8^2 = m_2^2 + n_2^2$ or $\rho = 3$ for $m_1^2 + n_1^2 = 1^2 + 7^2 = 5^2 + 5^2 = m_2^2 + m_2^2$. Since the solutions
of (1.2) are well known (cf. Isaacson/Keller [21]), we have

$$\dim N(F'(0,\lambda_0)) = 1 + \rho, \quad \rho \geq 1, \qquad (1.4)$$

where ρ is the multiplicity of the eigenvalue λ_0 of the Laplacian $-\Delta$, see (1.2).

To define the symmetries for (1.1), let D_4 be the symmetry group of the square $[0,1]^2$,
generated by the flips S_1 and S_2 with axes $x = 1/2$ and $y = x$ and let $Z_2 := \{+1,-1\}$.
The action of $\Gamma := Z_2 \times D_4$ on $u \in C(\Omega)$ is defined as

$$(\pm\delta u)(x,y) := \pm u(\delta(x,y)) \qquad \text{for } \pm 1 \in Z_2, \delta \in D_4. \qquad (1.5)$$

The problem (1.1) has the symmetry of Γ, namely

$$F(\gamma u,\lambda) = \gamma F(u,\lambda) \qquad \text{for all } \gamma \in \Gamma, (u,\lambda) \in C^2(\Omega) \times \mathbf{R} \qquad (1.6)$$

and F is called Γ-equivariant.

In the context with bifurcation theory the equivariance may be used and has been
extensively studied in recent years to reduce the complexity of bifurcation phenomena,
see e.g. Cicogna [12], Dellnitz/Werner [15], Golubitsky/Stewart/Schaeffer [16], Healey
[18, 19], Healey/Kielhöfer [20], Sattinger [26], Vanderbauwhede [27], Werner [28], All-
gower/Böhmer/Mei [1-3], etc. Since, for high dimensional null space $N(F'(0,\lambda_0))$, the

problem (1.1) is not generic in the usual sense of bifurcation theory, see e.g. Golubit-sky/Stewart/Schaeffer [16], though it occurs very often in applications, the equivariance does not seem to allow a full reduction of the problem. To study the bifurcation behaviour of (1.1) we will use a modified Lyapunov-Schmidt method, see e.g. Böhmer [7], Mei [23, 24], Allgower/Böhmer/Mei [1, 3], Böhmer/Mei [8].

2. Weak Form and Modified Lyapunov-Schmidt Method

In Section 1 we started with the Laplacian and Neumann boundary conditions, see (1.1). To simplify the discussions we would like to use the inner product in $L^2(\Omega)$ and write (1.1) in a weak form. For a discussion based upon the Laplacian, instead of the weak form (2.4), we refer to Golubitsky/Stewart/Schaeffer [16], Healey/Kielhöfer [20] and the references therein. Let

$$X := H^1(\Omega) \tag{2.1}$$

and

$$a(u,v) = \int_\Omega \nabla u \nabla v dx dy, \qquad \text{for } u,v \in X. \tag{2.2}$$

We define a linear operator $T : g \in L^2(\Omega) \mapsto Tg \in X$ by

$$a(Tg,v) = -(g,v), \qquad \forall\, v \in X, \quad g \in L^2(\Omega). \tag{2.3}$$

Here and in the sequel, $(\cdot,\cdot)$ represents the inner product in $L^2(\Omega)$. According to the theory of elliptic differential equations, the operator T is bounded, self-adjoint and compact from $L^2(\Omega)$ into X. A weak form of (1.1) is

$$G(u,\lambda) := u + \lambda T f(u) = 0. \tag{2.4}$$

The mapping $G : X \times \mathbf{R} \mapsto X$ is well defined and smooth. The singular set S_0 of (2.4) on the trivial solution curve $\{(0,\lambda), \lambda \in \mathbf{R}\}$ is given by the eigenvalues of $-\Delta$

$$S_0 := \{(0,(m^2+n^2)\pi^2) \mid m, n \in \mathbf{N} \cup \{0\}\}.$$

For $(0,\lambda_0) \in S_0$, let $G_0, D_u G_0, \ldots$ represent the evaluations of $G, D_u G, \ldots$ at $(0,\lambda_0)$. Then

$$\begin{cases} D_u G_0 = I + \lambda_0 T, \quad D_\lambda^l G_0 = 0,\, l = 1,2,\ldots \\ D_{uu} G_0 = 0, \quad D_{u\lambda} G_0 = T, \\ D_{uuu} G_0 = -\lambda_0 bT, \quad b := -f'''(0), \quad D_{uu\lambda} G_0 = 0, \quad D_{u\lambda\lambda} G_0 = 0. \end{cases} \tag{2.5}$$

Since T is self-adjoint and compact, $(D_u G_0)^* = D_u G_0 = I + \lambda_0 T$, the Leray-Schauder theory yields (cf. Crouzeix/Rappaz [13])

$$X = N(D_u G_0) \oplus R(D_u G_0), \tag{2.6}$$

where $\oplus$ represents the orthogonal sum under the inner product of $L^2(\Omega)$. For any $\lambda_0 \in S_0$ we have either

$$\lambda_0 = 0 \tag{2.7}$$

or

$$\lambda_0 = (m_1^2 + n_1^2)\pi^2 = \cdots = (m_p^2 + n_p^2)\pi^2 = m^2\pi^2, \quad m_i \cdot n_i \neq 0, i = 1, \ldots, p \tag{2.8}$$

or

$$\lambda_0 = (m_1^2 + n_1^2)\pi^2 = \cdots = (m_p^2 + n_p^2)\pi^2, \quad m_i \cdot n_i \neq 0, i = 1, \ldots, p, \tag{2.9}$$

where for each pair (m_i, n_i) with $m_i \neq n_i$ we have identified (n_i, m_i) with another pair (m_j, n_j) and $(m_i, n_i) \neq (m_j, n_j)$ for $i \neq j$. For the three different cases (2.7)-(2.9) we choose the following bases for $N(D_u G_0)$:

for (2.7): $$N((D_u G_0)^*) = N(D_u G_0) = \text{span}[\phi_1] \tag{2.10}$$

with

$$\phi_1 = 1; \tag{2.11}$$

for (2.8): $$N((D_u G_0)^*) = N(D_u G_0) = \text{span}[\phi_1, \ldots, \phi_p, \phi_{p+1}, \phi_{p+2}] \tag{2.12}$$

with

$$\phi_i = 2\cos m_i\pi x \cos n_i\pi y, i = 1, \ldots, p, \phi_{p+1} = \sqrt{2}\cos m\pi x, \phi_{p+2} = \sqrt{2}\cos m\pi y, \tag{2.13}$$

and, with the $\phi_i, i = 1, \ldots, p$ in (2.13),

for (2.9): $$N((D_u G_0)^*) = N(D_u G_0) = \text{span}[\phi_1, \ldots, \phi_p]. \tag{2.14}$$

To unify the discussion, we use the definition

$$\rho = \begin{cases} 1 & \text{if (2.7) holds;} \\ p+2 & \text{if (2.8) holds;} \\ p & \text{if (2.9) holds} \end{cases} \tag{2.15}$$

and rewrite (2.10), (2.12), (2.14) as follows

$$N((D_u G_0)^*) = N(D_u G_0) = \text{span}[\phi_1, \ldots, \phi_\rho]. \tag{2.16}$$

Based on (2.6), (2.16), we consider necessary conditions for the form of a solution curve $(u(t), \lambda(t)) \in X \times \mathbf{R}$ passing through the bifurcation point $(u_0(\equiv 0), \lambda_0)$. We may rescale the curve parameter t such that

$$\text{a)} \quad (u(0), \lambda(0)) = (u_0, \lambda_0) = (0, \lambda_0), \qquad \text{b)} \quad (\dot{u}(0), \dot{\lambda}(0)) \neq (0, 0). \tag{2.17}$$

Hence, there are $\alpha_i(t), \beta(t) \in \mathbf{R}, i = 1, \dots, \rho$, and $v(t) \in R(D_u G_0)$ such that for an interval $\mathcal{I} \subset \mathbf{R}, 0 \in \mathcal{I}$,

$$
\begin{cases}
u(t) = \displaystyle\sum_{i=1}^{\rho} t\alpha_i(t)\phi_i + tv(t), \\
\lambda(t) = \lambda_0 + t\beta(t), \quad \text{for } t \in \mathcal{I} \subset \mathbf{R}.
\end{cases}
\tag{2.18}
$$

Theorem 2.1: *A solution curve $\{(u(t), \lambda(t)), t \in \mathcal{I}\}$ of (2.4), satisfying (2.17), has necessarily either the form*

$$
\begin{cases}
u(t) = t\displaystyle\sum_{i=1}^{\rho} \alpha_i(t)\phi_i + t^3 v(t), \quad \displaystyle\sum_{i=1}^{\rho}|\alpha_i(0)| > 0, \\
\lambda(t) = \lambda_0 + t^2 \beta(t), \qquad \beta(0) \neq 0, \, t \in \mathcal{I}
\end{cases}
\tag{2.19}
$$

or the form

$$
\begin{cases}
u(t) = t^2 \displaystyle\sum_{i=1}^{\rho} \alpha_i(t)\phi_i + t^3 v(t), \\
\lambda(t) = \lambda_0 + t\beta(t), \qquad \beta(0) \neq 0, \, t \in \mathcal{I}.
\end{cases}
\tag{2.20}
$$

Proof: This can be proved directly by differentiating $G(u(t), \lambda(t)) = 0$ at $t = 0$ and using Lemma 2.1 below, see Mei [24], also Allgower/Böhmer/Mei [3]. ■

Using a modified Lyapunov-Schmidt method, one can show that (2.20) corresponds to the trivial solution curve of (2.4), see Mei [24]. Therefore, we are left with (2.19) and the assertion that all nontrivial solution curves of (2.4) bifurcating at $(0, \lambda_0)$ have locally this structure (2.19). Furthermore, the inequality

$$
\beta(0) \neq 0
\tag{2.21}
$$

in (2.19) indicates that (2.19) can be locally transformed into:

$$
\begin{cases}
u(t) = t\displaystyle\sum_{i=1}^{\rho} \alpha_i(t)\phi_i + t^3 v(t), \\
\lambda(t) = \lambda_0 + \text{sign}(b)t^2, \qquad t \in \mathcal{I}.
\end{cases}
\tag{2.22}
$$

If $f'''(0) < 0$, then sign(b) =-sign$(f'''(0))>0$ and solution branches bifurcate to the right at $(0, \lambda_0)$. Conversely, if $f'''(0) > 0$, they bifurcate to the left, see e.g. Fig. 4.1. To determine the functions $v(t)$, $\alpha_i(t)$, $i = 1, \dots, \rho$, we define a product space

$$
E := X \times \mathbf{R}^\rho \times \mathbf{R}
\tag{2.23}
$$

endowed with the product norm

$$
\|x\| = \|v\|_X + \sum_{i=1}^{\rho} |\alpha_i| + |t|, \qquad \forall\, x := (v, \alpha_1, \dots \alpha_\rho, t) \in E.
\tag{2.24}
$$

We consider the enlarged system in E:

$$\mathbf{F}(x) := \begin{pmatrix} G(t\sum_{i=1}^{\rho} \alpha_i \phi_i + t^3 v, \lambda_0 + \text{sign}(b)t^2)/t^3 \\ (\phi_1, v) \\ \vdots \\ (\phi_\rho, v) \end{pmatrix} = 0. \tag{2.25}$$

At $t = 0$, $\mathbf{F}(x)$ is defined by the limit of (2.25), i.e. for $x_0 := (v, \alpha_1, \ldots, \alpha_\rho, 0)$,

$$\mathbf{F}(x_0) := \begin{pmatrix} D_u G_0 v + T[\text{sign}(b)\sum_{i=1}^{\rho} \alpha_i \phi_i - \lambda_0 b(\sum_{j=1}^{\rho} \alpha_j \phi_j)^3]/6 \\ (\phi_1, v) \\ \vdots \\ (\phi_\rho, v) \end{pmatrix} = 0, \tag{2.26}$$

since

$$G(t\sum_{i=1}^{\rho} \alpha_i \phi_i + t^3 v, \lambda_0 + \text{sign}(b)t^2) =$$

$$= t^3 \{D_u G_0 v + T[\text{sign}(b)\sum_{i=1}^{\rho} \alpha_i \phi_i - \frac{1}{6}\lambda_0 b(\sum_{i=1}^{\rho} \alpha_i \phi_i)^3]\} + O(t^5). \tag{2.27}$$

Obviously, the mapping $\mathbf{F}$ is smooth in E. At the same time, the definition of $\mathbf{F}$ provides a one-to-one correspondence between the solution curves in (2.22) and the solution curves $(v(t), \alpha_1(t), \ldots, \alpha_\rho(t), t)$ of (2.25). In other words, we transform the problem (2.4) into the form (2.25)-(2.26) which will be solved in two steps: we first determine all solutions of (2.26) and prove that they are all nonsingular. Secondly, we use the implicit function theorem to get the solution curves for (2.25) in a neighborhood of the solutions of (2.26). These curves yield, via (2.22), the corresponding nontrivial solution branches of (2.4).

Because of (2.6), for $\lambda_0 \neq 0$ the system (2.26) is solvable if and only if there are $\alpha_i \in \mathbf{R}$, $i = 1, \ldots, \rho$ such that, with the self-adjointness of T,

$$\mathbf{f}(\alpha_1, \ldots, \alpha_\rho) := -\lambda_0 \, \text{sign}(b) \cdot \left((\phi_j, D_u G_0 v + T[\text{sign}(b)\sum_{i=1}^{\rho} \alpha_i \phi_i - \frac{\lambda_0 b}{6}(\sum_{i=1}^{\rho} \alpha_i \phi_i)^3])\right)_{j=1}^{\rho}$$

$$= (f_j(\alpha_1, \ldots, \alpha_\rho))_{j=1}^{\rho} = 0 \tag{2.28}$$

with

$$f_j(\alpha_1, \ldots, \alpha_\rho) := (\phi_j, [\sum_{i=1}^{\rho} \alpha_i \phi_i - \frac{\lambda_0 |b|}{6}(\sum_{i=1}^{\rho} \alpha_i \phi_i)^3]), \quad j = 1, \ldots, \rho. \tag{2.29}$$

We call (2.28) the **reduced bifurcation equations** for (2.4), see also Decker/Keller [14], Allgower/Böhmer/Mei [1, 3], Böhmer [7], Böhmer/Mei [8] and Mei [23, 24]. Compared to the bifurcation equations in the standard Lyapunov-Schmidt method, (2.28) has the advantage that it consists of a system of ρ cubic polynomials in the ρ unknowns α_i, independent of the parameter t. Modifications of Bezout's theorem show that if (2.28) is not degenerate, it has at most 3^ρ solutions in $\mathbf{C}^\rho$, see e.g. Bouligand [9], Allgower/Georg/Miranda [5]. Since we are only interested in solutions which lead to different solution curves of (2.25) (resp. (2.4)), we identify $\pm(\alpha_1, \ldots, \alpha_\rho)$ and exclude the trivial solution $\alpha = (0, \ldots, 0)$ which leads to the trivial solution branch of (2.24) (resp. (2.4)). We consider in the sequel merely non-vanishing solutions of (2.28).

Similarly, for the trivial case $\lambda_0 = 0$ in (2.7), we may consider (2.4) in the quotient space $H^1(\Omega)/\mathbf{R}$. Then $(0, \lambda_0) = (0,0)$ becomes a nonsingular solution of (2.4) and $\{(0, \lambda); \lambda \in \mathbf{R}\}$ is the unique solution curve of (2.4) which passes through $(0,0)$. We exclude this trivial case $\lambda_0 = 0$ in the following discussion.

First of all, based on the orthogonal properties of trigonmetric functions, the following lemma is a special case of a result in Allgower/Böhmer/Mei [3] and requires mainly technical calculations:

Lemma 2.1: *The basis functions ϕ_i, $i = 1, \ldots, p$ in (2.13) have the following properties*

$$(\phi_i^2, \phi_i^2) = \frac{9}{4}, \quad (\phi_i, \phi_i) = (\phi_i^2, \phi_j^2) = 1, \quad i \neq j \quad \text{and} \quad i, j = 1, \ldots, p,$$
$$(\phi_i, \phi_j \phi_k \phi_l) = 0, \quad \text{for} \quad j, k, l \in \{1, \ldots, p\} \setminus \{i\}, \; i = 1, \ldots, p. \tag{2.30}$$

Moreover, for the last two eigenfunctions in (2.13), ϕ_{p+1}, ϕ_{p+2}, we have

$$(\phi_{p+i}^2, \phi_{p+i}^2) = \frac{3}{2}, \quad (\phi_{p+i}, \phi_{p+i}) = 1, \; i = 1, 2,$$
$$(\phi_{p+1}^2, \phi_{p+2}^2) = (\phi_{p+i}^2, \phi_j^2) = 1, \; i = 1, 2, \; j = 1, \ldots, p$$
$$(\phi_{p+i}, \phi_{p+j}^3) = 0, \; i \neq j, \; i, j = 1, 2, \quad (\phi_{p+1}, \phi_{p+2}\phi_i\phi_j) = 0, \; i, j = 1, \ldots, p, \tag{2.31}$$
$$(\phi_{p+i}, \phi_j\phi_k\phi_l) = 0, \quad \text{for} \quad j, k, l \in \{1, \ldots, p\}, \; i = 1, 2.$$

We derive from (2.29) and Lemma 2.1

$$f_j(\alpha_1, \ldots, \alpha_\rho) = \alpha_j - \frac{\lambda_0|b|}{6}\left(\sum_{i=1}^\rho (\phi_j, \phi_i^3)\alpha_i^3 + 3\sum_{\substack{i,l=1 \\ i \neq l}}^\rho (\phi_j, \phi_i^2\phi_l)\alpha_i^2\alpha_l + \right.$$

$$\left. + 6\sum_{\substack{i,k,l=1 \\ i \neq l \neq k \neq i}}^\rho (\phi_j, \phi_i\phi_k\phi_l)\alpha_i\alpha_k\alpha_l\right)$$

$$= \alpha_j - \frac{\lambda_0|b|}{6}\left((\phi_j, \phi_j^3)\alpha_j^3 + \sum_{\substack{i=1 \\ i \neq j}}^\rho \alpha_i^2\alpha_j\right), j = 1, \ldots, \rho.$$

Hence, (2.28) becomes

$$f_j(\alpha_1, \ldots, \alpha_\rho) = \left[1 - \frac{\lambda_0 |b|}{2} \left(\frac{3}{4}\alpha_j^2 + \sum_{\substack{i=1 \\ i \neq j}}^{\rho} \alpha_i^2\right)\right]\alpha_j = 0, \quad j = 1, \ldots, p, \qquad (2.32a)$$

$$f_j(\alpha_1, \ldots, \alpha_\rho) = \left[1 - \frac{\lambda_0 |b|}{2} \left(\frac{1}{2}\alpha_j^2 + \sum_{\substack{i=1 \\ i \neq j}}^{\rho} \alpha_i^2\right)\right]\alpha_j = 0, \quad j = p+1, p+2. \qquad (2.32b)$$

Summarizing, we have reduced (2.26) to an algebraic system and each solution of (2.32) leads to a solution of (2.26) and vice versa. Consequently, we will show that this solution yields a solution branch of (2.25), and hence a solution curve of (2.4) passing through $(0, \lambda_0)$.

3. Solutions of the Reduced Bifurcation Equations

We consider in this section the solutions of (2.32) and their regularity.

First of all, one sees that in any solutions of (2.32) with two non-vanishing components α_i, α_j, the i-th and j-th equations of (2.32) become

$$1 - \frac{\lambda_0 |b|}{2}\left(b_i \alpha_i^2 + \sum_{\substack{l=1 \\ l \neq i}}^{\rho} \alpha_l^2\right) = 0, \qquad (3.1)$$

$$1 - \frac{\lambda_0 |b|}{2}\left(b_j \alpha_j^2 + \sum_{\substack{l=1 \\ l \neq j}}^{\rho} \alpha_l^2\right) = 0, \qquad (3.2)$$

where

$$b_i := \begin{cases} 3/4 & \text{for } i \in \{1, \ldots, p\}; \\ 1/2 & \text{for } i \in \{p+1, p+2\}. \end{cases} \qquad (3.3)$$

Therefore, if $i, j \in \{1, \ldots, p\}$ or $\{i, j\} = \{p+1, p+2\}$, the difference of (3.1) and (3.2) yields

$$\alpha_i^2 = \alpha_j^2. \qquad (3.4)$$

Otherwise, we have for all nontrivial components with indices $i \leq p$ and $j > p$

$$\alpha_i^2 = 2\alpha_j^2 \quad \text{for } i \leq p < j \leq p+2. \qquad (3.5)$$

Hence, whenever we have nontrivial components α_i for a solution of (2.32) with indices in $i \leq p$ and $j > p$, the statements (3.4) and (3.5) imply, for appropriate c, see (3.7),

$$\alpha_i = \pm c, \alpha_j = \pm c/\sqrt{2} \quad \text{for } i \leq p < j \leq p+2. \qquad (3.6)$$

Now we are able to prove the following.

Lemma 3.1: *For any $k = 1, 2, \ldots, \rho$, let $\alpha_{i_\nu} \neq 0$ for k different indices $i_\nu \in \{1, \ldots, \rho\}, \nu = 1, \ldots, k$ and $\alpha_j = 0$ for $j \neq i_\nu, \nu = 1, \ldots, k$. With $\bar{k}$, the number of nontrivial $\alpha_{p+1}, \alpha_{p+2}$, we define*

$$c := c(k, \bar{k}) := [\lambda_0 |b| (4k - 2\bar{k} - 1)/8]^{-\frac{1}{2}}. \tag{3.7}$$

Then (2.32) has $2^k \binom{\rho}{k}$ different solutions with k nontrivial components of the form

$$\alpha_{i_\nu} = \pm c \text{ for } i_\nu \leq p, \ \alpha_{i_\nu} = \pm c/\sqrt{2} \text{ for } i_\nu > p \text{ and } \quad \alpha_j = 0, j \neq i_\nu, \ \nu = 1, \ldots, k. \tag{3.8}$$

Proof: We start with the case $k = \bar{k} > 0$. Then (3.2) and (3.3) imply with $\alpha_j = \pm c/\sqrt{2}$ as in (3.8)

$$1 - \frac{\lambda_0 |b|}{2} [\frac{1}{2} + (\bar{k} - 1)] \frac{c^2}{2} = 0, \quad \text{hence } c = c(\bar{k}, \bar{k}).$$

Now, let $0 \leq \bar{k} < k$. Then we use (3.1) for an index $i = i_\nu \leq p$ and obtain with (3.3)

$$1 - \frac{\lambda_0 |b|}{2} [\frac{3}{4} c^2 + (k - \bar{k} - 1) c^2 + \bar{k} \frac{c^2}{2}] = 0, \quad \text{hence } c = c(k, \bar{k}).$$

Therefore, for any given k in $\{1, \ldots, p+2\}$ and $i_\nu, \nu = 1, \ldots, k$, the statement (3.8) yields all the solutions of (2.32) which have exactly k nontrivial components. The choices of $i_1, \ldots, i_k$ in $\{1, \ldots, \rho\}$ and the combinations of $\pm \alpha_{i_\nu}, \nu = 1, \ldots, k$ show that we have exactly $2^k \binom{\rho}{k}$ different nontrivial solutions of (2.32) of the form (3.8) with k nontrivial components. ∎

Theorem 3.2: *The system (2.32) has exactly $3^\rho - 1$ different and isolated nontrivial solutions. With $\mathbf{f} = (f_1, \ldots, f_\rho)^T$, for each of these solutions $(\alpha_1^0, \ldots, \alpha_\rho^0)$ the matrix $Df_0 := Df(\alpha_1^0, \ldots, \alpha_\rho^0)$ is strictly negative definite, i.e. has only negative eigenvalues.*

Proof: In Lemma 3.1 we have found $2^k \binom{\rho}{k}$ different nontrivial solutions of the form (3.8). The summation for $k = 1, \ldots, \rho$ shows that we have

$$\sum_{k=1}^{\rho} 2^k \binom{\rho}{k} = 3^\rho - 1$$

different nontrivial solutions of (2.32). We will show now, that these solutions are isolated. Hence, by Bezout's Theorem, we have found all (nontrivial) solutions for (2.32).

Let $(\alpha_1^0, \ldots, \alpha_\rho^0)$ be an arbitrary nontrivial solution of (2.32). Choose $k \geq 1$ indices $i_1, \ldots, i_k \in \{1, \ldots, \rho\}$ such that

$$\alpha_{i_1} \cdot \alpha_{i_2} \cdots \alpha_{i_k} \neq 0, \quad \alpha_j = 0, \ j \notin \{i_1, \ldots, i_k\}.$$

We consider the derivative of $\mathbf{f} = (f_1, \ldots, f_\rho)^T$ at this solution point $(\alpha_1^0, \ldots, \alpha_\rho^0)$

$$Df_0 = \begin{pmatrix} a_1 - \lambda_0 |b| (\alpha_1^0)^2 & -\lambda_0 |b| \alpha_2^0 \alpha_1^0 & \cdots & -\lambda_0 |b| \alpha_\rho^0 \alpha_1^0 \\ -\lambda_0 |b| \alpha_1^0 \alpha_2^0 & a_2 - \lambda_0 |b| (\alpha_2^0)^2 & \cdots & -\lambda_0 |b| \alpha_\rho^0 \alpha_2^0 \\ \vdots & \vdots & \ddots & \vdots \\ -\lambda_0 |b| \alpha_1^0 \alpha_\rho^0 & -\lambda_0 |b| \alpha_2^0 \alpha_\rho^0 & \cdots & a_\rho - \lambda_0 |b| (\alpha_\rho^0)^2 \end{pmatrix} =: A - \lambda_0 |b| \vec{\alpha} \vec{\alpha}^T,$$

where $A := diag(a_1, \ldots, a_p)$ and

$$a_j := \begin{cases} 1 - \frac{\lambda_0|b|}{8}[9(\alpha_j^0)^2 + 4\sum_{i \neq j}(\alpha_i^0)^2] & \text{for } j = 1, \ldots, p; \\ 1 - \frac{\lambda_0|b|}{4}[3(\alpha_j^0)^2 + 2\sum_{i \neq j}(\alpha_i^0)^2] & \text{for } j = p+1, p+2, \end{cases}$$

see (2.32) and

$$\vec{\alpha} := (\alpha_1^0, \ldots, \alpha_p^0)^T.$$

Since $\alpha_{i_1} \neq 0$, we get from the i_1-th equation in (2.32)

$$\sum_{i=1}^{p}(\alpha_i^0)^2 = \begin{cases} \frac{2}{\lambda_0|b|} + \frac{\alpha_{i_1}^2}{4} & \text{for } i_1 \in \{1, \ldots, p\}; \\ \frac{2}{\lambda_0|b|} + \frac{\alpha_{i_1}^2}{2} & \text{for } i_1 \in \{p+1, p+2\}. \end{cases}$$

Hence, if $0 \leq \bar{k} < k$ we may choose $i_1 \in \{1, \ldots, p\}$ and have

$$a_j = \begin{cases} -\frac{\lambda_0|b|}{8}[5(\alpha_j^0)^2 + (\alpha_{i_1}^0)^2] < 0 & \text{for } j = 1, \ldots, p; \\ -\frac{\lambda_0|b|}{8}[2(\alpha_j^0)^2 + (\alpha_{i_1}^0)^2] < 0 & \text{for } j = p+1, p+2. \end{cases}$$

Similarly, if $0 < \bar{k} = k$ we have $i_1 \in \{p+1, p+2\}$ and therefore

$$a_j = \begin{cases} -\frac{\lambda_0|b|}{8}[5(\alpha_j^0)^2 + 2(\alpha_{i_1}^0)^2] < 0 & \text{for } j = 1, \ldots, p, (\alpha_j^0 = 0); \\ -\frac{\lambda_0|b|}{4}[(\alpha_j^0)^2 + (\alpha_{i_1}^0)^2] < 0 & \text{for } j = p+1, p+2. \end{cases} \qquad (3.9)$$

Let, for an appropriately chosen $\alpha_{i_1}^0$, c_j and $c_{i_1} \in \{1, 2, 5\}$ denote the coefficients of α_j^0 and $\alpha_{i_1}^0$, respectively. Then

$$-\lambda_0|b|\vec{\alpha}^T A^{-1}\vec{\alpha} = 8\lambda_0 b \sum_{j=1}^{p} \frac{(\alpha_j^0)^2}{\lambda_0|b|(c_j(\alpha_j^0)^2 + c_{i_1}(\alpha_{i_1}^0)^2)}$$

$$= 8\sum_{l=1}^{k} \frac{(\alpha_{i_l}^0)^2}{(c_j(\alpha_{i_l}^0)^2 + c_{i_1}(\alpha_{i_1}^0)^2)}$$

$$\neq -1.$$

The Sherman/Morrison formula shows that $D\mathbf{f}_0$ is nonsingular (cf. Ortega/Rheinboldt [25]). The negative definiteness of $D\mathbf{f}_0$ is a consequence of

$$x^T D\mathbf{f}_0 \, x = x^T A x - \lambda_0|b|(x^T\vec{\alpha})^2 = -\sum_{j=1}^{p}|a_j|x_j^2 - \lambda_0|b|(x^T\vec{\alpha})^2 < 0$$

for every nontrivial $x \in \mathbf{R}^p$, since $a_j < 0$ for $j = 1, \ldots, p$, see (3.9).   ∎

4. Bifurcating Solution Branches

We state here the main results of our discussion.

Theorem 4.1: *At a corank ρ bifurcation point $(0, \lambda_0)$, (1.1) has exactly $(3^\rho - 1)/2$ different non-trivial solution curves $(u(t, \cdot), \lambda(t))$ of the form*

$$
\begin{cases}
u(t, x, y) = t \sum_{i=1}^{\rho} \alpha_i(t) \phi_i(x, y) + t^3 v(t, x, y), \\
\lambda(t) = \lambda_0 - \mathrm{sign}(f'''(0)) t^2, \quad t \in [-t_0, t_0],
\end{cases}
\tag{4.1}
$$

bifurcating from $\{(0, \lambda), \lambda \in \mathbf{R}\}$, where $(\alpha_1(0), \ldots, \alpha_\rho(0))$ satisfies (2.32). Moreover, if $f'''(0) > 0$, these solutions are stable. Conversely, for $f'''(0) < 0$, these solutions are unstable.

Proof: Obviously, a solution $(v, \alpha_1, \ldots, \alpha_\rho)$ of (2.26) corresponds uniquely to a solution $(\alpha_1, \ldots, \alpha_m)$ of (2.32). By Theorem 3.2, (2.32) has exactly $3^\rho - 1$ nontrivial solutions. For each solution $(\alpha_1^0, \ldots, \alpha_\rho^0)$ of (2.32), there exists a unique $v^0 \in R(D_u G_0)$, such that

$$
D_u G_0 v^0 + \mathrm{sign}(b) T [\sum_{i=1}^{\rho} \alpha_i^0 \phi_i - \lambda_0 |b| (\sum_{i=1}^{\rho} \alpha_i^0 \phi_i)^3 / 6] = 0.
$$

Hence, $(v^0, \alpha_1^0, \ldots, \alpha_\rho^0, 0)$ satisfies (2.26). At the same time, consider the equation

$$
D_{(v, \alpha_1, \ldots, \alpha_\rho)} F(v^0, \alpha_1^0, \ldots, \alpha_\rho^0, 0)(w, \beta_1, \ldots, \beta_\rho) = (g, c_1, \ldots, c_\rho).
$$

in $E \times \mathbf{R}^\rho$. More precisely

$$
D_u G_0 w + \mathrm{sign}(b) \sum_{i=1}^{\rho} \beta_i T [\phi_i - \frac{\lambda_0 |b|}{2} (\sum_{j=1}^{\rho} \alpha_j^0 \phi_j)^2 \phi_i] = g,
\tag{4.2a}
$$

$$
(\phi_k, w) = c_k, \quad k = 1, \ldots, \rho.
\tag{4.2b}
$$

According to (2.6), $\beta_1, \ldots, \beta_\rho$ in (4.2a) can be computed by

$$
(\phi_k, \mathrm{sign}(b) \sum_{i=1}^{\rho} \beta_i T [\phi_i - \frac{\lambda_0 |b|}{2} (\sum_{j=1}^{\rho} \alpha_j^0 \phi_j)^2 \phi_i]) = (\phi_k, g), \quad k = 1, \ldots, \rho,
$$

or in matrix form, see the proof of Theorem 3.2, (2.28) and (2.29)

$$
-\frac{\mathrm{sign}(b)}{\lambda_0} Df_0 (\beta_1, \ldots, \beta_\rho)^T = ((\phi_1, g), \ldots, (\phi_\rho, g))^T.
\tag{4.3}
$$

Since Df_0 is nonsingular, the system (4.3) is uniquely solvable. Substituting its solution into (4.2a), we obtain a system for w in $R(D_u G_0)$ which is also uniquely solvable. Thus the operator $D_{(v,\alpha_1,\ldots,\alpha_\rho)}\mathbf{F}(v^0,\alpha_1^0,\ldots,\alpha_\rho^0,0)$ is nonsingular. Applying the implicit function theorem to $\mathbf{F}$ at $(v^0,\alpha_1^0,\ldots,\alpha_\rho^0,0)$, we get a constant $t_0 > 0$ and a unique mapping $\{(v(t),\alpha_1(t),\ldots,\alpha_\rho(t),t),\ t \in (-t_0,t_0)\}$, such that

$$\begin{cases} \mathbf{F}(v(t),\alpha_1(t),\ldots,\alpha_\rho(t),t) = 0, \\ (v(0),\alpha_1(0),\ldots,\alpha_\rho(0)) = (v^0,\alpha_1^0,\ldots,\alpha_\rho^0). \end{cases}$$

Substituting $v(t),\alpha_1(t),\ldots,\alpha_\rho(t)$ into (2.22), one obtains a solution curve (4.1) of (1.1). Due to the oddness of G (resp. $\mathbf{F}$), we see that $\pm(v^0,\alpha_1^0,\ldots,\alpha_\rho^0,0)$ results in the same solution curve of (2.25) (resp. (1.1)). Therefore, $3^\rho - 1$ different solutions of (2.32) lead to exactly $(3^\rho - 1)/2$ different solution curves of (1.1).

Since $D\mathbf{f}_0$ has strictly negative eigenvalues, observing the special definition (2.28), (2.29) of $\mathbf{f}$, we see if $f'''(0) > 0$, then $-\lambda_0 \operatorname{sign}(b) > 0$ and $-\frac{\operatorname{sign}(b)}{\lambda_0} D\mathbf{f}_0$ has also strictly negative eigenvalues. Hence, the bifurcating solutions at $(0,\lambda_0)$ are stable, see Sattinger [26], Golubitsky/Stewart/Schaeffer [16]. We recall that in this case all bifurcations occur to the left, i.e. subcritical, see (2.22) ff. On the other hand, if $f'''(0) < 0$, then $-\lambda_0 \operatorname{sign}(b) < 0$ and $-\frac{\operatorname{sign}(b)}{\lambda_0} D\mathbf{f}_0$ has strictly positive eigenvalues. Hence, these solutions are unstable. In this case all bifurcations occur to the right, i.e. supercritical, see (2.22) ff.   ∎

Summarizing, a solution curve $(u(t),\lambda(t))$ in (4.1) of (1.1) is determined by computing $v(t),\alpha_1(t),\ldots,\alpha_m(t)$ as a nonsingular solution curve of (2.25) passing through the solution point $(v^0,\alpha_1^0,\ldots,\alpha_\rho^0,0)$ of (2.26). The different "tangents" of the solution curves (4.1) at $(0,\lambda_0)$ are determined by the isolated solutions $(v^0,\alpha_1^0,\ldots,\alpha_\rho^0)$ of (2.26). In other words, each solution branch in (4.1) has a cone shaped δ-neighborhood of $(0,\lambda_0)$. There are $(3^\rho - 1)/2$ such cones in a neighborhood of $(0,\lambda_0)$. If we choose $\delta < \delta_0$ sufficiently small, the correspondence between the nonsingular solutions of (2.25), (2.26) and the solution branches of (1.1) implies that these cones do not overlap, see e.g. Fig. 4.1. That means, at least in a small neighborhood of $(0,\lambda_0)$, (1.1) has no secondary bifurcations. The stability of the bifurcating solution branches depends upon the sign of $f'''(0)$, as indicated in Theorem 4.1. Finally we mention that a similar bifurcating scenario holds for other boundary conditions, e.g. Dirichlet, Dirichlet and Neumann along different sides of Ω as well as some other special cases related to hidden symmetries of (1.1) with periodic boundary conditions, see e.g. Gomes [17], Healey/Kielhöfer [20], Allgower/Böhmer/Mei [3].

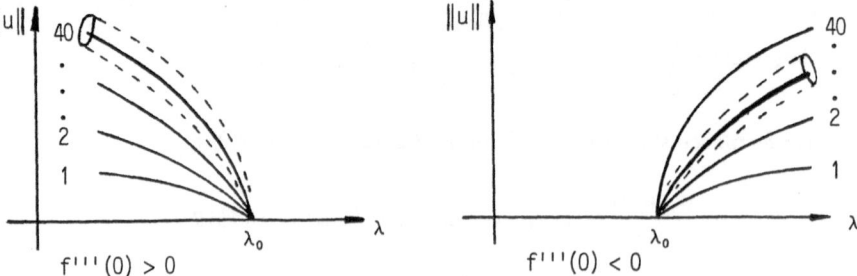

Fig. 4.1

5. A Simple Example

We consider the simplified buckling problem

$$
\begin{cases}
\Delta u + \lambda \sin(u) = 0 & \text{in} \quad \Omega := [0,1] \times [0,1], \\[2mm]
\dfrac{\partial u}{\partial n} = 0 & \text{on} \quad \partial \Omega
\end{cases}
\tag{5.1}
$$

at its corank 4 bifurcation point $(0, \lambda_0 = 25\pi^2)$. For (2.8)-(2.12), we have

$$
\phi_1 = 2\cos 3\pi x \cos 4\pi y, \quad \phi_2 = 2\cos 4\pi x \cos 3\pi y, \quad \phi_3 = \sqrt{2}\cos 5\pi x, \quad \phi_4 = \sqrt{2}\cos 5\pi y
$$

and the reduced bifurcation equations are

$$
\begin{cases}
\alpha_1 - \dfrac{\lambda_0}{2}[\dfrac{3}{4}\alpha_1^2 + \alpha_2^2 + \alpha_3^2 + \alpha_4^2]\alpha_1 = 0, \\[3mm]
\alpha_2 - \dfrac{\lambda_0}{2}[\dfrac{3}{4}\alpha_2^2 + \alpha_1^2 + \alpha_3^2 + \alpha_4^2]\alpha_2 = 0, \\[3mm]
\alpha_3 - \dfrac{\lambda_0}{2}[\dfrac{1}{2}\alpha_3^2 + \alpha_1^2 + \alpha_2^2 + \alpha_4^2]\alpha_3 = 0, \\[3mm]
\alpha_4 - \dfrac{\lambda_0}{2}[\dfrac{1}{2}\alpha_4^2 + \alpha_1^2 + \alpha_2^2 + \alpha_3^2]\alpha_4 = 0.
\end{cases}
\tag{5.2}
$$

Solving (5.2) directly, we obtain 40 different isolated nontrivial solutions. The other 40 nontrivial solutions follow from the symmetry that both $\pm(\alpha_1, \ldots, \alpha_4)$ are solutions of (5.2). Substituting back into (2.26) with $f(u) = \sin(u)$, one gets 40 nonsingular solutions of (2.26). At each solution point of (2.26), the implicit function theorem yields a unique solution curve passing through $(0, 25\pi^2)$. Altogether, we get 40 different solution curves of the corresponding (2.25) and consequently, 40 different solution branches of (5.1) bifurcating from $\{(0, \lambda); \lambda \in \mathbf{R}\}$ at $(0, \lambda_0 = 25\pi^2)$.

The 14 figures below show nodal lines of the $u_i'(0, x, y) = \frac{\partial u_i}{\partial t}(0, x, y), i = 1, 40$. The total number of 40 nodal lines is obtained by collecting all elements of the representive orbits, indicated in these figures.

References

[1] Allgower, E. L., Böhmer, K. and Mei, Z. (1990) A generalized equibranching lemma with application to $D_4 \times Z_2$ symmetric elliptic problems, Bericht Nr. 9, Fachbereich Mathematik, University of Marburg

[2] Allgower, E. L., Böhmer, K. and Mei, Z. (1990) On new bifurcation results for semi-linear elliptic equations wit symmetries, to appear in: MAFELAP, Brunel 1990

[3] Allgower, E. L., Böhmer, K. and Mei, Z. (1990) Bifurcation analysis for the 2-d nonlinear Laplacian with two orthogonal periodic boundary conditions, Preprint

[4] Allgower, E. L., Chien, C.-S. and Georg, K. (1989) Large sparse continuation problems, J. Comp. Appl. Math. **26**, 3-22

[5] Allgower, E. L., Georg, K. and Miranda, R. (1990) Computing real real solutions of polynomial systems, Preprint Colorado State University, Fort Collins, CO.

[6] Berger, M. S. (1969) On one parameter families of real solutions of nonlinear operator equations, *Bull. Amer. Math. Soc.* **75**, 456-459

[7] Böhmer, K. (1990) Developing a numerical Lyapunov-Schmidt method, Bericht zur Fachbereich Mathematik der Philipps-Universität Marburg, to appear in: *Computing*

[8] Böhmer, K. and Mei, Z. (1990) On a numerical Lyapunov-Schmidt method, In: Allgower, E. L. & Georg, K. (eds.): *Computational Solution of Nonlinear Systems of Equations*, Lectures in Applied Mathematics, **26** Providence, 1990, 79-88

[9] Bouligand, G. (1948) Revue Gener.Sc.Pur.Appl., Bull. Soc philomath. 55

[10] Budden, P. and Norbury, J. (1979) A nonlinear elliptic eigenvalue problem, *J. Inst. Math. Appl.* **24**, 9-33

[11] Budden, P. and Norbury, J. (1982) Solution branches for nonlinear equilibrium problems - bifurcation and domain perturbations, *IMA J. Appl. Math.* **28**, 109-129

[12] Cicogna, G. (1981) Symmetry breakdown from bifurcation, *Lett. Nuovo Cimento* **31**, 600-602

[13] Crouzeix, M. and Rappaz, J. (1989) On Numerical Approxiamtion in Bifurcation Theory, Springer-Verlag, 1990

[14] Decker, D. W. and Keller, H. B. (1980) Multiple limit point bifurcation, *J. Math. Anal. Appl.* **75**, 417-430

[15] Dellnitz, M., & Werner, B. (1989) Computational methods for bifurcation problems with symmetries-with special attention to steady state and Hopf bifurcation points, *J. Comp. Appl. Math.* **26**, 97-123

[16] Golubitsky, M., Stewart, I. and Schaeffer, D. G. (1988) Singularities and Groups in Bifurcation Theory, Vol. II, Springer-Verlag, Heidelberg Berlin New York

[17] Gomes, M. (1989) Steady-state mode interaction in rectangular domain, Dissertation, Mathematics Institute, University of Warwick

[18] Healey, T. J. (1988) Global bifurcation and continuation in the presence of symmetry with an application to solid mechanics, *SIAM J. Math. Anal.* **19**, 824-840

[19] Healey, T. J. (1988) A group-theoretic approach to computational bifurcation problems with symmetry, *Comput. Meths. in Appl. Mech. Engrg.* **67**, 257-295

[20] Healey, T. J. and Kielhöfer, H. (1990): Symmetry and nodal properites in global bifurcation analysis of quasi-linear elliptic equations, *Arch. Rat. Mech. Anal.*

[21] Isaacson, E. and Keller, H. B. (1969) Analysis of Numerical Methods, John Wiley & Sons, Inc., New York London Sydney

[22] Lions, P. L. (1982) On the existence of positive solutions of semilinear elliptic equations, *SIAM Review* **24**, 441-467

[23] Mei, Z. (1989) Numerical approximations of corank-2 bifurcation problems, Dissertation, Department of Mathematics, University of Marburg

[24] Mei, Z. (1990) Bifurcations of a simplified buckling problem and its discretizations, Preprint, Department of Mathematics, University of Marburg

[25] Ortega, J. M. and Rheinboldt, W. C. (1970) Iterative Solution of Nonlinear Equations in Several Variables, Academic Press, New York

[26] Sattinger, D. H. (1979) Group Theoretic Methods in Bifurcation Theory, Lecture Notes in Math. 762, Springer-Verlag, Berlin Heidelberg New York

[27] Vanderbauwhede, A. (1982) Local Bifurcation Theory and Symmetry, Pitman London

[28] Werner, B. (1988) Computational methods for bifurcation problems with symmetries and applications to steady states of n-box reaction-diffusion models, in *Numerical Analysis 1987*, D. F. Griffiths, & G. A. Watson (eds.) 279-293

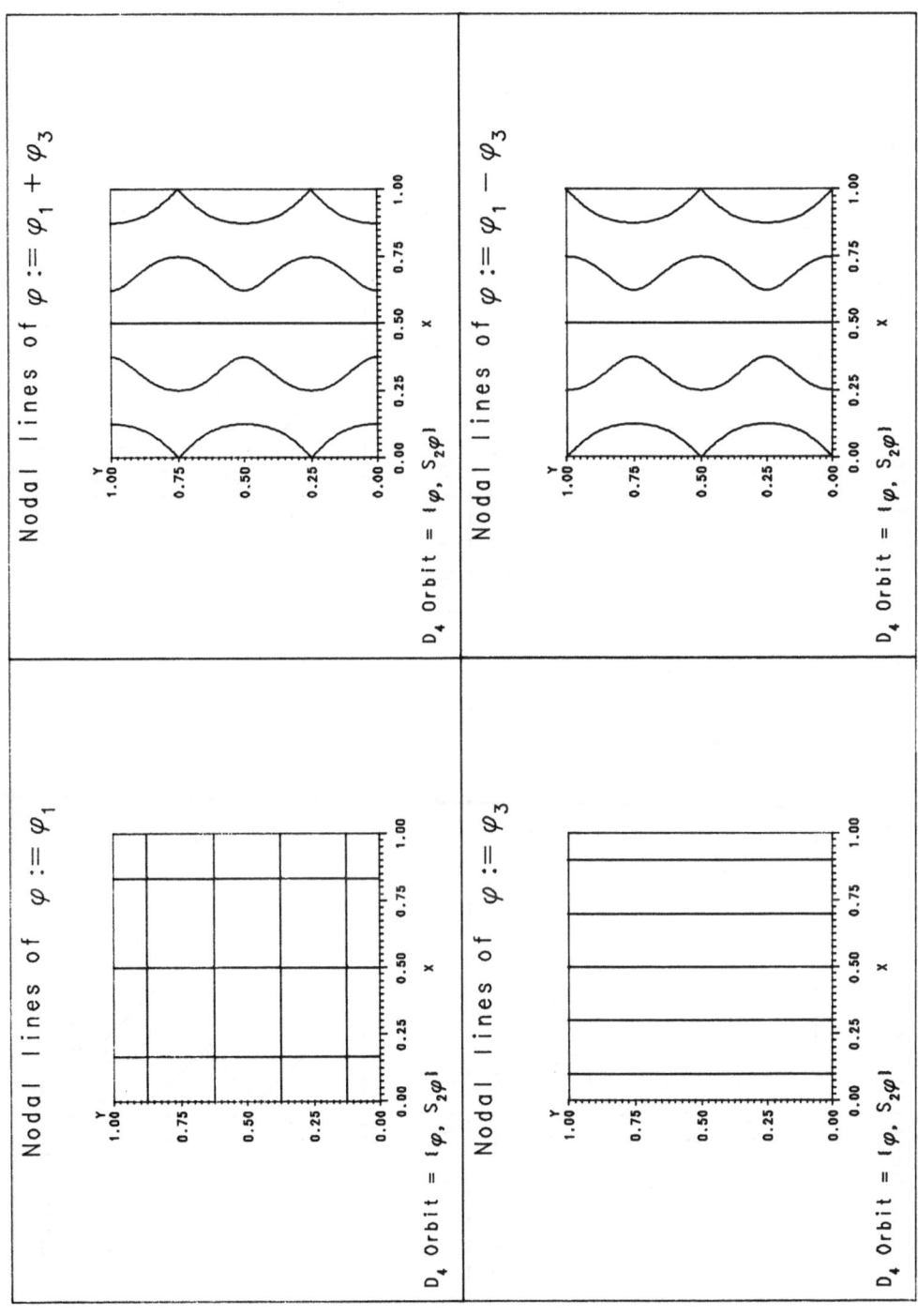

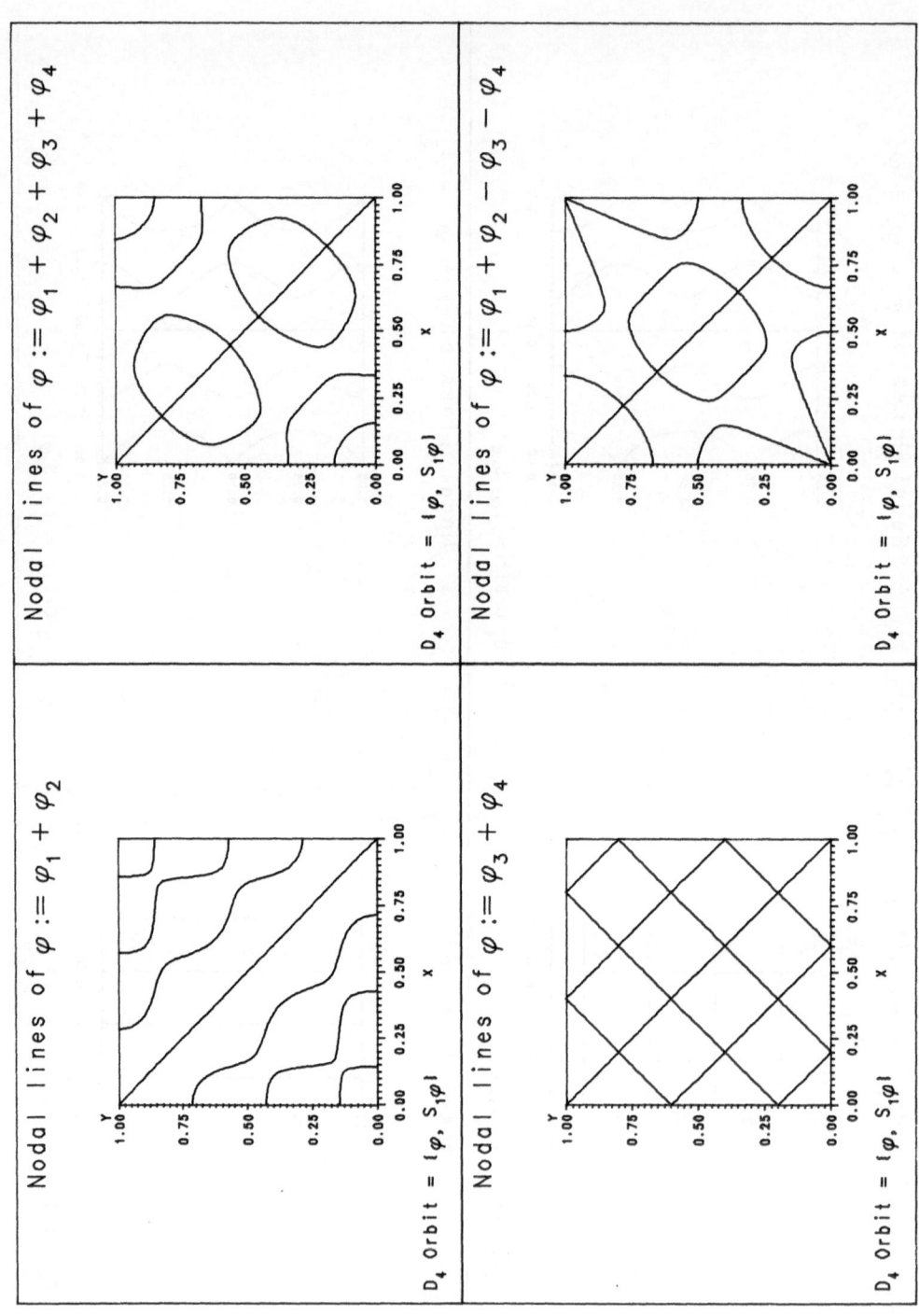

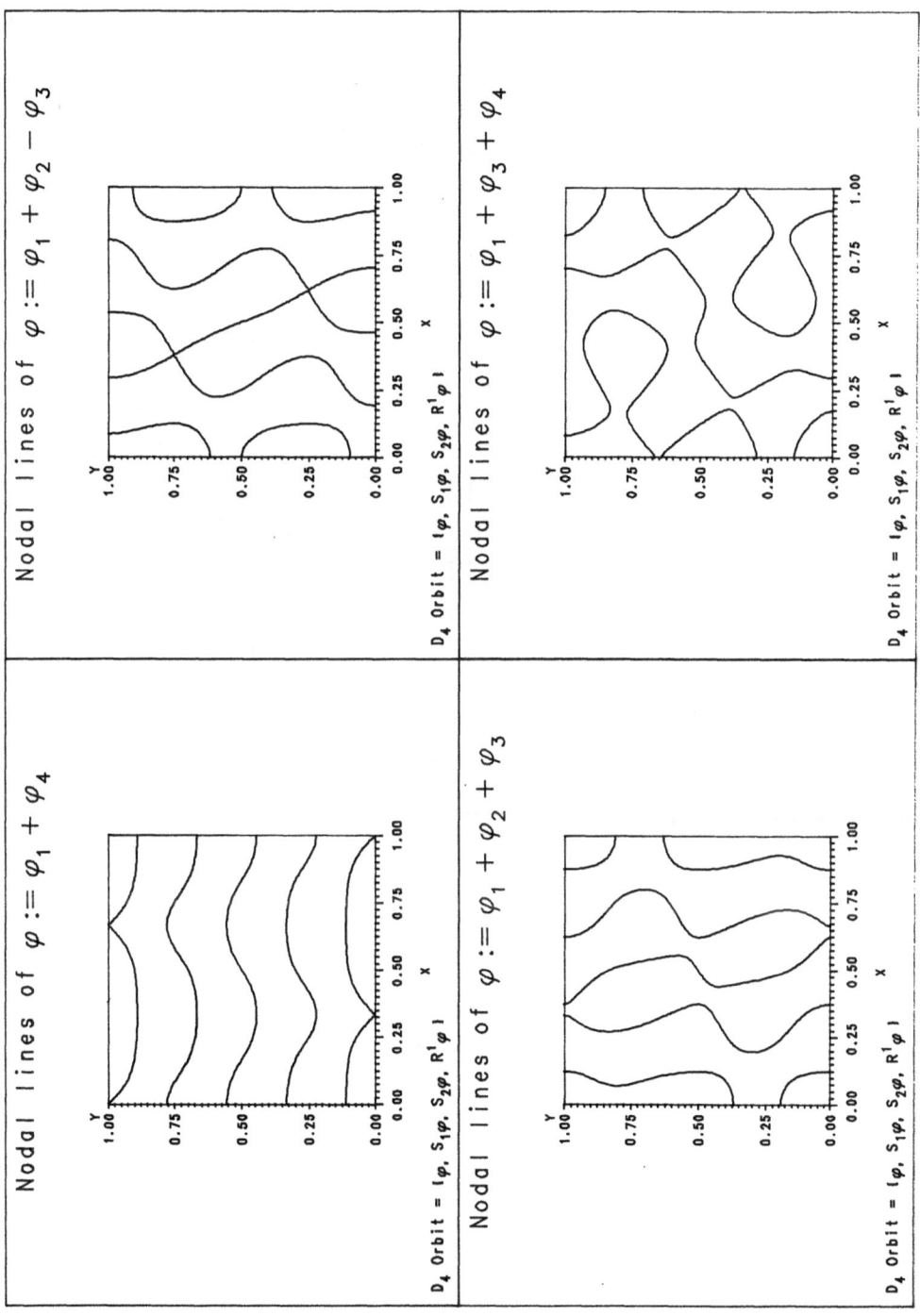

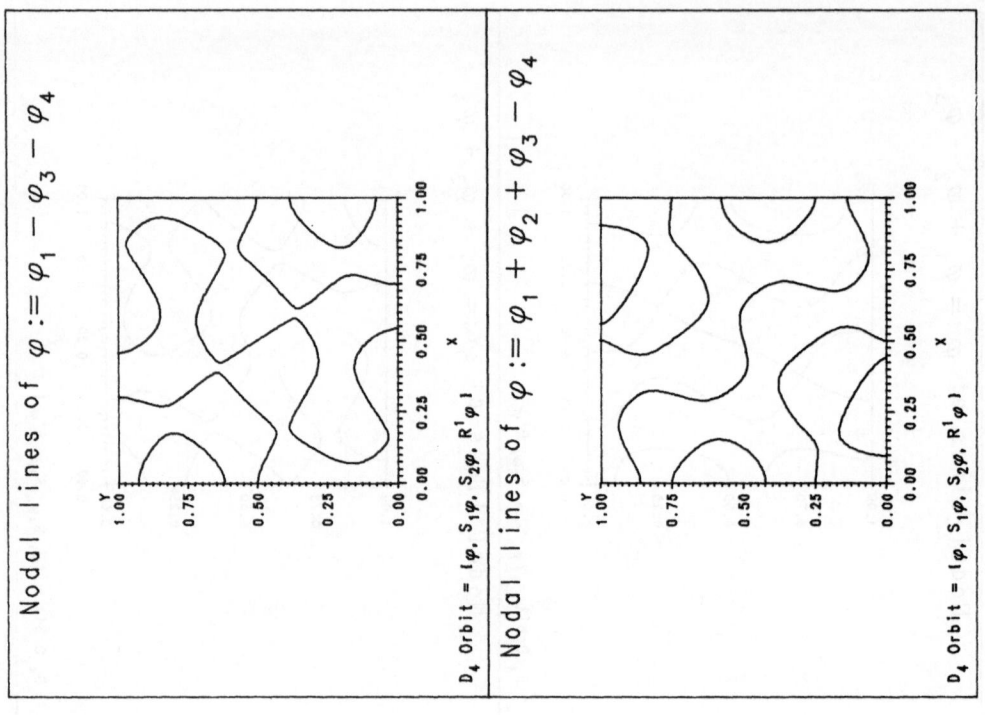

Nodal lines of $\varphi := \varphi_1 - \varphi_3 - \varphi_4$

D_4 Orbit = { φ, $S_1\varphi$, $S_2\varphi$, $R^1\varphi$ }

Nodal lines of $\varphi := \varphi_1 + \varphi_2 + \varphi_3 - \varphi_4$

D_4 Orbit = { φ, $S_1\varphi$, $S_2\varphi$, $R^1\varphi$ }

International Series of Numerical Mathematics, Vol. 97, © 1991 Birkhäuser Verlag Basel

The Effect of Fluctuations on the Transition Behavior of a Nonlinear Chemical Oscillator

J. Amrehn, Th.-M. Kruel, F.W. Schneider

Institut für Physikalische Chemie der Universität Würzburg
Marcusstr. 9–11, D–8700 Würzburg, Germany

F. Buchholtz

Department of Chemistry, Brandeis University, Waltham, MA

We present the effect of Gaussian distributed noise on the secondary periodic bifurcation of a nonlinear chemical oscillator, namely the driven Brusselator model under CSTR boundary conditions. This system contains a cubic autocatalysis and shows a strong amplification of noise which is imposed on the variables of the model at discrete time intervals ("shot noise"). We show that the location of the secondary periodic bifurcation between large and small amplitude oscillations changes dramatically as a function of the noise frequency at constant noise amplitude. The simulated behavior is explained on the basis of a bifurcation analysis. The present calculations serve as an interpretation for the dramatic stirring effects observed in the oscillatory chemiluminescence of luminol in a CSTR.

1 Introduction

The effect of noise on nonlinear mechanisms is of particular interest in the field of modelling nonlinear chemical reactions. Inclusion of noise into nonlinear equations is necessary to compare the unavoidably noisy experiment with the theoretical model. Noise is usually treated by the Langevin equation and its corresponding Fokker–Planck equation if an analytical expression is available. On the other hand, chemical reaction mechanisms are usually extremely complex and may consist of a large number of individual steps. Closed analytical expressions are not available in general. Thus one has to resort to computer simulations in order to study the effect of fluctuations on the nonlinear model.

The effect of fluctuations is particularly dramatic for mechanisms which contain cubic autocatalysis [1–3] such as the Brusselator model [4]. We present the effect of fluctuations (shot noise) on the Brusselator model with the boundary conditions of a Continuous Flow Stirred Tank Reactor (CSTR). A CSTR is generally used in experiments on chemical oscillations in an open system. Of particular interest in this work is the dependence of the transition between high and low amplitude oscillations upon the frequency of the fluctuations. This transition is characterized as a secondary periodic bifurcation. We will be able to show that this secondary periodic bifurcation point is significantly blurred and shifted on the parameter axis in the presence of noise.

Our numerical calculations simulate a real chemical system, namely the oscillating chemiluminescence of luminol [5,6] in a CSTR. Fluctuations are introduced in the experiment by the flow and stirring processes which cause local inhomogeneities in all species concentrations due to finite mixing times.

In general we distinguish between two types of noise [7] : variations in the bifurcation parameters (type P (parameter) noise) and fluctuations in the intermediate concentrations (type V (variable) noise). Stirring effects mainly correspond to type V noise whereas type P noise is caused by variations of the flow through the reactor or by temperature fluctuations. In this work we use type V noise for the simulation of stirring effects by imposing fluctuations on the concentration variables.

2   The Model

We choose the highly nonlinear Brusselator model [4], driven by a linear oscillator [8], as an object of our numerical investigations. The proposed mechanism is motivated by our experiments with the luminol oscillator [5,6,9]. In the latter a chemical oscillator, the socalled Orbán–oscillator [10], drives the luminol oxidation, which, as a result, displays sharp pulses of regular chemiluminescent oscillations. The sharpness of the peaks suggests the presence of a high order nonlinearity.

The Brusselator describes the conversion $A + B \longrightarrow D + E$ which occurs via the intermediates X and Y. The latter undergo a cubic autocatalysis:

$$
\begin{aligned}
A &\xrightarrow{k_1} X \\
X + B &\xrightarrow{k_2} Y + D \\
2X + Y &\xrightarrow{k_3} 3X \\
X &\xrightarrow{k_4} E
\end{aligned}
$$

The driving oscillator is represented by the species B where the influx changes sinusoidally whereas the influx of A is kept constant

$$B^{in} = B_0^{in}(1 + \alpha \cos \omega t) \tag{1}$$

$$A^{in} = const. \tag{2}$$

This situation essentially represents a coupling of a linear with a nonlinear oscillator in a CSTR. The linear oscillator is a simplification of the Orbán–oscillator [10] whereas the nonlinear oscillator is an approximate description of the luminol oscillator.

The differential equations are:

$$\frac{dA}{dt} = -k_1 A + k_f(A^{in} - A) \tag{3}$$

$$\frac{dB}{dt} = -k_2 BX + k_f(B^{in} - B) \tag{4}$$

$$\frac{dX}{dt} = k_1 A - k_2 BX + k_3 X^2 Y - k_4 X - k_f X \tag{5}$$

$$\frac{dY}{dt} = k_2 BX - k_3 X^2 Y - k_f Y \tag{6}$$

where: $k_1 = 2.0$, $k_2 = 4.15$, $k_3 = 1100$, $k_4 = 1.85$, $A^{in} = 6.0$, $B_0^{in} = 80.0$, $\alpha = 0.5$, $\omega = 0.2\,rad/sec$.

The numerical values have been chosen to fit our experimental results. The intensity of the chemiluminescence is represented by species X. The products D and E are of no further interest. k_f is the flow rate of species through the CSTR.

3 Simulation of Gaussian distributed shot noise

As carried out in previous work [7, 11–13] we simulate fluctuations by the addition of random variables to each state variable of the system (A, B, X, Y) at constant time intervals Δt^{fluc} (i.e. a constant number of integration steps). The perturbed elements $X_i^{fluc} = X_i(1 + \xi_i)$ enter the next integration step. These fluctuations are interactive in contrast to instrumental noise which does not affect the system itself. The type of noise used here is also called "shot noise" since the amplitude of the noise changes abruptly in a rectangular fashion.

The noise ξ itself is Gaussian white noise, i.e. with uniform power spectral density. The amplitude of the Gaussian fluctuations is chosen to be 1% (specified by the mean variance of the distribution). Variation in the time interval Δt^{fluc} between two successive fluctuations enables us to vary the frequency of the fluctuations in a well defined way.

4 Results

Fig.1 displays the simultaneous plot of the time series for B^{in}, X and Y for the above set of parameters in the absence of any fluctuations. It is seen that the oscillations in X display sharp pulses followed by two quasi–steady states (SS III and SS I). The time dependent behavior can be understood from the investigation of the stability of the non–driven system, i.e. when $B^{in} = const.$, or, equivalently, when the driving amplitude $\alpha = 0$.

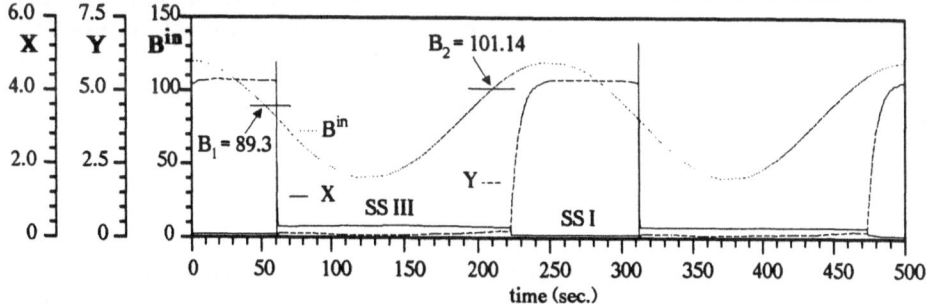

Figure 1: *Plot of the time series of species X, Y and B^{in} for $k_f = 0.18s^{-1}$ without any noise. At $t \approx 60sec$ B^{in} falls below the turning point B_1 (Fig.2) and triggers the transition from SS I (low X) to SS III (high X) via a sharp overshoot. At $t \approx 222sec$ B^{in} rises above the Hopf point B_2 (Fig.2) thereby triggering the jump from SS III back to SS I.*

The transitions in the time series of X are delayed with respect to the critical values (B_1 and B_2) of the oscillating B^{in}. This is easily explained by the fact that B^{in} couples with X via the flow term $k_f(B^{in} - B)$ which is rate determining in that particular case.

A study of the stability of the non–driven system is shown in Fig.2. The steady state branches are drawn for B^{in} as a bifurcation parameter. The upper steady state SS III becomes unstable for increasing B^{in} via a supercritical Hopf bifurcation. The unstable branch of SS II connects SS III and SS I (stable) via two turning points. This situation of stable and unstable branches is drawn for two different values of the flow rate, $k_f = 0.36s^{-1}$ (upper branch) and $k_f = 0.18s^{-1}$ (lower branch). The range of the oscillations of B^{in} in the driven system is indicated by the horizontal arrow.

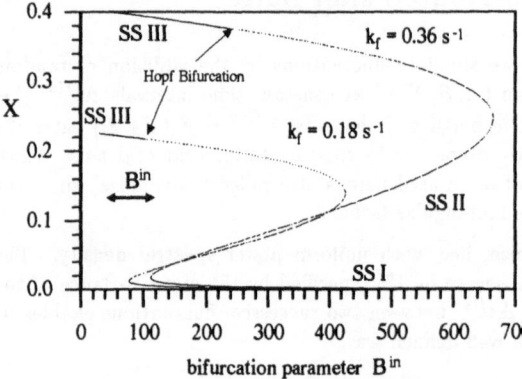

Figure 2: *Stability analysis for the non–driven system. Solid lines denote stable steady states, dotted lines unstable ones. Branches SS I and SS III are connected via the unstable branch SS II. Stability is lost via a Hopf bifurcation and gained again at a turning point (SS II → SS I). The horizontal arrow shows the range in which B^{in} oscillates. For $k_f = 0.18s^{-1}$ B^{in} crosses the Hopf point, but not for $k_f = 0.36s^{-1}$.*

The oscillation of B^{in} switches the system between the two steady states SS III and SS I. Switching from SS I to SS III occurs when B^{in} falls below the left turning point which connects SS I and SS II; switching from SS III to SS I takes place when B^{in} surpasses the Hopf point (HB) at which SS III becomes unstable.

These trigger–type oscillations of X stop when one of the instabilities (turning point or Hopf bifurcation) are no longer crossed by the independent oscillations of B^{in}. In this particular case the large amplitude oscillations cease to exist when a critical flow rate is exceeded. Fig.2 shows the situation where $k_f = 0.36s^{-1}$ is above the critical flow rate ($k_f^{crit} = 0.203s^{-1}$, compare Fig.4). Here the oscillations of B^{in} no longer cross the Hopf bifurcation of SS III. The point of a critical flow rate represents a secondary periodic bifurcation which corresponds to a Hopf bifurcation for the underlying non–driven system.

Critical flow rates are observed also in our experiments with the luminol oscillator [9] indicating the physical relevance of the model.

In the numerical simulations of the time series (Fig.3) we use a linear increase (ramp) of k_f from 0.15 s^{-1} to 0.30 s^{-1} within 5000 seconds in accordance with our experimental ramping. The increase of k_f has been chosen to be as slow as possible in the calculations as well as in experiments in order to reduce the delay which occurs between B_1 and the burst in X as well as between B_2 and the transition SS III → SS I.

Fig.3a) shows the simulation without any noise. The large amplitude oscillations cease at $t \approx 1020sec$ corresponding to $k_f^{crit} = 0.181s^{-1}$. The discrepancy between k_f^{crit} from the simulation and the value of 0.203 s^{-1} from the static bifurcation analysis (Fig.4) is easily explained by the delay effect caused by the flow term $k_f(B^{in} - B)$. Although B^{in} crosses the Hopf bifurcation, the system is too slow to cross the critical threshold in X before B^{in} oscillates back below the Hopf bifurcation. k_f^{crit} is thus also a function of the driving frequency ω and it approaches the static value of 0.203 s^{-1} for $\omega \to 0$ (not shown).

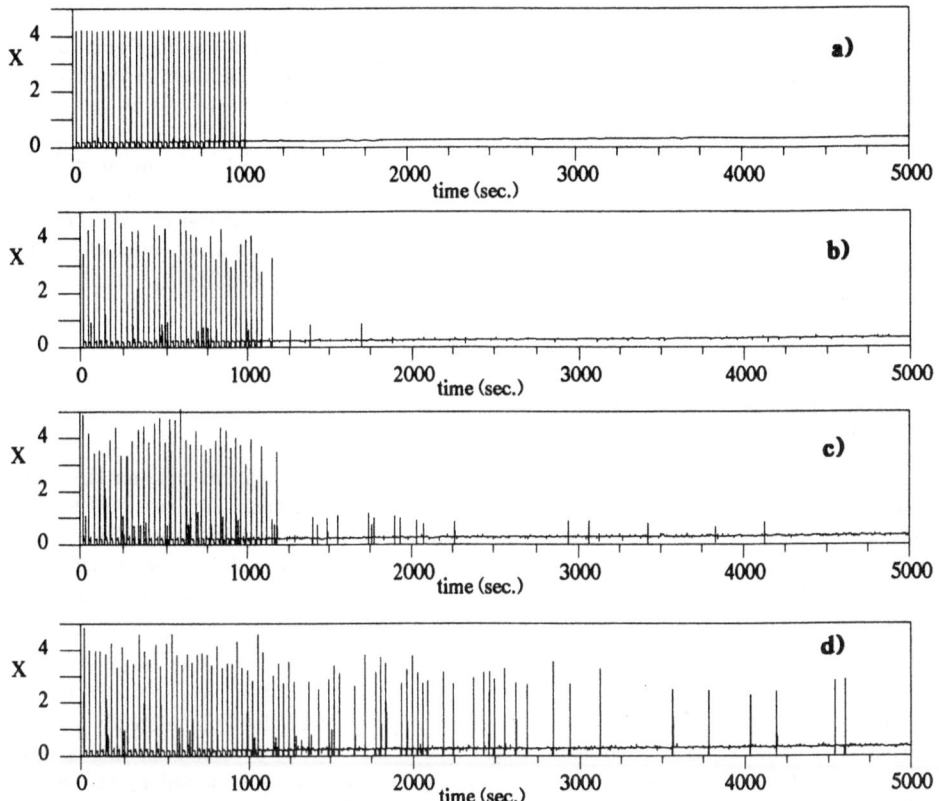

Figure 3: *Time series of X, obtained by integration with a Runge–Kutta–scheme with small stepsize.*
a) *Transition between large amplitude and small amplitude oscillations without superimposed noise. The transition occurs at $t \approx 1020sec$ corresponding to $k_f^{crit} \approx 0.181s^{-1}$. The small amplitude oscillations are too small to be seen in the plot.*
b–d) *Transition scenario at $\omega^{fluc} = \omega$, $\omega^{fluc} = 2\omega$, and $\omega^{fluc} = 3\omega$, respectively. k_f^{crit} is significantly blurred and shifted towards higher values.*

We present the time series for various noise frequencies ranging from $\omega^{fluc} = \omega$ to $\omega^{fluc} = 3\omega$ (Fig.3b–d), where $\omega^{fluc} = 2\pi/\Delta t^{fluc}$. For increasing fluctuation frequencies it is seen that the transition (periodic bifurcation) from large to small amplitude oscillations is blurred and shifted towards higher values of k_f^{crit}, i.e. longer reaction times. One obtains a transition region rather than a well defined transition point.

Even though the noise amplitude is rather small (1 %), the shift of k_f^{crit} towards higher values is dramatic when the frequency of the fluctuations is increased by a factor of 3. The exact value of k_f^{crit} cannot be determined precisely at high fluctuation frequencies (Fig.3d). The trend of increasing

k_f^{crit} with increasing fluctuation frequency is in agreement with the experimental dependence of k_f^{crit} on the stirring rate [9].

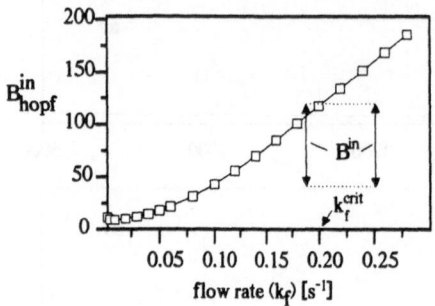

Figure 4: *Dependence of the Hopf point on the flow rate. The independent oscillator in B^{in} is indicated by the vertical arrows at two different flow rates. At $k_f = 0.185s^{-1}$ the oscillations in B^{in} cross the Hopf bifurcation and trigger the transition from SS III to SS I. Above $k_f^{crit} = 0.203s^{-1}$ B^{in} no longer crosses the Hopf bifurcation: the large amplitude oscillations disappear (shown for $k_f = 0.25s^{-1}$).*

5   Discussion

The highly nonlinear Brusselator model in a CSTR provides an excellent example for the effect of random fluctuations upon the temporal behavior of a nonlinear chemical oscillator [7, 12–14]. We show that the interaction of a highly nonlinear bistable system and an external perturbation on B^{in} may lead to periodic oscillations of burst–like appearance. One may explain their occurrence and disappearence by investigating the stability of the non–driven system as a function of the control parameter k_f.

We have found that the imposition of low amplitude (1 %) Gaussian distributed noise upon the variables of the system leads to significant shifts in the critical flow rate at which the large amplitude oscillations disappear. Oscillations occur at parameter values where steady states are observed in the absence of external noise. The range of noise–induced oscillations increases with the frequency of the fluctuations. The frequency dependence may be explained by the fact that the more often the system is perturbed during one cycle of B^{in} , the greater is the chance that the fluctuation occurs exactly when B^{in} is close to one of the critical values of B_1 (turning point) or B_2 (Hopf bifurcation). This is also demonstrated in Fig.4, where the Hopf bifurcation of the non–driven system is plotted as a function of the control parameter k_f (without noise). At $k_f^{crit} = 0.203s^{-1}$ the driving oscillations of B^{in} can no longer cross the bifurcation line. This scenario is substantially altered upon the inclusion of noise: the well defined bifurcation line is replaced by a bifurcation region whose precise width cannot be determined. This gives B^{in} the chance to cross the instability at even higher flow rates thereby triggering noise induced oscillations.

Nicolis and Puhl [2] have demonstrated the occurence of shifts (advancement or postponement) of bifurcation lines upon the inclusion of inhomogeneities in a reactor. Their analytical results have been obtained for a cubic autocatalysis in a CSTR. The displacement of a Hopf bifurcation as a result of external noise has also been shown for the non–driven Brusselator model by Lefever and Turner [14].

In the experimental luminol oscillator [5] sharp pulses of chemiluminescence are observed in a CSTR experiment. It is assumed that the chemiluminescence is driven by an internal species (e.g. Cu(I)OOH) [15], which oscillates periodically. The shape of the narrow light pulses and the appearance of two steady states is well reproduced by the simulated time series. The trend in the disappearance of oscillations also conforms with the experimental behavior: a significant increase of k_f^{crit} is observed upon an increase of the stirring rate in the experiment [9] which is analogous to an increase in the frequency of the fluctuations. Therefore we conclude that the effective fluctuation frequency is proportional to the stirring rate in the CSTR.

We also performed computer calculations at a constant fluctuation frequency and increasing fluctuation amplitude. However the observed effects are much smaller than those observed in the experiment and in the present calculations. Thus the experimentally observed phenomena can be modelled by a highly nonlinear trigger mechanism with the additional inclusion of fluctuations of variable frequencies at constant amplitude.

Acknowledgement: This work was supported by the Volkswagenstiftung and the Fonds der chemischen Industrie. We thank Prof. R. Seydel for generous support with his program package BIFPACK which was used for these calculations.

References

[1] A. Puhl and G. Nicolis, *Chem.Engng.Sci.* **41**, 3111 (1986).

[2] A. Puhl and G. Nicolis, *J.Chem.Phys.* **87**, 1070 (1987).

[3] M. Frankowicz, in *"Spatial Inhomogeneities and Transient Behavior in Chemical Kinetics"*, P. Gray, G. Nicolis, F. Baras, P. Borckmans and S.K. Scott (eds.), Manchester University Press, Manchester 1990, p. 705.

[4] I. Prigogine and R. Lefever, *J.Chem.Phys.* **48**, 1695 (1968).

[5] J. Amrehn, P. Resch, and F.W. Schneider, *J.Phys.Chem.* **92**, 3318 (1988).

[6] J. Amrehn, P. Resch and F.W. Schneider, in *"Spatial Inhomogeneities and Transient Behavior in Chemical Kinetics"*, P. Gray, G. Nicolis, F. Baras, P. Borckmans and S.K. Scott (eds.), Manchester University Press, Manchester 1990 p. 671.

[7] Th.–M.Kruel, A. Freund, and F.W. Schneider, *J.Chem.Phys.* **93**, 416 (1990).

[8] T. Kai and K. Tomita, *Prog.Theor.Phys.* **61**, 54 (1979).

[9] J. Amrehn, Th.–M. Kruel, and F.W. Schneider, to be published.

[10] M. Orbán, *J.Am.Chem.Soc.* **108**, 6893 (1986).

[11] A. Freund, Th.–M.Kruel, and F.W. Schneider, *Ber.Bunsenges. Phys.Chem.* **90**, 1079 (1986).

[12] A. Freund, Th.–M.Kruel, and F.W. Schneider, in *"Proceedings from MIDIT 1986 Workshop"* , edited by R.D. Parmentier and P.L. Christiansen (Manchester University, Manchester, 1987), p. 601.

[13] A. Freund, Th.–M.Kruel, and F.W. Schneider, in *"From Chemical to Biological Organization"*, edited by M. Markus, S.C. Müller, and G. Nicolis (Springer, Berlin, 1988) p. 49.

[14] R. Lefever and J.Wm. Turner, *Phys.Rev.Lett.* **56**, 1631 (1986).

[15] Y. Luo, M. Orbán, K. Kustin and I.R. Epstein, *J.Am.Chem.Soc.* **111** (13), 4541 (1989).

International Series of Numerical Mathematics, Vol. 97, © 1991 Birkhäuser Verlag Basel   27

EXAMPLES OF BOUNDARY CRISIS PHENOMENON
IN STRUCTURAL DYNAMICS

A. K. Bajaj

School of Mechanical Engineering

Purdue University

West Lafayette, IN 47907, U.S.A.

1. INTRODUCTION

The term "crisis" was introduced by Grebogi et al. [1] to describe certain sudden qualitative changes in chaotic dynamics of nonlinear dynamical systems as some control parameter is varied. These changes occur when a chaotic attractor collides with a coexisting unstable fixed point or periodic orbit, or its stable manifold. This collision may result in i) sudden changes in the size of chaotic attractors, ii) sudden appearances of chaotic attractors, or iii) sudden destructions of chaotic attractors along with their basins. As an example, consider the one-dimensional map

$$y_{n+1} = c - y_n^2 \tag{1}$$

where c is the control parameter. This map exhibits two different types of crisis as the control parameter c is varied. It is well known [1,2] that at c = -0.25, a tangent bifurcation results in the creation of an unstable and a stable fixed point (denoted $\bar{y}_u$ and $\bar{y}_s$, respectively). The stable fixed point or period-1 solution $\bar{y}_s$ undergoes a period-doubling bifurcation cascade which accumulates at $\bar{c} = 1.40095...$ The unstable period-1 solution $\bar{y}_u$ forms the boundary of the basin of attraction for the stable solutions branch. As c is increased further, the chaotic attractor, arising at $c = \bar{c}$, undergoes a crisis at $c = c^* = 2$ where it collides with the unstable period-1 solution $\bar{y}_u$ on its basin boundary. This is called a boundary or exterior crisis. Beyond c*, the chaotic attractor is unstable and for all initial states, the system ultimately diverges to minus infinity. However, for (c - c*) << 1, if an orbit is started with initial conditions in the region where previously the chaotic attractor existed, the orbit may persist for a long time near the former chaotic attractor and this phenomenon is called transient chaos.

The one-dimensional map (1) exhibits also another type of crisis which arises in the region of well known periodic windows. The periodic windows are generated by a tangent bifurcation and undergo a period-doubling cascade resulting in a multiband chaotic attractor. Beyond some critical value $c = \hat{c}$, each multiband chaotic attractor enlarges to a single, broadband attractor.

The phenomenon of sudden enlargement of a chaotic attractor is called interior crisis and results when an unstable periodic solution collides with a coexisting chaotic attractor.

Crises are very common in nonlinear systems and can occur whenever a chaotic attractor competes with another attractor. This attractor can be a fixed point, a limit cycle, or another chaotic attractor. For systems governed by nonlinear differential equations, the maps are usually higher dimensional than one and then the crises arise due to heteroclinic or homoclinic tangencies [3]. Thus, a chaotic attractor becomes tangent to the stable manifold of a saddle-type fixed point or periodic orbit, which defines the basin boundary of the chaotic attractor. In many cases, the tangency occurs simultaneously with the collision of the attractor with the saddle itself [2].

Many examples of systems exhibiting crises have been reported and studied in the literature. These include forced damped pendulum [3], Lorenz equations [4], Josephson junction [5], Duffing's equation [6], and Belousov - Zhabotinskii reaction [7]. There are very few examples of systems of dimension > 3 because of the obvious difficulties associated with constructing stable manifolds of saddle points.

In the present work, we report on the occurrence of "boundary" crisis in examples from resonant motions of weakly nonlinear structural systems and give numerical evidence for it. One problem considered is that of the harmonically excited nonplanar vibrations of stretched strings [8]. A one-mode truncation of the partial differential equations, represented by two nonlinear coupled oscillators, is studied using the method of averaging as well as by direct numerical integration. The averaged equations, with different nonlinear coefficients, also govern the resonant dynamics of beams [9], spherical pendulum [10], as well as axisymmetric shells [11]. The averaged system is found to possess complicated dynamics including isolated and Hopf bifurcating limit cycle branches, chaotic attractors, and the phenomenon of "boundary" crisis. A qualitative discussion of the stable and unstable manifolds of singular points along with the computation of chaotic attractors is used to show the occurrence of crisis whereby, at lower damping levels, the chaotic attractors suddenly disappear and do not exist over a wide excitation frequency interval. This phenomenon is shown to also exist for the other physical systems governed by the averaged equations. "Crisis" is also verified for the truncated string equations.

2. EQUATIONS OF MOTION AND AVERAGING

The nondimensional equations for one-mode truncation of the partial differential equations governing the dynamics of the string can be shown [8] to be given by

$$z_1{}'' + z_1 = \hat{\varepsilon}\{2\cos\tau - 2\alpha z_1{}' - 2\beta z_1{}'' - 4z_1(z_1^2 + z_2^2)\}\,, \tag{2}$$

$$z_2{}'' + z_2 = \hat{\varepsilon}\{-2\alpha z_2{}' - 2\beta z_2{}'' - 4z_2(z_1^2 + z_2^2)\}\,, \tag{3}$$

where z_1 and z_2 represent, respectively, the amplitude of the first spatial mode in the plane of external harmonic forcing, and in the plane orthogonal to it. The normalized parameters α and β represent, respectively, the damping, and the frequency detuning between the excitation and the linear modal frequency of the first mode. These parameters are normalized with respect to the amplitude of harmonic forcing, and $\hat{\varepsilon}$, the small parameter for asymptotic analysis, is a measure of this forcing. For a complete definition of the parameters and the derivation of equations (1) - (2), the reader should refer to [8].

The coupled oscillators (2) - (3) can be studied by the method of averaging for $0 < \hat{\varepsilon} \ll 1$. Let

$$z_i(\tau) = p_i(\tau)\cos\tau + q_i(\tau)\sin\tau\,, \tag{4}$$
$$z_i'(\tau) = -p_i(\tau)\sin\tau + q_i(\tau)\cos\tau\,, \quad i = 1,2\,.$$

Substituting (4) in (2)-(3) and averaging the resulting equations over the period 2π, gives the averaged equations

$$
\begin{aligned}
\dot{p}_1 &= -\alpha p_1 - (\beta + EA/2)\,q_1 + BMp_2\,, \\
\dot{q}_1 &= -\alpha q_1 + (\beta + EA/2)p_1 + BMq_2 + 1\,, \\
\dot{p}_2 &= -\alpha p_2 - (\beta + EA/2)q_2 - BMp_1\,, \\
\dot{q}_2 &= -\alpha q_2 + (\beta + EA/2)\,p_2 - BMq_1\,,
\end{aligned}
\tag{5}
$$

where $E = (p_1^2 + q_1^2 + p_2^2 + q_2^2)$, $M = p_1q_2 - p_2q_1$, and where A and B denote the nonlinear coefficients characterizing the different physical systems. For example, A and B take values (A,B) = (-3, 1), (0.25, -0.75) and (3.491, -5.958) for the string [8], the spherical pendulum [10] and the beam [9], respectively.

3. SOLUTIONS OF THE AVERAGED EQUATIONS

In order to trace the sequence of bifurcations which ultimately leads to a crisis, we present the solutions of the averaged equations in the form of amplitude response curves as a function of the frequency β, for various fixed values of the damping α. We concentrate here on parameter values representing the string system; the qualitative picture, however, applies to the whole family of systems. Note that constant solutions of the averaged equations represent harmonic motion for the string. Dynamic (limit cycles etc.) solutions of the averaged equations imply amplitude-and phase-modulated motions for the string.

For large damping ($\alpha > 0.991$), only the constant solution with $p_2 = q_2 = 0$ exists and it is single-valued for all frequencies β. The string therefore has a unique steady-state periodic response in the plane of excitation and this is called the planar response. For $\alpha < 0.991$, the planar response becomes multi-valued in the frequency interval (β_2, β_6). Two stable steady-state motions are now possible for the same excitation frequency. The response curve for $\alpha = 0.7$ and the qualitative phase portrait for β between β_2 and β_6 are shown in Figure 1. The stable manifold of the middle saddle-type constant solution determines the domains of attraction of the upper and the lower stable solutions.

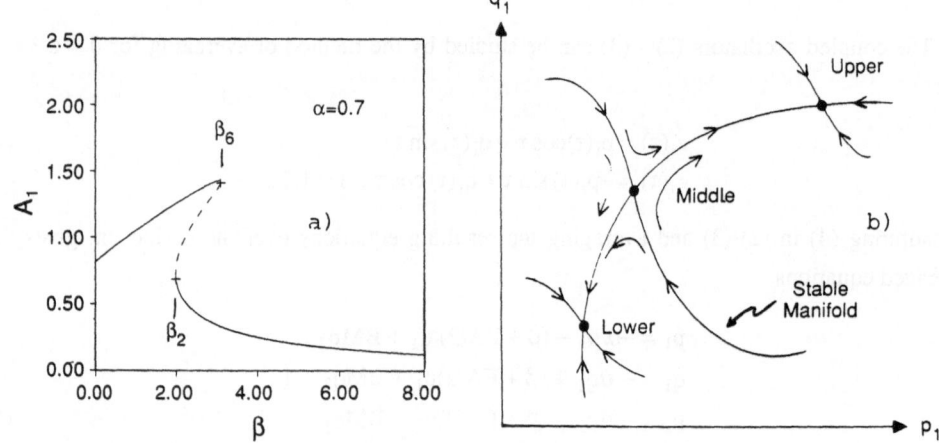

Figure 1. a) Constant amplitude response curve for $\alpha = 0.7$. b) Qualitative phase
portrait in (p_1, q_1) plane for $\beta_2 < \beta < \beta_6$.

The non-planar branch of solutions ($p_2, q_2 \neq 0$) is known [8] to exist provided $\alpha < 0.687$. This branch arises from the single mode solutions due to pitchfork bifurcations at frequencies β_1 and β_5. In fact, there are two non-planar solution branches due to symmetry of the governing equations (4). For sufficiently large damping ($\alpha > 0.477$), the non-planar branch is single valued although the non-planar constant solution undergoes a Hopf bifurcation at frequencies β_{1*} and β_{2*} for damping $\alpha < 0.577$. Thus, the averaged equations are expected to possess limit cycle solutions near β_{1*} and β_{2*}. Figure 2 shows the amplitude response curve for $\alpha = 0.513$ along with the qualitative phase portrait in the (p_2, q_2) plane, for a frequency β lying between β_{1*} and β_{2*}. Note that the origin in the (p_2, q_2) plane represents all the three planar solutions, although the unstable manifolds asymptotic to the limit cycles around non-planar constant solutions correspond to those for the upper planar solution. The stable manifold of the upper planar solution determines, in part, the domains of attraction for the stable limit cycles. In fact, as the frequency β is varied, the limit cycles around the non-planar constant solutions undergo period-doubling bifurcations resulting ultimately in chaotic attractors. Then, the unstable manifolds for the upper planar solution are asymptotic to the chaotic attractors.

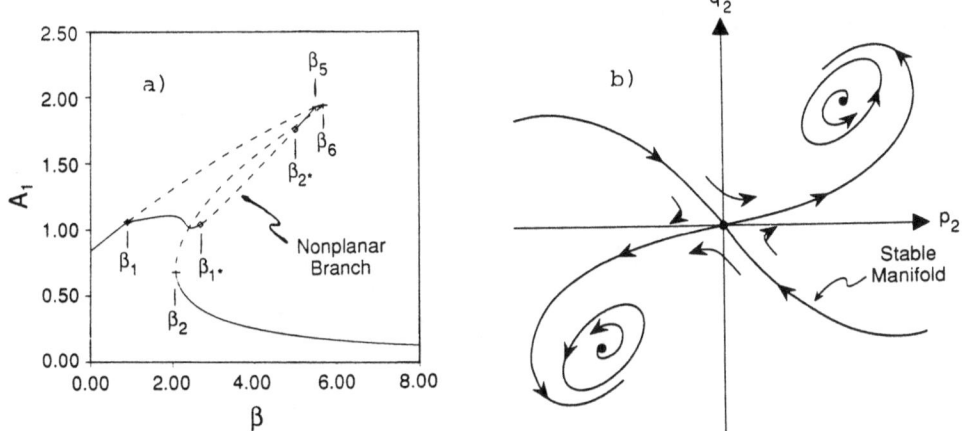

Figure 2. a) Constant amplitude response curve for α = 0.513. b) Qualitative phase portrait in (p_2, q_2) plane for $\beta_{1*} < \beta < \beta_{2*}$.

On decreasing the damping slightly, the dynamic solutions around each of the unstable non-planar solutions merge via a homoclinic orbit and now result in a chaotic attractor of the Lorenz type, encircling the two non-planar solutions and the upper planar solution. Recall that the domain of attraction of this solution is determined by the stable manifold of the middle saddle-type planar solution and the other coexisting stable solution is the lower planar constant solution. An example of the chaotic attractor of the Lorenz type for damping α = 0.50 is shown in Figure 3. As the damping is reduced further, it turns out that the Lorenz type attractors (periodic, chaotic etc.) abruptly dissappear over a frequency interval. The only stable solution found in this frequency interval is the lower planar constant solution. This phenomenon is first observed for α = 0.495 and can be seen to be a "boundary" crisis or a heteroclinic bifurcation. As already discussed in the introduction, a crisis occurs when the chaotic attractor touches its basin boundary. In the present context, this should mean that the Lorenz type chaotic attractor comes into contact with the stable manifold of the middle planar constant solution. In fact, in the present case, the attractor touches the saddle point itself. After a crisis has occured and the attractor destroyed, trajectories that are started in the region will remain, for a finite time, in the vicinity where the attractor existed (ghost of the chaotic attractor) but will eventually lead to the lower planar fixed point. Figure 4a shows the Lorenz type chaotic attractor and the other relevant singular points at a frequency β, $\beta = 3.80$, just prior to the "boundary" crisis. The attractor comes very close to the middle planar saddle point. At a slightly higher freqeuncy $\beta = 3.89$, the attractor has grown sufficiently to collide with the saddle so that, as shown in Figure 4b, the attractor like object is only a transient and the solution ultimately converges to the lower planar constant solution.

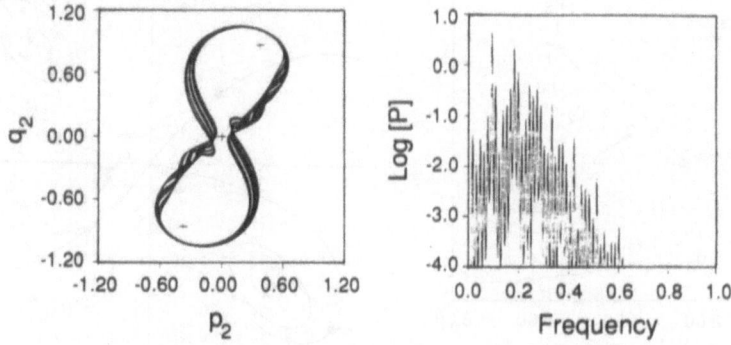

Figure 3. Lorenz type chaotic attractor and its spectrum for $\alpha = 0.5$, $\beta = 3.80$.

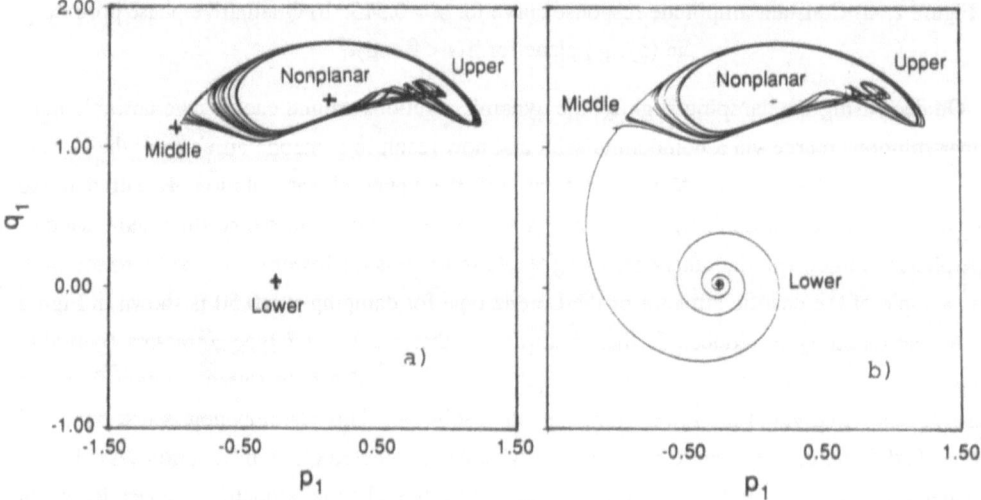

Figure 4. a) The chaotic attractor ($\beta = 3.80$), and b) the transient chaos ($\beta = 3.89$), immediately before and after the "boundary" crisis in the string system for $\alpha = 0.495$.

4. CRISIS IN OTHER SYSTEMS

As noted earlier, the averaged equations for the string differ only in the nonlinear coefficients A and B from the averaged equations of many other physical systems, including the spherical pendulum, and a beam. Many of the important features of the solutions for the string, including the crisis phenomenon, are found for every member of the family.

Typical constant amplitude response curves for the spherical pendulum are shown in Figure 5. The value of damping α, in Figure 5, has been chosen to correspond with a damping value used by Miles [10]. There are several important differences between the response curve for the spherical pendulum and the response curves for the string system. One clear difference is in the nature of non-linear coefficients, A and B, such that for the pendulum, the backbone curve for the planar branch and the backbone curve for the non-planar branch bend in different directions. As a result, when the non-planar solutions become unstable by Hopf bifurcation, there is an interval in frequency, $(\beta_2 - \beta_{1*})$, where there are no stable constant solutions. In this interval, the only attractors that exist are limit cycles and chaotic attractors. Almost every choice of initial conditions for frequencies in the interval $(\beta_2 - \beta_{1*})$ leads to periodic or chaotic solutions.

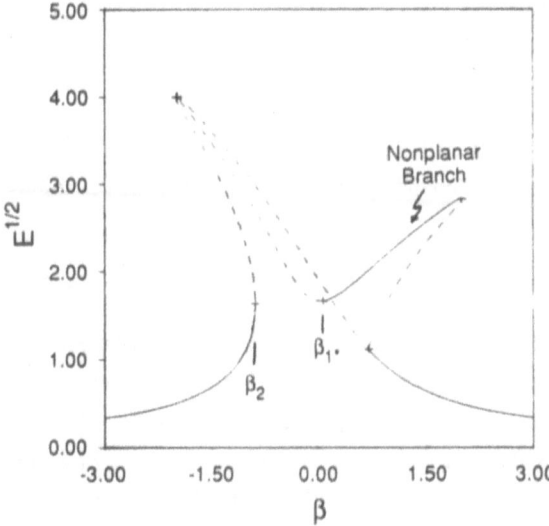

Figure 5. Constant amplitude response curve for $\alpha = 0.25$ for the spherical pendulum.

For the values of α chosen, the solution branch emanating from the Hopf bifurcation at β_{1*} quickly gives rise to a chaotic attractor as β is reduced. This attractor grows rapidly as the turning point β_2 is approached (see Figure 5). Reducing β past β_2, the lower and the middle planar fixed points are created. The crisis occurs very near the turning point in the constant amplitude response curves. Figure 6a shows the Lorenz type attractor and relevant fixed points for Miles' pendulum equations just prior to the "boundary" crisis. Figure 6b shows the transient behavior immediately after the chaotic attractor is destroyed by a crisis.

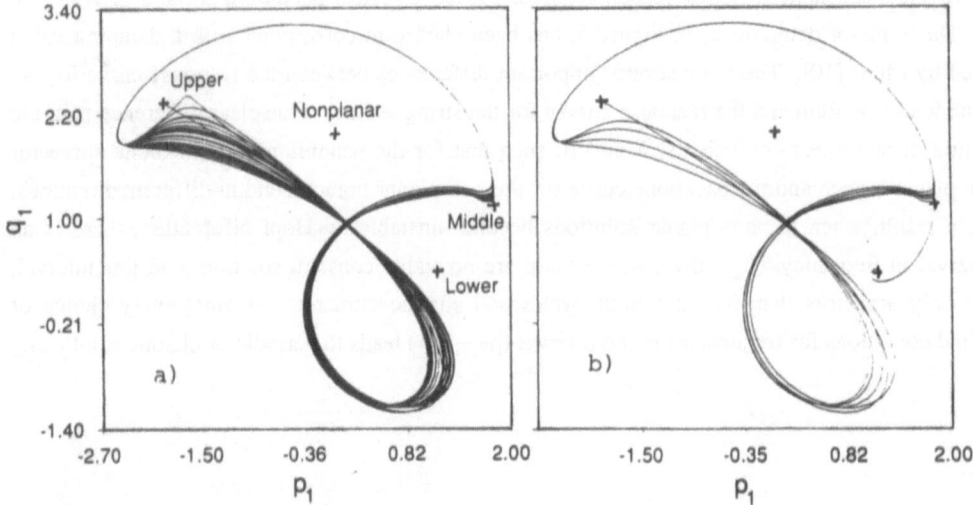

Figure 6. a) The chaotic attractor ($\beta = -0.967$), and b) the transient chaos ($\beta = -0.969$), immediately before and after the "boundary" crisis for the spherical pendulum for $\alpha = 0.25$.

5. CRISIS IN COUPLED OSCILLATORS AND CONCLUDING REMARKS

The chaotic attractor in the averaged equations was shown to undergo a "boundary" crisis, in which the attractor touches its basin boundary and is destroyed. For $\varepsilon = 0.1$, where $\hat{\varepsilon} = (10\varepsilon)^{1/3}/4$, the nonautonomous system governed by equations (2) - (3) also exhibits a "boundary" crisis. Figure 7a shows, for $\beta = 4.5$, the Poincare' section of a Lorenz type chaotic attractor along with the relevant fixed points (i.e. the 2π-period solutions) of the Poincare' map just prior to the crisis. Figure 7b shows for only a slightly different β, $\beta = 4.49$, the Poincare' section of the transient chaos once the chaotic attractor has been destroyed. Here the crisis arises due to the collision of the chaotic attractor with the period-1 saddle-type middle planar solution as well as its stable manifold. Figure 8 shows the corresponding time history of the non-planar component, z_2, of the motion of the string. It clearly demonstrates the dramatic effect that a crisis has on the dynamics of the system.

Summarizing, we have investigated the dynamic behavior of whirling motions of a nonlinear string forced harmonically in a plane. Strong numerical evidence is provided to suggest that the amplitude-modulated chaotic motions predicted by the averaged equations are destroyed by a "crisis". Very similar behavior is found in the original coupled oscillators, testifying to the usefulness of asymptotic analysis even for very complex dynamical motions.

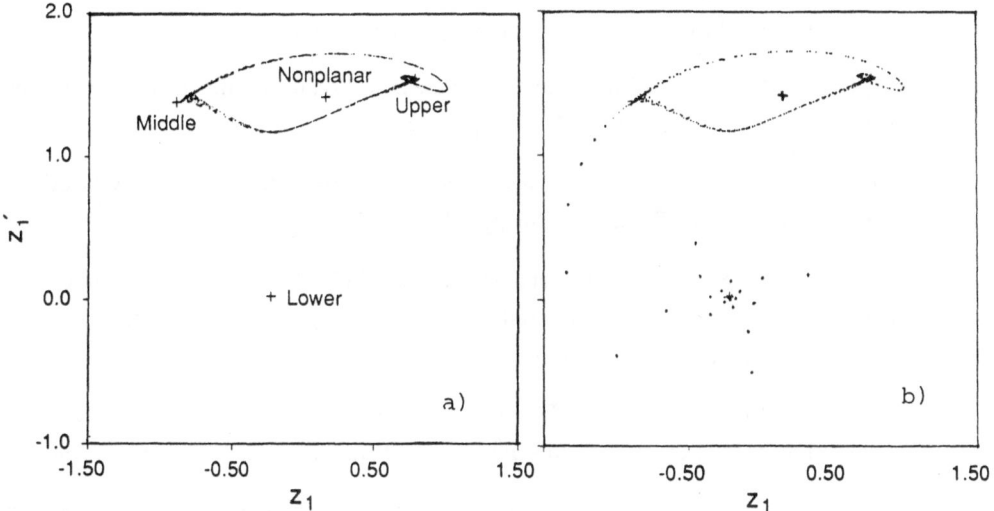

Figure 7. "Crisis" in forced coupled osicllators for $\alpha = 0.5$. a) Poincare' section of the chaotic attractor ($\beta = 4.50$), b) Poincare' section displaying transient chaos ($\beta = 4.49$).

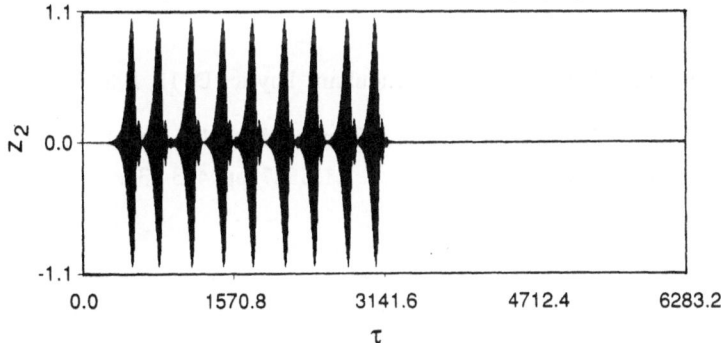

Figure 8. Time response of the out-of-plane component displaying transient chaos for
$\alpha = 0.5, \beta = 4.49$.

REFERENCES

1. Grebogi, C., Ott, E. and Yorke, J. A., Crises, sudden changes in chaotic attractors, and transient chaos. Physica D, 7, 1983, pp. 181-200.

2. Thompson, J.M.T. and Stewart, H.B., Nonlinear dynamics and chaos. John Wiley, New York, 1986.

3. Grebogi, C., Ott, E. and Yorke, J.A., Critical exponent of chaotic transients in nonlinear dynamical systems. Phys. Rev. Lett., 57, 1986, pp. 1284-1287.

4. Yorke, J.A. and Yorke, E.D., Matastable chaos: the transition to sustained chaotic behavior in the Lorenz model. J. Stat. Phys., 21, 1979, pp. 263-277.

5. Gwinn, E.G. and Westervelt, R.M., Fractal basin boundaries and intermittency in driven damped pendulum. Phys. Rev., 33, 1986, pp. 4143-4155.

6. Ueda, Y., Yoshida, S., Stewart, H.B. and Thompson, J.M.T., Basin explosions and escape phenomena in the twin-well Duffing oscillator: compound global bifurcations organizing behavior. Phil. Trans. R. Soc. Lond. A., 332, 1990, pp. 169-186.

7. Richetti, P., DeKepper, P., Roux, J.C. and Swinney, Harry L., A crisis in Belousov-Zhabotinskii reaction: experiment and simulation. J. Stat. Phys., 48, 1987, pp. 977-990.

8. Johnson, J.M. and Bajaj, A.K., Amplitude modulated and chaotic dynamics in resonant motion of strings. J. Sound and Vib., 128, 1989, pp. 87-107.

9. Maewal, A., Chaos in a harmonically excited elastic beam. J. Appl. Mech., 53, 1986, pp. 625-632.

10. Miles, J., Rosonant motion of a spherical pendulum. Physica D, 11, 1984, pp. 309-323.

11. Maewal, A., Miles' evolution equations for axisymmetric shells: simple strange attractors in structural dynamics. Int. J. Non-Lin. Mech., 21, 1987, pp. 433-438.

International Series of Numerical Mathematics, Vol. 97, © 1991 Birkhäuser Verlag Basel

Bifurcation, Pattern Formation, and Transition to Chaos in Combustion

A. Bayliss and B. J. Matkowsky

Dept. of Engineering Sciences and Applied Mathematics
Northwestern University
Evanston, IL 60208, U. S. A.

1. Introduction. We study problems in the dynamics and pattern formation of both solid and gaseous fuel combustion. Our goal is to understand the basic mechanisms affecting the combustion process, which is a necessary prerequisite to controlling it. A related goal is to describe the spatio-temporal patterns that are observed experimentally as well as to predict new dynamical behavior, as yet unobserved. Thus we wish to determine the mechanisms of successive transitions to increasingly complex patterns of dynamical behavior, as parameters of the problem are varied. The problems are formulated as initial boundary-value problems for systems of highly nonlinear partial differential equations modeling the relevant combustion process.

We employ both analytical and numerical approaches in our investigations, as well as the interaction of the two approaches. In combustion problem, the activation energies of the chemical reactions that occur, are typically large. As a result, the spatial region in which the chemical reactions occur to a significant extent, are thin, layer type regions termed reaction zones. In the limit of infinite activation energy, the reaction zone shrinks to a (in general moving) surface termed a flame front (flame sheet model) [10, 29]. In this case, the distributed Arrhenius reaction rates become localized reaction rates on the surface, given by a surface δ function whose strength is calculated by the method of matched asymptotic expansions [10, 11, 29]. The analytical investigations are carried out for the case of infinite activation energy, while the numerical investigations are carried out for finite, though large, activation energies. A high resolution of the reaction zones is necessary to accurately describe the structure and dynamics of the solution. Since the location of the reaction zone is not know a-priori, but in fact moves during the course of combustion, in a possibly oscillatory manner, the task of its accurate resolution is a challenge to numerical methods. Inaccurate resolution in the calculation can lead to numerical instabilities, and more perniciously, to spurious and incorrect predictions of dynamical behavior. To meet this challenge, we introduced an adaptive pseudo-spectral method [2] - [8]. It has been successful in describing the experimentally observed, and analytically described patterns appearing during combustion. In addition, it has enabled us to predict new, as yet unobserved dynamical patterns.

2. Numerical Method. We employ the Chebyshev pseudo-spectral method in non-periodic directions and the Fourier pseudo-spectral method in periodic (e.g.

angular) directions. In spectral methods (see e.g. [12, 19]) the solution, u, is expanded as a finite sum of basis functions

$$u \simeq u_J \equiv \sum_{j=0}^{J} a_j T_j(x)$$

where T_j is the jth Chebyshev polynomial. The expansion coefficients a_j are obtained from collocation, that is the function u_J is forced to solve the equation at a set of $J + 1$ points x_j, the collocation points.

The major advantage of pseudo-spectral methods over finite difference methods is enhanced accuracy for a fixed discretization size. In fact pseudo-spectral methods exhibit infinite order accuracy when used to approximate infinitely differentiable functions, that is, the error can be shown to decrease faster than any inverse power of J. However, although these methods are highly accurate when used to approximate functions which exhibit relatively gradual spatial variation, they have difficulties in approximating functions exhibiting localized regions of rapid variation, such as the temperature rise across the narrow reaction zone. Severe spatial oscillations can occur in approximating rapidly varying functions which are not well resolved. These oscillations can affect the dynamics and in certain cases lead to the computation of spurious dynamics (i.e. inadequately resolved computations can indicate chaos whereas the exact solution is in fact periodic).

If the location of the reaction zone is known in advance, the oscillations can be eliminated by introducing a suitable change of coordinates so that in the new coordinate system the solution has a more gradual variation. However, typically the location of the reaction zone is not known in advance and is one of the objects of the computation. In order to realize the benefits of the pseudo-spectral method in computing rapidly varying solutions to combustion problems, we developed an adaptive pseudo-spectral method which has proven to be particularly effective in computing rapidly varying functions. We introduce a family of coordinate transformations of the form

(1) $x = q(s, \vec{\alpha})$

where s is the new computational coordinate and $\vec{\alpha}$ represents a parameter vector which is typically of dimension one or two. We choose $\vec{\alpha}$ so that in the new coordinate system the solution exhibits a more gradual variation and thus is better approximated by a small number of basis functions. Since the behavior of the solution changes during the course of the computation, appropriate values of $\vec{\alpha}$ must be chosen dynamically so as to adapt to changes in the solution [2, 4].

In order to adaptively choose a coordinate transformation in which the spectral approximation is more accurate it is necessary to develop error measures. The error measures are computed for each value of $\vec{\alpha}$ until a minimum is found. For spectral methods based on expansions in Chebyshev polynomials we have derived and employed a functional which measures the numerical error, which is an integral functional of the solution and its derivatives [2]. This adaptive procedure has been extended to higher dimensional transformations [3].

We now describe some of the results we have obtained.

3. Gasless Condensed Phase Combustion.

Gasless condensed phase combustion is being studied as a new and innovative method for the fabrication of high tech ceramic and metallic materials. Thus the goal of combustion in this case is not merely the production of heat, but rather the fabrication of special materials which have desirable characteristics such as being especially hard, especially pure and being impervious to very high temperatures.

In this method, pioneered by A. G. Merzhanov and colleagues, a cylindrical sample consisting of a compacted powder mixture is ignited at one end. A thermal wave then propagates through the sample, converting unburned reactants to products. The solid reactant is converted directly to a solid product, without the prior formation of a gaseous phase.

We consider a reaction front, sometimes referred to as a solid flame, propagating through a cylindrical sample. Experiments have revealed a variety of modes of propagation of the thermal wave (see e.g. [30] and the references therein). In addition to the uniformly propagating planar front, there have been observations of (i) pulsating combustion, in which a planar front propagates with an oscillatory velocity, (ii) spin combustion, in which one or more hot spots (luminous points) are observed to move in a helical motion along the surface of the sample, and (iii) multiple point combustion, in which the hot spots appear, disappear, and reappear repeatedly. Experiments indicate that the burning may occur either throughout the sample, or only on the surface and not in its interior, and that sometimes, though not always, melting of one or more of the reactants occurs.

The nondimensional model we consider is given by

$$
(2) \qquad \Theta_t = \phi_t \Theta_z + \nabla^2 \Theta + \left(\frac{1}{\alpha(1+\gamma)} \right) \Lambda Y \exp\left(-\frac{N(1-\sigma)(\Theta-1)}{\sigma+(1-\sigma)\Theta} \right),
$$

$$
Y_t = \phi_t Y_z - \left(\frac{1}{\alpha} \right) \Lambda Y \exp\left(-\frac{N(1-\sigma)(\Theta-1)}{\sigma+(1-\sigma)\Theta} \right),
$$

where

$$
\left(\begin{array}{c} a \\ b \end{array} \right) = \left\{ \begin{array}{ll} a, & z < 0 \\ b, & z > 0. \end{array} \right.
$$

Here Θ and Y denote nondimensional temperature and mass fraction respectively, N is the nondimensional activation energy, σ is the ratio of the unburned to burned temperature, $z = x_3 - \phi(x_1, x_2, t)$ is a coordinate moving with the front ϕ, and Λ, the solid flame speed eigenvalue, is determined from the solution corresponding to the uniformly propagating planar front. The model accounts for melting, which causes the reaction rate to increase by the factor $\alpha > 1$, due to the increased surface to surface contact caused by melting. The condition at the melting surface is

$$(1 + \nabla \phi^2)[\Theta_z] + \gamma Y(t, z = 0)\phi_t = 0.$$

where $[q]$ denotes the jump in the quantity q across the melting surface, and γ is the heat of fusion. The boundary conditions are

$$Y \to 1, \quad \Theta \to 0 \quad \text{as } z \to -\infty,$$
(3)
$$\Theta \to 1 \quad \text{as } z \to \infty.$$

We considered this problem in both one dimension and on the two-dimensional surface of a cylinder. In the surface combustion problem we impose the periodicity conditions

$$g(\psi + 2\pi) = g(\psi)$$

where g denotes Θ, Y, and ϕ, and ψ is the angular coordinate.

Matkowsky and Sivashinsky [28] modeled and analyzed the one-dimensional case, employing a flame sheet model without melting. They showed that the pulsating mode arose as a Hopf bifurcation from the uniformly propagating planar front, as a critical value μ_c of the parameter $\mu = \Delta/2$ where $\Delta = N(1 - \sigma)$, was exceeded. Above μ_c the uniform solution is unstable and perturbations evolve to the bifurcated state, i.e. to the pulsating propagating state. The analysis is in qualitative agreement with experimental results, which showed that mixtures with low activation energies ($\mu < \mu_c$) propagated in the uniformly propagating mode, while mixtures with higher activation energies ($\mu > \mu_c$) propagated in the pulsating mode, and that the average velocity in the pulsating mode was less than that of the uniformly propagating mode.

To determine the global behavior of the system, beyond the local description given by bifurcation theory, we employed our adaptive pseudo-spectral method [4, 6, 7]. This method has been employed for models both with [21], and without melting [28]. In both cases we found the numerically predicted critical value μ_c in very good agreement with the value determined analytically from the flame sheet model. For both models we showed how the sinusoidal oscillations predicted by bifurcation theory developed into relaxation oscillations as μ was increased. For the melting model, the definition of μ is modified to $\mu = \Delta/(2(1 - M))$ where the melting parameter M is defined as $M = (1 - (1 + \gamma)/\alpha)\exp(\Delta(\Theta_m - 1))$, and Θ_m denotes the melting temperature [21]. The nonmelting model corresponds to $M = 0$ ($\alpha = 1, \gamma = 0$).

Transitions to chaos occur for both the melting and nonmelting models. However the route to chaos is different for each of the models. For the melting model, as μ was increased, we found three distinct windows of both singly (T) and doubly ($2T$) periodic solutions (see Fig. 1). Beyond the third $2T$ window solutions exhibit intermittency, where a gradual drift away from the doubly periodic solution is interrupted by randomly occurring bursts. Upon further increasing μ the bursts occur more frequently until apparently fully chaotic solutions occur. These solutions are illustrated in Figs. 2, 3, and 4 respectively. The power spectral density (PSD) has been computed for these solutions, and exhibits broad band spectral content.

The behavior of the nonmelting model is quite different. As μ was increased, we observed a sequence of successive period doublings. Upon increasing μ yet further, apparently fully chaotic solutions were found.

Analytical studies of the problem in which burning occurs on the surface of a cylindrical sample, and throughout the sample, appeared in [9, 16, 22, 23]. In these papers, with melting taken into account, solutions describing pulsating combustion, spinning combustion and multiple point combustion were presented.

We have studied the problem of surface combustion numerically, employing a two-dimensional adaptive pseudo-spectral method [3]. We have computed a standing wave solution branch describing multiple point combustion, for the melting model. As μ is increased the solution evolves from nearly sinusoidal behavior to one exhibiting relaxation oscillations in time and a strong localization in the angular variable, describing the luminous point. Upon increasing μ further a low frequency envelope appears and the solution appears to be quasi-periodic.

4. Gaseous Combustion.
There have been numerous experimental observations of spatio/temporal patterns in gaseous fuel combustion (see e.g. [14, 17, 18, 24]). Our objective is to determine the different patterns which can occur and to describe transitions between such patterns.

We first consider the structure and dynamics of flames stabilized on a line source (corresponding to a point source in two dimensions) of fuel. The diffusional-thermal model is described by a system of transport equations in polar coordinates r and ϕ. The source of fuel is located at $r = 0$. The unknowns are the non-dimensionalized temperature Θ, and non-dimensionalized concentration C of a deficient component of the fuel mixture. Upon suitably non-dimensionalizing, the model is given by [27]

$$(4) \qquad \Theta_t = \nabla^2\Theta - \frac{\kappa\Theta_r}{r} + C\Lambda \exp\left(\frac{N(1-\sigma)(\Theta-1)}{\sigma+(1-\sigma)\Theta}\right),$$

$$C_t = \frac{\nabla^2 C}{L} - \frac{\kappa C_r}{r} - C\Lambda \exp\left(\frac{N(1-\sigma)(\Theta-1)}{\sigma+(1-\sigma)\Theta}\right).$$

The Lewis number L is the ratio of thermal diffusivity to mass diffusivity, the strength of the source is $2\pi\kappa$, σ is the ratio of the unburned to burned temperature, N is a nondimensional activation energy, and $\Lambda = M^2/(2L)$, where $M = N(1-\sigma)$. We note that Λ, which is referred to as the flame speed eigenvalue, depends on the non-dimensionalization. The boundary conditions are

$$C \to 1, \qquad \Theta \to 0 \text{ as } r \to 0,$$
$$C \to 0, \qquad \Theta \to 1 \text{ as } r \to \infty.$$

The solution to (4) has been studied analytically in the limit $M \to \infty$ [15, 27]. In this limit the reaction zone shrinks to a surface $r = \Psi(\phi)$, called the flame front.

The following stationary, axisymmetric solution exists:

$$\Theta = \begin{cases} (\frac{r}{\kappa})^{\kappa} + O(\frac{1}{M}), & r \leq \kappa, \\ 1, & r \geq \kappa, \end{cases}$$

$$C = 1 - \Theta + O(\frac{1}{M}),$$

$$\Psi = \kappa,$$

and is referred to as the basic solution.

Linear stability analysis of this problem, as well as other problems in gaseous combustion, indicates that two types of instabilities can occur. The first, referred to as the cellular instability, occurs if L is less than a critical value $L_c < 1$. Here a single real eigenvalue crosses into the right half plane. Crossing this stability threshold leads to a transition to a stationary cellular flame, in which peaks and troughs appear on the front. The peaks point toward the combustion products, and the troughs are convex towards the fresh fuel mixture. The flame appears brighter at the troughs (which are hotter) and darker at the peaks. This regime describes, e.g. the combustion of rich hydrocarbon/air mixtures.

The second instability, referred to as the pulsating instability, occurs if L is greater than a critical value $L_{c_2} > 1$. Here a pair of complex conjugate eigenvalues crosses into the right half plane. Crossing this stability threshold in the axisymmetric case, leads to a transition to a pulsating flame, in which the flame position oscillates in time about a mean position given by the basic solution. In the nonaxisymmetric case, cells appear on the front as its position oscillates in time leading e.g. to traveling waves, standing waves or quasi-periodic waves. The regime $L > 1$ describes, e.g. the combustion of lean hydrocarbon/air mixtures.

A linear stability analysis of this problem [15], indicated that the effect of increasing κ, which is inversely proportional to the curvature of the flame front, was destabilizing. In [27], a nonlinear analysis for fixed $L < 1$, employing κ as a bifurcation parameter was carried out, and a local description of stationary cellular flames in a neighborhood of the bifurcation point was given.

In [8] we numerically determined a cascade of bifurcations from a stationary axisymmetric solution to stationary cellular solutions with increasing angular mode number, as the bifurcation parameter κ was increased, with L fixed at a value < 1. We also observed regions of bistability in which two solutions co-existed, each with its own domain of attraction. A summary of these results is shown in Fig. 5 where a bifurcation diagram of stationary cellular solution branches is plotted.

As L is decreased with κ held fixed we found dynamic behavior in the cellular regime. In particular we found transitions from a stationary axisymmetric flame, to a stationary four mode cellular flame, to a four mode cellular flame with a very slowly traveling wave (TW) along the flame front [5]. The latter transition appears to arise from an infinite period, symmetry breaking bifurcation from the stationary cellular solution branch at $L = L_*$. That is, the stationary cellular flame enjoys symmetry with respect to both rotations and reflections, while the TW solution breaks the reflection symmetry. In Fig. 6, we plot the speed S, of the slowly traveling wave vs. L and note that S decays to zero as $(L_* - L)^{1/2}$. We note that some of the four mode solutions in the vicinity of L_* appear to be unstable, while further away from L_* they are stable. We present contour plots of Θ for the case of a stable stationary four mode cellular solution in Fig. 7 and for a stable spinning (TW) four mode cellular solution in Fig. 8. The symmetry breaking is apparent. Such bifurcations are sometimes referred to as bifurcations from a group orbit (see e.g. [13]).

Upon decreasing L further we found a transition to mixed mode solutions exhibiting apparently quasi-periodic dynamics. In these solutions the transverse behavior of the solution is composed of a combination of mode numbers 4 and 3. The predominant mode number is 4, which is also the mode number of the pure traveling waves found for slightly larger values of L. A plot of Θ for the mixed mode solution is shown in Fig. 9. We note that a predominant 4 mode pattern is visible, however the presence of additional incommensurate modes is clear. The PSD of the temperature at one fixed point in space is shown in Fig. 10. indicating the presence of the two incommensurate frequencies and multiples of their sums and differences.

In the regime $L > 1$ we considered both the axisymmetric and the nonaxisymmetric problem. In the axisymmetric case we employed σ as the bifurcation parameter. Upon increasing σ we found several period doublings, apparently describing a period doubling cascade to chaos.

In the non-axisymmetric case we studied the effect of varying κ and N. Upon increasing κ we found a sequence of modal transitions between traveling and standing wave solutions (see Fig. 11). We found transitions from a seven cell standing wave to an eight cell traveling wave to a nine cell traveling wave. These transitions are accompanied by intervals of bistability. As N is increased, for a fixed value of κ, we found transitions between standing waves and traveling wave solutions. In addition the cells become more sharply defined and localized in space. A plot of Θ for one such value is shown in Fig. 12. This solution corresponds to a seven cell spinning (TW) flame.

We now consider freely propagating flames in the parameter regime $L > 1$. In [25, 26] we considered the problem in one and two dimensions respectively. We identified a Hopf bifurcation point at $L = L_c > 1$, and described the bifurcation of a pulsating propagating solution (a solution with a planar front whose propagation velocity was oscillatory in time) in one dimension, and a pulsating propagating solution with a traveling wave along its front in two dimension. The latter analysis however was restricted to consideration of the nonlinear evolution of the single most dangerous mode k_0 (the mode whose growth rate was maximum). No consideration was given to the continuous band of wave numbers to which the basic solution (the planar front propagating with constant velocity) was unstable.

When these, as well as the effects of heat losses, are taken into account, we find that a system of modified complex Ginzburg-Landau equations results. More specifically the flame front ϕ is given by

$$\phi = -mt + \epsilon[R_1(\tau_1, \tau_2, \eta)e^{i(\omega_0 t + k_0 y)} + R_2(\tau_1, \tau_2, \eta)e^{i(\omega_0 t - k_0 y)} + \text{c.c.} - m\psi] + O(\epsilon^2),$$

where m is a solution of $m = e^{-H/m^2}$, H denotes the heat loss coefficient, $\omega_0 = \omega(k_0)$ is the frequency of the most dangerous mode (calculated from the dispersion relation $\omega = \omega(k)$), and ϵ measures the amount by which L exceeds the bifurcation point L_c. The complex amplitudes R_i of the traveling waves along the front, are functions of the slow time variables $\tau_j = \epsilon^j t$ $(j = 1, 2)$, and the slow spatial variable $\eta = \epsilon y$. In terms of the coordinates $\eta_{1,2} = \eta \pm \omega_0' \tau_1$, where $\omega_0' = \omega'(k) \mid_{k=k_0}$, the amplitudes $R_{1,2}(\tau_2, \eta_1, \eta_2)$ satisfy

(5)
$$\frac{\partial R_1}{\partial \eta_2} = \frac{\partial R_2}{\partial \eta_1} = 0,$$

and the modified complex Ginzburg Landau equations

(6)
$$\frac{\partial R_1}{\partial \tau_2} = AR_1 + B\frac{\partial^2 R_1}{\partial \eta_1^2} + (C + ik_0 D)R_1 \mid R_1 \mid^2$$
$$+ (E - ik_0 D)R_1 \langle \mid R_2 \mid^2 \rangle,$$

(7)
$$\frac{\partial R_2}{\partial \tau_2} = AR_2 + B\frac{\partial^2 R_2}{\partial \eta_2^2} + (C + ik_0 D)R_2 \mid R_2 \mid^2$$
$$+ (E - ik_0 D)R_2 \langle \mid R_1 \mid^2 \rangle,$$

where the averaged terms are defined by

$$\langle f \rangle = \lim_{T \to \infty} \frac{1}{2T} \int_{-T}^{T} f(\tau_2, z) \, dz,$$

and ψ, which describes the shift in the position of the front, is determined from $R_{1,2}$.

The dependence of the coefficients A, B, C, D, E on the physico-chemical parameters of the problem has been determined in [31] where the equations are derived and then analyzed to uncover interesting dynamical behavior. We remark that variants of equations (5)-(7) were obtained by Alvarez-Pereira and Vega for the case $H = 0$, and by Knobloch and DeLuca who assumed that $R_{1,2}$ were periodic, and used symmetry arguments to obtain the form of the equations, without determining the coefficients in terms of the parameters of a specific problem [20].

Acknowledgement. We are pleased to acknowledge helpful conversations with J. M. Vega and C. Alvarez-Pereira who directed our attention to the significance of the averaged terms in (6)-(7). This paper summarizes BJM's invited lecture at the Conference on Bifurcation and Chaos: Analysis, Algorithms, Applications, held in Wurzburg, FRG, August 1990. This research was supported by NSF grant ASC 87-19573 and DOE grant DEFG02-87ER25027.

References

[1] C. Alvarez-Pereira and J. M. Vega, to appear.

[2] A. Bayliss, D. Gottlieb, B. J. Matkowsky and M. Minkoff, J. Comput. Phys. **81** (1989), 421-443.

[3] A. Bayliss, R. Kuske and B. J. Matkowsky, J. Comput. Phys. **91** (1990), 174-196.

[4] A. Bayliss and B. J. Matkowsky, J. Comput. Phys. **71** (1987), 147-168.

[5] A. Bayliss and B. J. Matkowsky, Appl. Math. Letters. **3** (1990), 75-79.

[6] A. Bayliss and B. J. Matkowsky, SIAM J. Appl. Math. **50** (1990), 437-459.

[7] A. Bayliss, B. J. Matkowsky and M. Minkoff, SIAM J. Appl. Math. **49** (1989), 1047-1063.

[8] A. Bayliss, B. J. Matkowsky and M. Minkoff, SIAM J. Appl. Math. **49** (1989), 1421-1432.

[9] M. R. Booty, S. B. Margolis and B. J. Matkowsky, SIAM J. Appl. Math. **48** (1988), 828-853.

[10] J. D. Buckmaster and G. S. S. Ludford, *Theory of Laminar Flames,* Cambridge University Press, Cambridge, U.K. (1983).

[11] J. D. Buckmaster and G. S. S. Ludford, *Lectures on Mathematical Combustion,* CBMS-NSF Regional Conference Series in Applied Mathematics, No. 43, SIAM, Philadelphia, PA (1983).

[12] C. Canuto, M. Y. Hussaini, A. Quarteroni and T. A. Zang, *Spectral Methods in Fluid Dynamics,* Springer-Verlag, NY (1987).

[13] J. D. Crawford and E. Knobloch, Physics Letters A 128 (1988), 327-331.

[14] M. El-Hamdi, M. Gorman, J. W. Mapp and J. I. Blackshear Jr., Combust. Sci. and Tech. **55** (1987), 33-40.

[15] M. Garbey, H. A. Kaper, G. K. Leaf and B. J. Matkowsky, Quarterly of Appl. Math. **47** (1989), 691-704.

[16] M. Garbey, H. A. Kaper, G. K. Leaf and B. J. Matkowsky, European J. Appl. Math. **1** (1990), 73-89.

[17] I. M. Gololobov, E. A. Granovskii and Yu. A. Gostintsev, Combustion, Explosion and Shock Waves **17** (1981), 22-26.

[18] M. Gorman, M. el-Hamdi and K. A. Robbins, Forum on Chaos, ASME V. 90, ed. K. N. Ghia (1990).

[19] D. Gottlieb and S. A. Orszag, *Numerical Analysis of Spectral Methods: Theory and Applications*, CBMS-NSF Regional Conference Series in Applied Mathematics, SIAM, Philadelphia (1977).

[20] E. Knobloch and J. DeLuca, preprint.

[21] S. B. Margolis, SIAM J. Appl. Math. **43** (1983), 331-369.

[22] S. B. Margolis, H. G. Kaper, G. K. Leaf, and B. J. Matkowsky, Combust. Sci. and Tech. **43** (1985), 127-165.

[23] S. B. Margolis and B. J. Matkowsky, SIAM J. Appl. Math. **48** (1988), 828-853.

[24] G. H. Markstein, ed., "Nonsteady Flame Propagation", Pergamon Press, Elmsford, NY (1967).

[25] B. J. Matkowsky and D. O. Olagunju, SIAM J. Appl. Math. **39** (1980), 290-300.

[26] B. J. Matkowsky and D. O. Olagunju, SIAM J. Appl. Math. **42** (1982), 486-501.

[27] B. J. Matkowsky, L. J. Putnick and G. I. Sivashinsky, SIAM J. Appl. Math. **38** (1980), 489-504.

[28] B. J. Matkowsky and G. I. Sivashinsky, SIAM J. Appl. Math. **35** (1978), 230-255.

[29] B. J. Matkowsky and G. I. Sivashinsky, SIAM J. Appl. Math. **37** (1979), 686-699.

[30] A. G. Merzhanov, Arch. Combustionis **1** (1981), 23-48.

[31] D. O. Olagunju and B. J. Matkowsky, to appear.

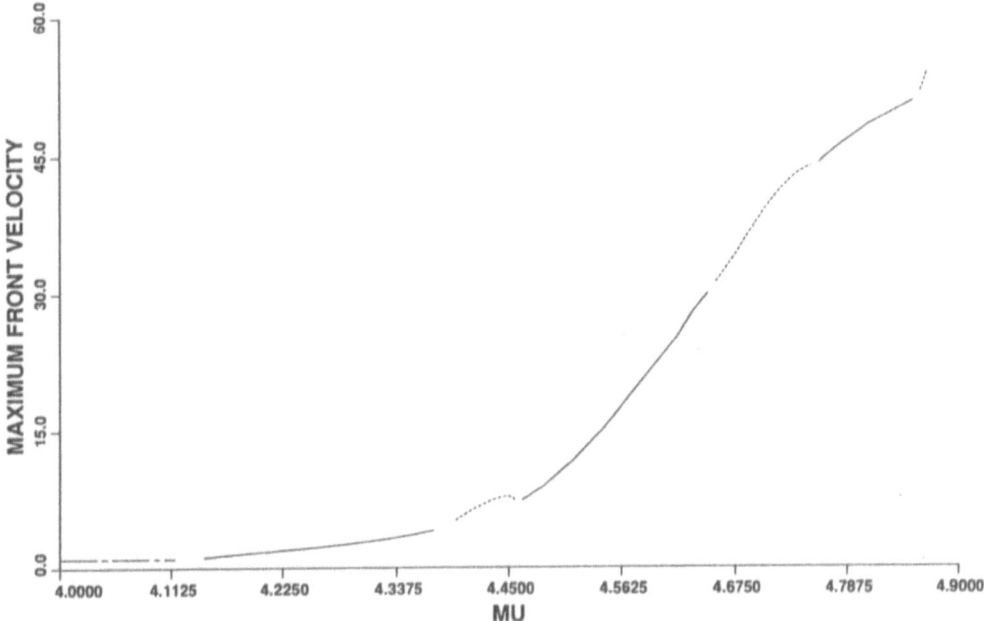

Figure 1. Maximum $|\phi_t|$ for various solution branches. Uniformly propagating solution branch $(- \cdot -\cdot)$, T branches (———), $2T$ branches ($\cdots\cdots$).

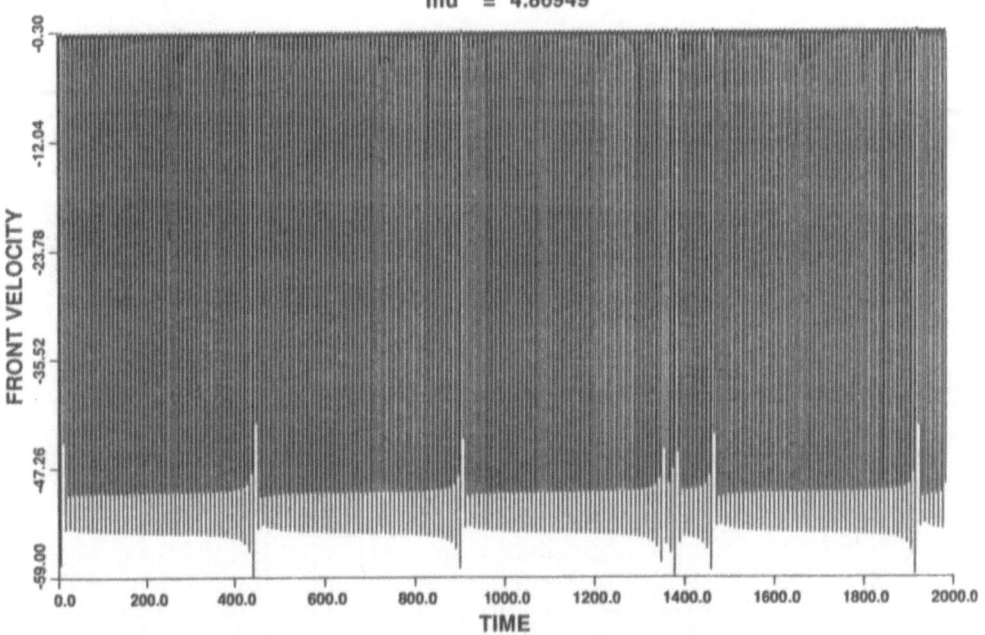

Figure 2. ϕ_t exhibiting intermittency with long laminar regions.

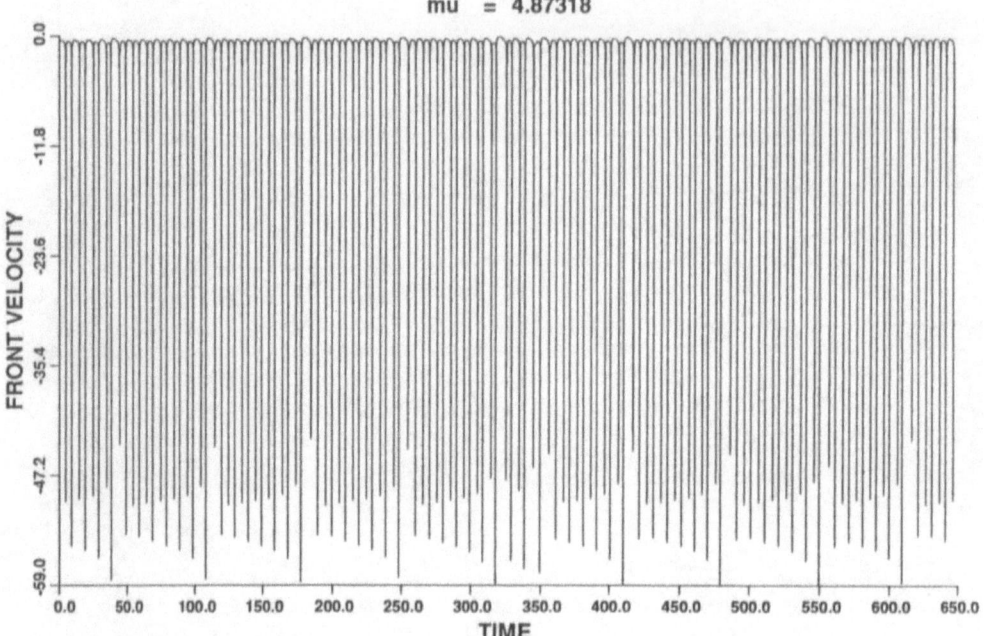

Figure 3. ϕ_t exhibiting intermittency with shorter laminar regions.

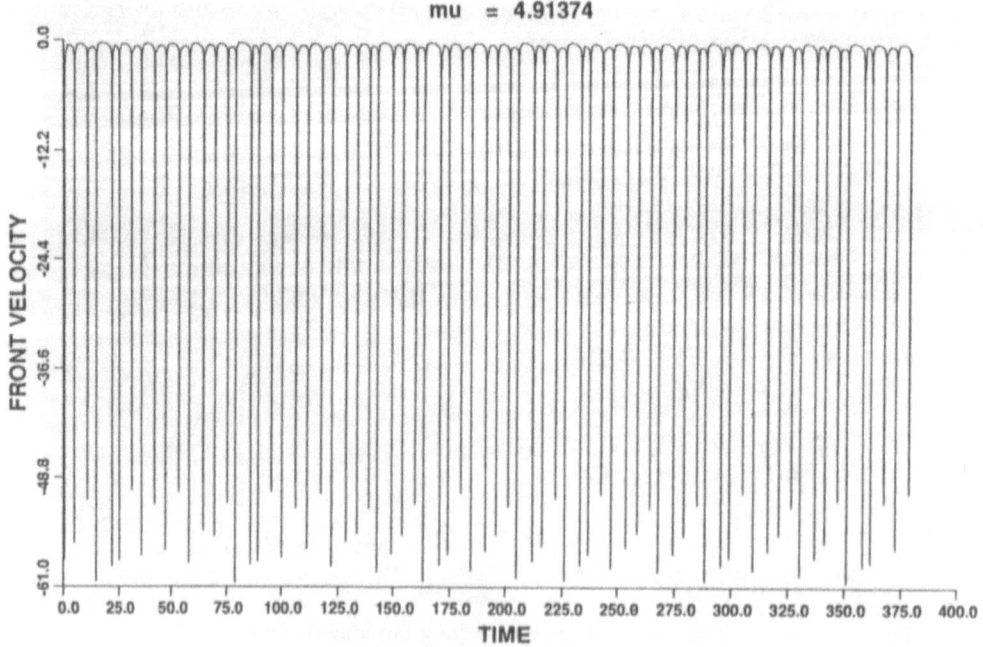

Figure 4. ϕ_t exhibiting apparently chaotic behavior.

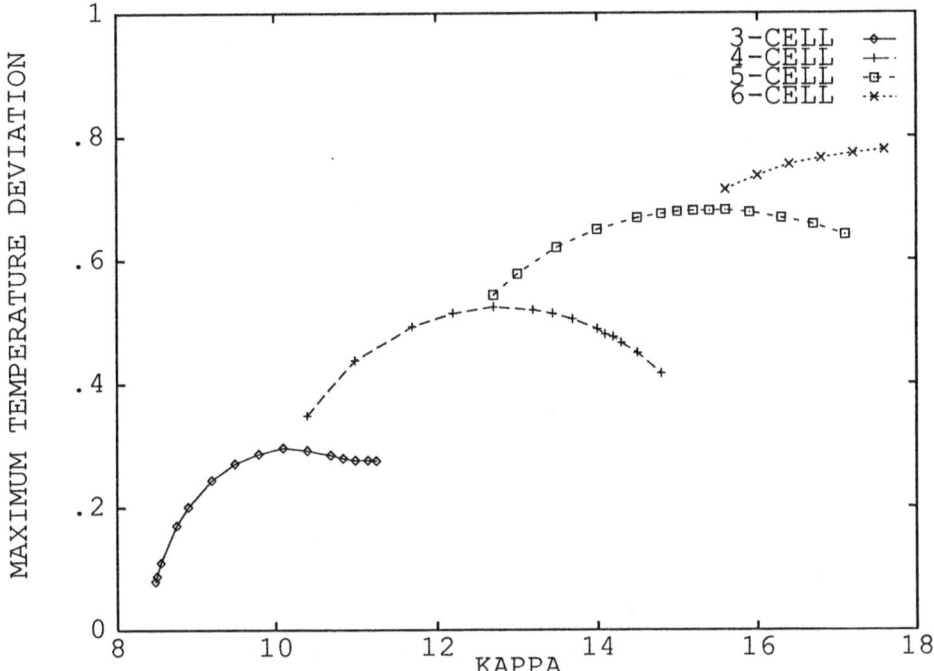

Figure 5. Stationary cellular solution branches.

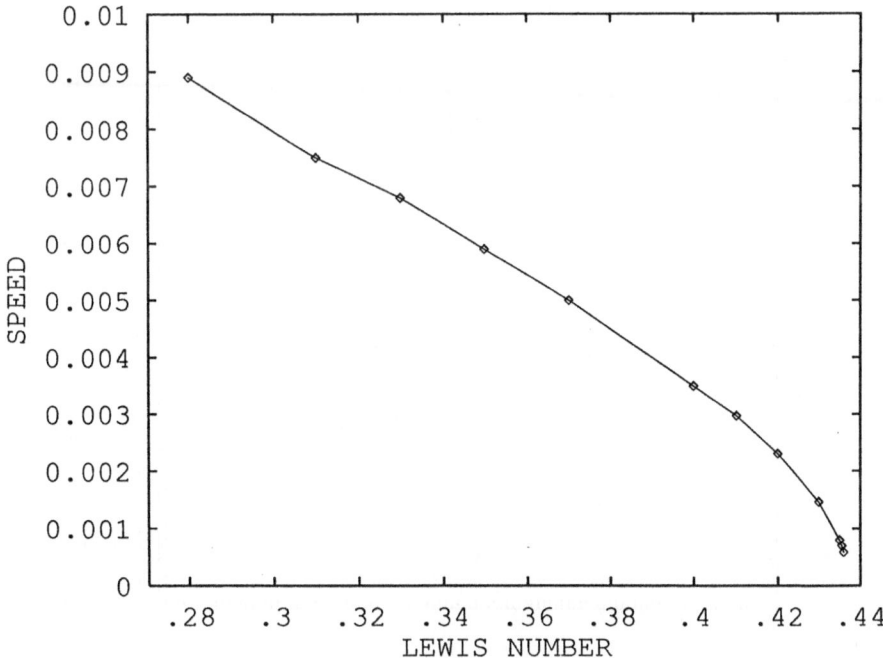

Figure 6. Speed of spinning cellular solutions.

A. Bayliss and B. J. Matkowsky

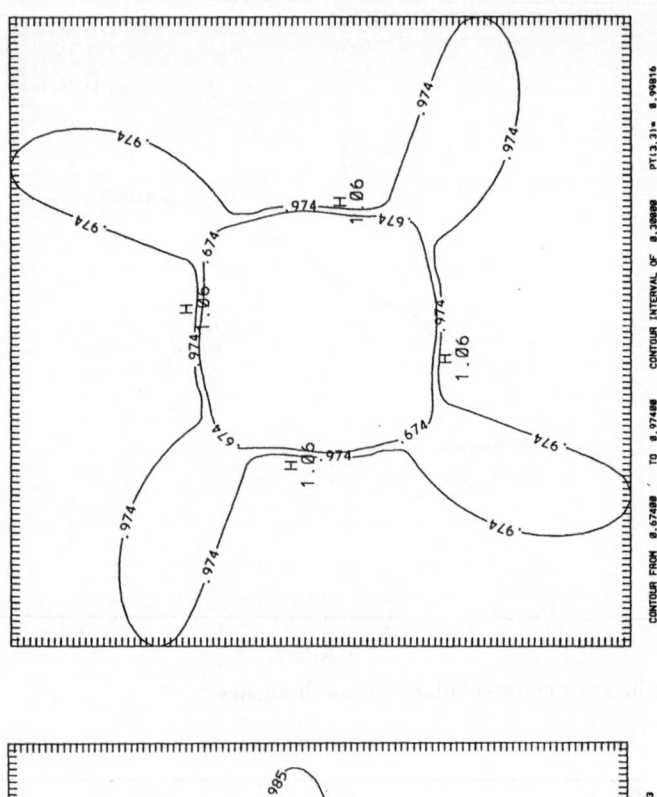

CONTOUR FROM 0.67400 TO 0.97400 CONTOUR INTERVAL OF 0.30000 PT(3,3)= 0.99916

Figure 8. Contour plot of spinning cellular solution, $L = .37$.

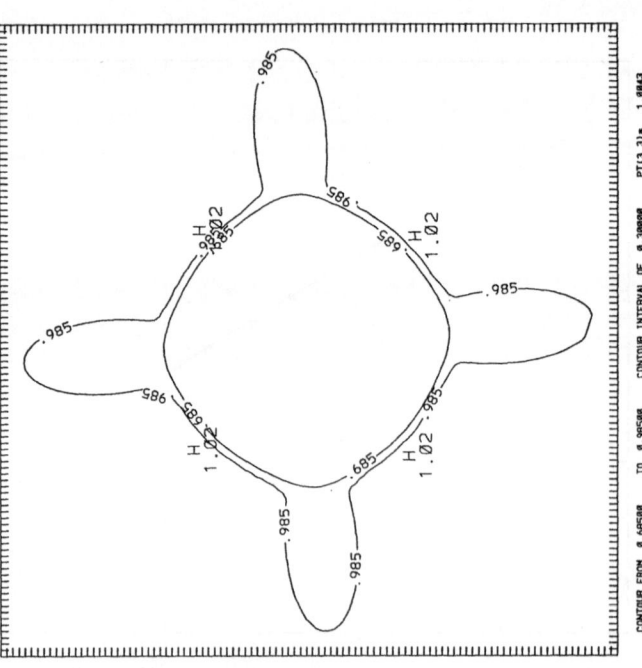

CONTOUR FROM 0.68500 TO 0.98500 CONTOUR INTERVAL OF 0.30000 PT(3,3)= 1.0043

Figure 7. Contour plot of stationary cellular solution, $L = .47$.

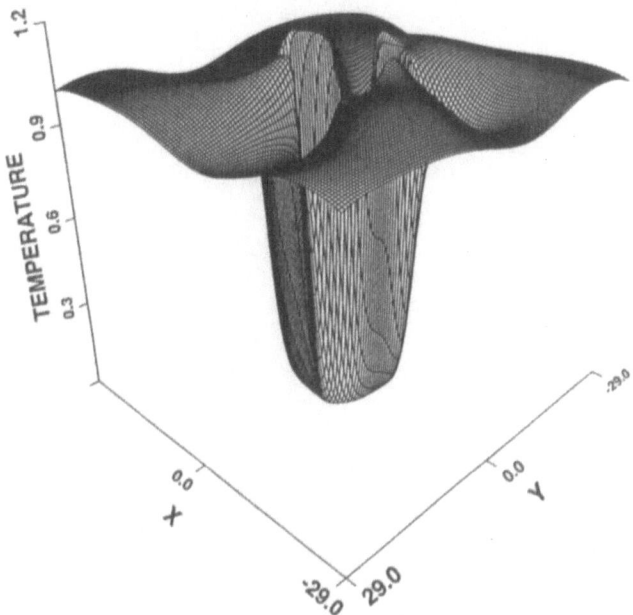

Figure 9. Θ for mixed mode (quasi-periodic) solution.

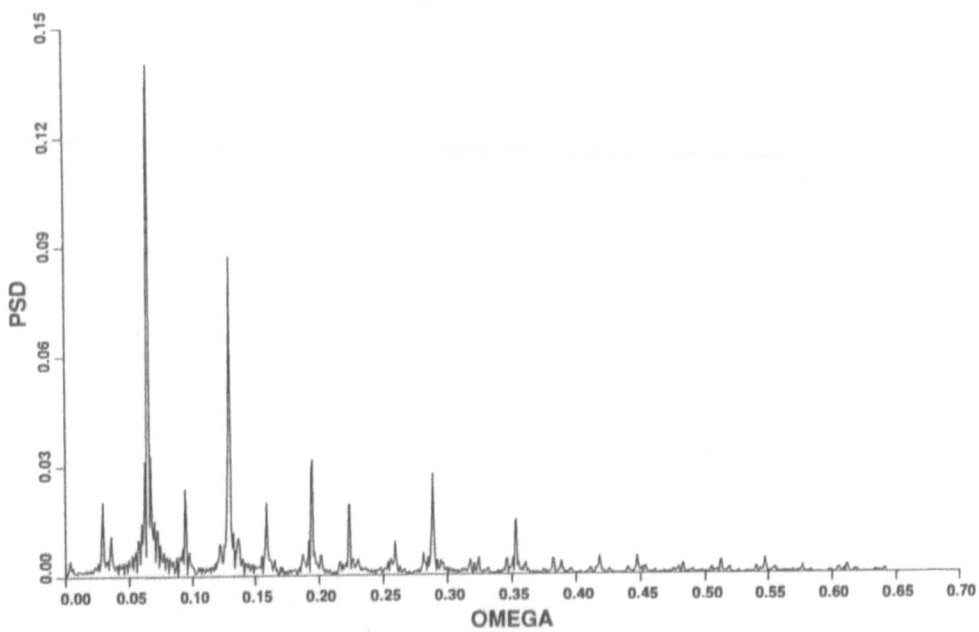

Figure 10. PSD of Θ at a fixed spatial point for mixed mode solution.

A. Bayliss and B.J. Matkowsky

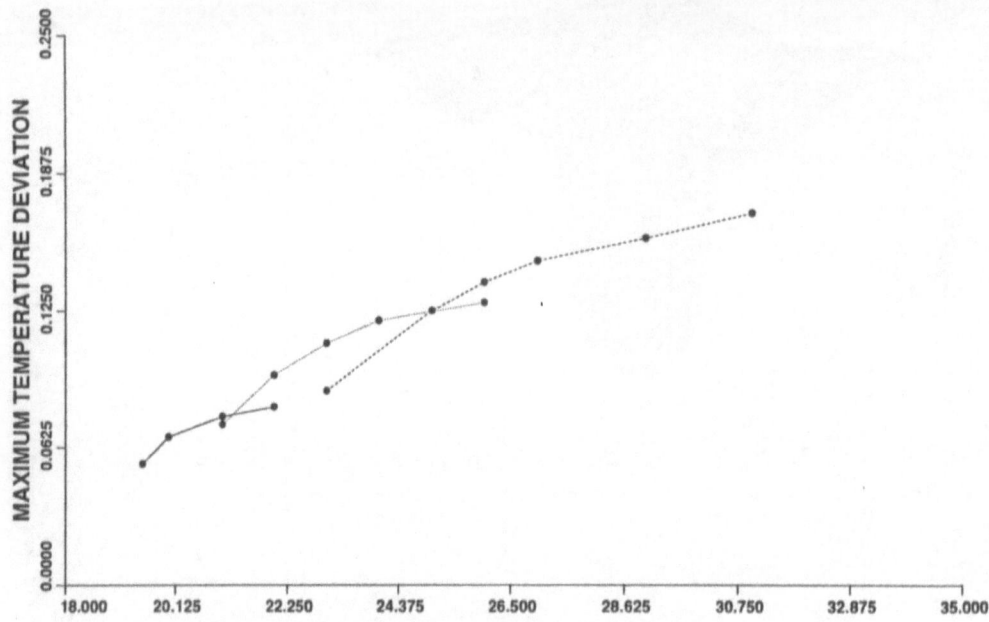

Figure 11. Unsteady solution branches for $L > 1$. Seven mode standing wave
(———), eight mode TW(⋯⋯), nine mode TW (– – –).

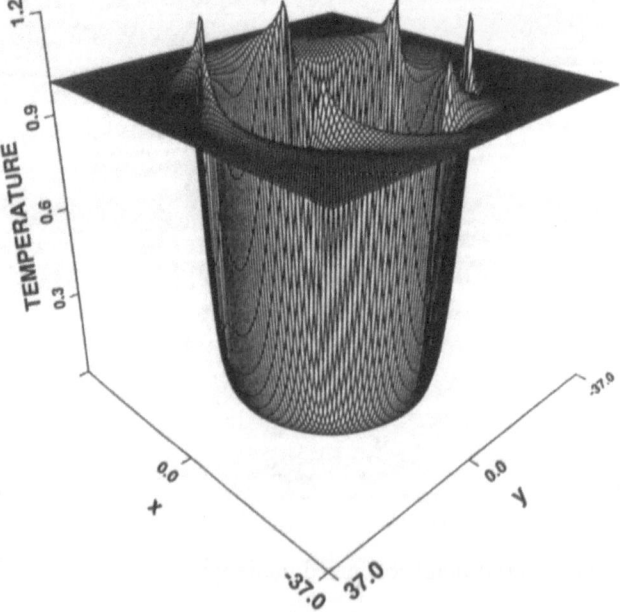

Figure 12. Θ for seven-cell spinning wave, exhibiting sharply defined cells.

International Series of Numerical Mathematics, Vol. 97, © 1991 Birkhäuser Verlag Basel

ON THE PRIMARY AND SECONDARY BIFURCATION OF EQUATIONS INVOLVING SCALAR NONLINEARITIES

N.W.Bazley and N.X.Tan

Institute of Mathematics, University of Cologne, West Germany

Introduction

It is well-known that many problems of quantum mechanics and chemical physics etc. can be described by eigenvalue equations involving scalar nonlinearities of the form

$$g\{\tfrac{1}{2}\langle Bu, u\rangle\}Bu = \lambda u, (\lambda, u) \in R \times X. \tag{1}$$

Here , g is a differentiable function with g^{-1} continuous and $g^{-1}(1) = 0, g^{-1}(x) > 0$ for $x > 1$ (or $g^{-1}(x) > 0$ for $x < 1$) , B is a self-adjoint linear continuous operator and $\langle .,.\rangle$ is the inner product in a Hilbert space X , R is the space of real numbers . The operator $N(u) = g\{\tfrac{1}{2}\langle Bu, u\rangle\}Bu$ is called a scalar nonlinearity . This notion was introduced by Bazley and Kupper [1] and independently by Medeiros [4] . In what follows we consider the primary and secondary bifurcation, the stability of primary bifurcation solutions from a simple eigenvalue and the unstability of secondary bifurcation solutions from a double eigenvalue of the above equation. Let $\bar{\lambda} > 0$ be an eigenvalue of the operator B with multiplicity p . We suppose that the nullspace of the operator $B - \bar{\lambda}id$:

$$Ker(B - \bar{\lambda}id) = [v^1, ..., v^p]$$

with $v^1, ..., v^p$ linearly independent vectors in X and $[v^1, ..., v^p]$ denoting the space spanned by $v^1, ..., v^p$. Without loss of generality we may assume $\langle v^i, v^j\rangle = \delta_{ij}, i, j = 1, ..., p$, with δ_{ij} denoting the Kronecker delta . First , we shall show that $(\bar{\lambda}, 0)$ is always a primary bifurcation solution of (1) and there exist at least two distinct parameter families in the case $p = 1$ and a continuum of parameter families of nontrivial solutions in a neighborhood of $(\bar{\lambda}, 0)$ in the case $p \geq 2$. Furthermore, we can describe these parameter families in an analytical form $(\lambda, u(\lambda)), \lambda > \bar{\lambda}$, (or $\lambda < \bar{\lambda}$) with $u(\lambda) = \sum_{j=1}^{p} \alpha_j v^j$, where $\alpha_j = \alpha_j(\lambda), j = 1, ..., p$ and $\sum_{j=1}^{p} \alpha_j^2 = \tfrac{2}{\lambda} g^{-1}(\tfrac{\lambda}{\bar{\lambda}})$. Let $(\lambda, u(\lambda)), \lambda > \bar{\lambda}$, be such a family . We shall prove that if $p = 1$, then for any $\lambda > \bar{\lambda}, (\lambda, u(\lambda))$ is not a secondary bifurcation solution of (1). However , if $p \geq 2$, then for any $\lambda_1 > \bar{\lambda}, (\lambda_1, u(\lambda_1))$ is a secondary bifurcation solution of Equation (1). Moreover, we can also describe parameter families of nontrivial solutions in a neighborhood of $(\lambda_1, u(\lambda_1))$. Further , we shall consider the stability of the primary bifurcation solutions $(\lambda, \pm u(\lambda)) = (\lambda, \pm(\tfrac{2}{\lambda} g^{-1}(\tfrac{\lambda}{\bar{\lambda}}))^{\frac{1}{2}} v^1)$ in the case $p = 1$. These solutions are stable if $g'\{g^{-1}(\tfrac{\lambda}{\bar{\lambda}})\} < 0$ when λ near $\bar{\lambda}$. The unstability of secondary bifurcation solutions $(\lambda, \pm v(\lambda))$ with

$$v(\lambda) = u(\lambda_1) + (\tfrac{2}{\lambda}(g^{-1}(\tfrac{\lambda}{\bar{\lambda}}) - g^{-1}(\tfrac{\lambda_1}{\bar{\lambda}}))^{\frac{1}{2}} w^1), \lambda \in R_{\lambda_1},$$

in the case $p = 2$ will be investigated in the conclusion of Section 2.

In Section 3 , we shall consider an example in elliptic partial differential equations concerning scalar nonlinearities , where our results obtained in Section 2 are very effective applicable.

2. The main results

In this section we shall prove some new results on the primary and secondary bifurcation, the stability of primary bifurcation solutions and on the unstability of secondary bifurcation solutions of Equation (1). For the sake of simplicity of notation , we only investigate the case

$g^{-1}(x) > 0$ for $x > 1$ (the other case would be studied similarly). Now , let $\bar{\lambda} \in R, \bar{\lambda} > 0$ be an eigenvalue of the operator B with multiplicity $p \geq 1$ and corresponding eigenvectors $v^1, ..., v^p, \langle v^i, v^j \rangle = \delta_{ij}, i, j = 1, ..., p$, i.e;

$$Ker(B - \bar{\lambda}id) = [v^1, ..., v^p].$$

Further , we define the multivalued mappings $I : R \mapsto R^p$ and $C : R \mapsto X$ by

$$I(\lambda) = \{\alpha = (\alpha_1, ..., \alpha_p) \in R^p / \textstyle\sum_{j=1}^{p} \alpha_j^2 = \frac{2}{\lambda} g^{-1}(\frac{\lambda}{\lambda})\}$$

and

$$C(\lambda) = \{u \in X / \exists \alpha = (\alpha_1, ..., \alpha_p) \in I(\alpha) : u = \textstyle\sum_{j=1}^{p} \alpha_j v^j \}.$$

For an arbitrary set D we denote by $|D|$ the number of elements in D and by R_β the set $\{\lambda \in R, \lambda > \beta\}$.

We have : for all $\lambda \in R_{\bar{\lambda}}$

$$|I(\lambda)| = |C(\lambda)| = \kappa = \begin{cases} 2, & \text{if } p = 1; \\ +\infty, & \text{if } p \geq 2. \end{cases}$$

Next , the result on the primary bifurcation of Equation (1) can be formulated as follows :

Theorem 1. Let $\bar{\lambda}, v^1, ..., v^p$ and κ be as above . Then $(\bar{\lambda}, 0)$ is a primary bifurcation solution of Equation (1) and there exist at least κ distinct parameter families of nontrivial solutions in a neighborhood of $(\bar{\lambda}, 0)$. Moreover , for any $\lambda \in R_{\bar{\lambda}}$, we can find a nontrivial solution $(\lambda, u(\lambda))$ with $u(\lambda) = \sum_{j=1}^{p} \alpha_j v^j$ and $\alpha = (\alpha_1, ..., \alpha_p) \in I(\lambda)$.

Further, let $\lambda \in R_{\bar{\lambda}}$ and $u(\lambda) \in C(\lambda)$. As was been shown in the proof of Theorem 2 , $(\lambda, u(\lambda))$ satisfies Equation (1) . We now consider the equation

$$F(\lambda, v) = g\{\tfrac{1}{2}\langle B(u(\lambda) + v), u(\lambda) + v \rangle\}B(u(\lambda) + v) - \lambda(u(\lambda) + v) = 0,$$

$$(\lambda, v) \in R_{\bar{\lambda}} \times X. \tag{2}$$

It is clear that $F(\lambda, 0) = 0$ for all $\lambda \in R_{\bar{\lambda}}$. If $(\lambda_1, 0)$, for some $\lambda_1 \in R_{\bar{\lambda}}$ is a primary bifurcation solution of Equation (2) , we say that $(\lambda_1, u(\lambda_1))$ is a secondary bifurcation solution of Equation (1). Any point $(\lambda, u(\lambda)), \lambda \in R_{\bar{\lambda}}$, is called a trivial solution and the other points (λ, v) satisfying Equation (2) are called nontrivial solutions of Equation (1) .

We have :

Theorem 2 . If $p = 1$ and $(\lambda, \pm u(\lambda)) = (\lambda, \pm(\frac{2}{\lambda}g^{-1}(\frac{\lambda}{\lambda}))^{\frac{1}{2}}v^1), \lambda \in R_{\bar{\lambda}}$, then $(\lambda, \pm u(\lambda))$, for any $\lambda \in R_{\bar{\lambda}}$ with $g'\{g^{-1}(\frac{\lambda}{\lambda})\} \neq 0$, is not a secondary bifurcation solution of Equation (1) .

Now, we consider Equation (2) in the case $p \geq 2$. Let $(\lambda, u(\lambda)), \lambda \in R_{\bar{\lambda}}$, be such that $u(\lambda) \in C(\lambda)$ for all $\lambda \in R_{\bar{\lambda}}$. Then $(\lambda, u(\lambda))$ satisfies Equation (1) for any $\lambda \in R_{\bar{\lambda}}$. Let $\lambda_1 \in R_{\bar{\lambda}}$ and $u(\lambda_1) = \sum_{j=1}^{p} \alpha_j^1 v^1 \in C(\lambda_1)$, with $\alpha_j^1 = \alpha_j^1(\lambda_1), j = 1, ..., p$, and let $\beta^i, i = 1, ..., p - 1$, be indenpendent vectors in R^p, with $\beta^i = (\beta_1^i, ..., \beta_p^i)$, and $\sum_{j=1}^{p} \alpha_j^1 \beta_j^i = 0, i = 1, ..., p-1$. We define the multivalued mappings $J : R_{\lambda_1} \mapsto R^{p-1}$ and $D : R_{\lambda_1} \mapsto X$ by

$$J(\lambda) = \{\delta = (\delta_1, ..., \delta_{p-1}) \in R^{p-1} / \textstyle\sum_{j=1}^{p-1} \delta_k^2 = \frac{2}{\lambda}\{g^{-1}(\frac{\lambda}{\lambda}) - g^1(\frac{\lambda_1}{\lambda})\}$$

and

$$D(\lambda) = \{v \in X / \exists \delta = (\delta_1, ..., \delta_{p-1}) \in J(\lambda), v = u(\lambda_1) + \textstyle\sum_{k=1}^{p-1} \delta_k w^k \}.$$

We have : for all $\lambda \in R_{\lambda_1}$

$$|J(\lambda)| = |D(\lambda)| = \chi = \begin{cases} 2, & \text{if } p = 2 ; \\ +\infty, & \text{if } p \geq 3. \end{cases}$$

Then, the result on the secondary bifurcation of Equation (1) in the case $p \geq 2$ can be statted as follows :

Theorem 3. Let g^{-1} be an increasing function and $(\lambda, u(\lambda))$, $\lambda \in R_{\bar{\lambda}}$, be as above . Then for any $\lambda_1 \in R_{\bar{\lambda}}$ with $g'\{g^{-1}(\frac{\lambda_1}{\bar{\lambda}})\} \neq 0$, $(\lambda_1, u(\lambda_1))$ is a secondary bifurcation solution of Equation (1) and there exist at least χ distinct parameter families of nontrivial solutions of (1) in a neighborhood of $(\lambda_1, u(\lambda_1))$. Moreover , for any $\lambda \in R_{\lambda_1}$ we can find a nontrivial solution $(\lambda, v(\lambda))$ in an analytical form $v(\lambda) = u(\lambda_1) + \sum_{j=1}^{p} \delta_j w^j$ with $\delta = \delta(\lambda) = (\delta_1(\lambda), ..., \delta_{p-1}(\lambda)) \in J(\lambda)$.

Next , we consider the stability of solutions $(\lambda, \pm u(\lambda))$, $\lambda \in R_{\bar{\lambda}}$, in the case $p = 1$ with

$$\pm u(\lambda) = \pm(\tfrac{2}{\bar{\lambda}} g^{-1}(\tfrac{\lambda}{\bar{\lambda}}))^{\frac{1}{2}} v^1)$$

and the unstability of solutions $(\lambda, \pm v(\lambda))$, $\lambda \in R_{\lambda_1}$, with

$$v(\lambda) = u(\lambda_1) \pm (\tfrac{2}{\bar{\lambda}}(g^{-1}(\tfrac{\lambda}{\bar{\lambda}}) - g^{-1}(\tfrac{\lambda_1}{\bar{\lambda}})))^{\frac{1}{2}} w^1$$

in the case $p = 2$. As were been shown, they are parameter families of nontrivial solutions in neighborhoods of $(\bar{\lambda}, 0)$ and of $(\lambda_1, u(\lambda_1))$, respectively. For the sake of simplicity of notation, we only investigate the solutions $(\lambda, u(\lambda))$ and $(\lambda, v(\lambda))$.

Next, we consider the equation

$$u' = f(\lambda, u), \text{ with } f(\lambda, 0) = 0 \text{ for all } \lambda \in R.$$

Suppose that $(\bar{\lambda}, 0)$ is a primary bifurcation solution of the equation $f(\lambda, v) = 0$ and $(\lambda, u(\lambda))$ is a parameter family of nontrivial solutions in a neighborhood of $(\bar{\lambda}, 0)$. We say that $(\lambda, u(\lambda))$ is a stable solution if the spectrum of $D_v f(\lambda, u(\lambda))$ lies in the left-hand plane , when λ near $\bar{\lambda}$. Otherwise , $(\lambda, u(\lambda))$ is said to be unstable .

We have :
Theorem 4 . Let $p = 1$ and $(\lambda, u(\lambda)) = (\lambda, (\tfrac{2}{\bar{\lambda}} g^{-1}(\tfrac{\lambda}{\bar{\lambda}}))^{\frac{1}{2}} v^1)$, $\lambda \in R_{\bar{\lambda}}$. Then $(\lambda, u(\lambda))$ is stable if $g'\{g^{-1}(\tfrac{\lambda}{\bar{\lambda}})\} < 0$ for λ near $\bar{\lambda}$.

Further , in the case $p = 2$, we have just proved in Theoren 2 that for any $\lambda \in R_{\bar{\lambda}}$, $(\lambda, u(\lambda))$ with $u(\lambda) \in C(\lambda)$, satisfies Equation (1) . Thus , $(\lambda, u(\lambda))$, $\lambda \in R_{\bar{\lambda}}$ is a parameter family of nontrivial solutions in a neighborhood of $(\bar{\lambda}, 0)$. Theorem 6 shows that if g^{-1} is increasing, then for any λ_1 with $g'\{g^{-1}(\tfrac{\lambda_1}{\bar{\lambda}})\} \neq 0$, $(\lambda_1, u(\lambda_1))$ is a secondary bifurcation solution of (1). Furthermore, we can find two distinct parameter families $(\lambda, \pm v(\lambda))$, $\lambda \in R_{\lambda_1}$, of nontrivial solutions in a neighborhood of $(\lambda_1, u(\lambda_1))$, where

$$(\lambda, \pm v(\lambda)) = (\lambda, u(\lambda_1) \pm \delta_1(\lambda) w^1) \tag{3}$$

and

$$\delta_1(\lambda) = (\tfrac{2}{\bar{\lambda}}(g^{-1}(\tfrac{\lambda}{\bar{\lambda}}) - g^{-1}(\tfrac{\lambda_1}{\bar{\lambda}}))^{\frac{1}{2}}.$$

Remarking that δ_1 is a continuous function in λ and $\delta_1(\lambda) \to 0$ as $\lambda \to \lambda_1$, we conclude that $\pm v$ are continuous mappings in λ, $\pm v(\lambda) \to u(\lambda_1)$ as $\lambda \to \lambda_1$. In what follows we only consider the unstability of $(\lambda, v(\lambda))$, (the unstability of $(\lambda, -v(\lambda))$ would be investigated similarly). We obtain the following theorem on the unstability of $(\lambda, v(\lambda))$.

Theorem 5. Let $p = 2$ and $(\lambda, v(\lambda))$ be as in (3) with $g'\{g^{-1}(\tfrac{\lambda}{\bar{\lambda}})\} \neq 0$. In addition , assume that $\bar{\lambda}$ is an isolated eigenvalue of the operator B. Then $(\lambda, v(\lambda))$ is unstable .

3. Example

Let us consider the primary and secondary bifurcation and the stability of boundary value problems of nonlinear elliptic partial differential equations of the form

$$\lambda A v = g(\tfrac{1}{2} b(v, v)) B v \quad \text{in } \Omega$$

$$(4)$$
$$\mathbf{B}_k v = 0, 0 \le k \le m - 1, \text{ in } \partial\Omega.$$

Here , Ω denotes a bounded domain in the n-dimensional Euclidean space R^n with the infinittely differentiable boundary $\partial\Omega$ which is a $(n - 1)$-dimensional linear manifold and Ω lies locally in one side of $\partial\Omega$, $\partial\Omega \in C^\infty$; $\mathbf{A}$ is a uniform elliptic differential operator with oder $2m$ of the form

$$Av = \sum_{|\alpha||\beta|\le m} (-1)^{|\alpha|} D^\alpha (a_{\alpha\beta}(x) D^\beta v)$$

with $a_{\alpha\beta} \in C^\infty(\Omega)$; $\mathbf{B}$ is a uniform elliptic differential operator but at most with order $2m - 2$ and given by

$$Bv = \sum_{|\alpha||\beta|\le m-1} (-1)^{|\alpha|} D^\alpha (b_{\alpha\beta}(x) D^\beta v)$$

with $b_{\alpha\beta} \in C^\infty(\Omega)$, and

$$b(v, v) = \sum_{|\alpha||\beta|\le m-1} \int_\Omega b_{\alpha\beta}(x) D^\alpha v D^\beta v) d\Omega;$$

g is a differential function with g^{-1} continuous , $g(0) = 1, g(t) > 0$ for $t > 1$ and $\mathbf{B}_k$, for any $k, 0 \le k \le m - 1$, is a linear homogeneous differential operator, which is all defined in a neighborhood of $\partial\Omega$. For simplicity we assume that the order of $\mathbf{B}_k$ is less than m and with the boundary operators the elliptic differential operators $\mathbf{A}$ and $\mathbf{B}$ of such forms that we may associate with $\mathbf{A}$ and $\mathbf{B}$ symmetric bilinear forms

$$a(u, v) = \int_\Omega uAv d\Omega = \sum_{|\alpha||\beta|\le m} \int_\Omega a_{\alpha\beta} D^\alpha u D^\beta v d\Omega,$$

$$b(u, v) = \int_\Omega uBv d\Omega = \sum_{|\alpha||\beta|\le m-1} \int_\Omega b_{\alpha\beta} D^\alpha u D^\beta v d\Omega$$

for all $u, v \in V = \{f \in C^\infty(\bar\Omega)/\mathbf{B}_k f = 0 \text{ on } \partial\Omega, 0 \le k \le m-1\}$. We provide V with the standard $\|.\|_m$-topology of the Sobolev space $H^m(\Omega)$, and the completion of $(V, \|.\|_m)$ becomes a linear and closed subspace of $H^m(\Omega)$ and is denoted by X. The restricted norm of $L_2(\Omega)$ to X is denoted by $\|.\|_{L2}$. Further , we make the following hypotheses on $a(.,.)$ and $b(.,.)$.

Hypothesis 6. There is a constant $K > 0, a(v, v) \ge K\|v\|_{L_2}^2$ for all $v \in X$.

Hypothesis 7. $b(v, v) \ge 0$ for all $v \in X$ and $b(v, v) = 0$ implies $v = 0$.

In what follows we say that $(\lambda, v) \in R \times X$ is a weak solution of the boundary value problem (4) if

$$\lambda a(v, u) = g(\tfrac{1}{2}b(v, v))b(v, u) \qquad (5)$$

holds for all $u \in X$.

Next , for any $u, v \in X$ we define the inner product $\langle .,. \rangle$ by

$$\langle u, v \rangle = a(u, v).$$

The norm $\|.\|$ corresponding to this inner product is given as usual by

$$\|u\| = (\langle u, v \rangle)^{\frac{1}{2}}.$$

one can easily verify that X together with this inner product and this norm becomes a Hilbert space . We denote it by X again . By the Riesz representation Theorem we conclude that there exists a continuous linear self-adjoint operator B such that

$$\langle Bv, u \rangle = b(v, u)$$

holds for all $u, v \in X$. It then follows that Euqality (14) can be rewritten as
$$\langle u, v \rangle = g(\tfrac{1}{2}\langle Bv, v \rangle)\langle Bv, u \rangle$$

for all $u, v \in X$. This is equivalent to the equation

$$\lambda v = g\{\tfrac{1}{2}\langle Bv, v \rangle\}Bv, \qquad (\lambda, v) \in R \times X. \tag{6}$$

By Lemma 2.1 in [5] the operator B has only positive enumerable eigenvalues. Each eigenvalue has a finite multiplicity ,i.e; they can be ordered $0 < \lambda_1 \le \lambda_2 \le ... < +\infty$. We apply Theorem 1 to show that for any $n = 1, 2, ..., (\lambda_n, 0)$ is a primary bifurcation solution of Equation (6) . Let $\bar{\lambda} = \lambda_n$ for some n and $Ker(B - \bar{\lambda}id) = [v^1, ..., v^p]$. We define the multivalued mappings I, C as in Section 2. Let $(\lambda, u(\lambda)), \lambda \in R_{\bar{\lambda}}$, be such that $u(\lambda) \in C(\lambda)$ for all $\lambda \in R_{\bar{\lambda}}$ and $\lambda_1 \in R_{\bar{\lambda}}$ with $g'\{g^{-1}(\frac{\lambda_1}{\bar{\lambda}})\} \ne 0$. We apply Theorem 3 to show that $(\lambda_1, u(\lambda_1))$ is a secondary bifurcation solution of (6) in the case $p \ge 2$. Further , we can also use Theorem 4 to prove that in the case $p = 1$ the solutions $(\lambda, \pm u(\lambda)) = (\lambda, \pm(\frac{2}{\bar{\lambda}}g^{-1}(\frac{\lambda}{\bar{\lambda}}))^{\frac{1}{2}}v^1)$ are stable if $g'\{g^{-1}(\frac{\lambda}{\bar{\lambda}})\} < 0$ for λ near $\bar{\lambda}$. In the case $p = 2$,we apply Theorem 5 to show that the secondary bifurcation solutions $(\lambda, \pm v(\lambda)) = (\lambda, \pm(\frac{2}{\bar{\lambda}}(g^{-1}(\frac{\lambda}{\bar{\lambda}}) - g^{-1}(\frac{\lambda_1}{\bar{\lambda}}))^{\frac{1}{2}}w^1)$ are unstable .

References

1 N.W.Bazley and T. Kuppper , Branches of solutions in nonlinear eigenvalue problems ;
     Appl.of Nonlin. Anal.in Phys.Sci. (H.Amann , N.Bazley, K. Kirchgassner,eds).London : Pitman 1981.

2 N.W.Bazley and R.J.Weinacht , A class of explicitly resolvable evolution equations ;
     Math.Meth.in the Appl.Sci., 7 ,1985.

3 N.W.Bazley and R.J.Weinacht , Bifurcation of periodic solution for wave equations with scalar nonlinearities;
     Math.Meth.in the Appl.Sci., 6 ,1984.

4 L.A. Medeiros , On a new class of of nonlinear wave equations ;
     J.Math.Anal.Appl. 69 , 1979 .

5 G.G.Raef , Asymmptotic solutions for a class of nonlinear vibration problems ;
     Ph.D.Thesis ,University of Utrecht Holland , 1982 .

6 J.Smoller , Shock waves and reaction-diffusion equations ;
     Springer-Verlag , New York , Heidelberg ,Berlin , 1982 .

International Series of Numerical Mathematics, Vol. 97, © 1991 Birkhäuser Verlag Basel   59

PERIODIC SOLUTIONS LEADING TO CHAOS IN AN OSCILLATOR
WITH QUADRATIC AND CUBIC NONLINEARITIES

F. Benedettini and G. Rega

Dipartimento di Ingegneria delle Strutture, delle Acque e del Terreno

Università di L'Aquila, L'Aquila, Italy

Abstract: Regions of periodic and chaotic response, types of bifurcation and strange attractors of an unsymmetric oscillator of interest in structural dynamics are analyzed. The bifurcation predictive capability of the stability analysis of simple approximate solutions is discussed.

1.   INTRODUCTION

Among the systems of structural dynamics which are likely to exhibit chaotic behaviour, the forced oscillator with quadratic and cubic nonlinearities plays a meaningful rôle both for being a rather accurate onedegree-of-freedom model to describe the planar finite dynamics of elastic structures with initial curvature and for containing an unsymmetric term.

In this paper the strong nonlinear oscillator

$$\ddot{q} + \mu\dot{q} + q + 35.95\ q^2 + 534.53\ q^3 = P\cos\Omega t, \qquad (1)$$

where the values of coefficients of nonlinear terms refer to a suspended cable of mechanical interest, is considered. The influence of even and odd nonlinearities in the regular dynamics of the system and some results concerning its chaotic response were presented in previous works [1]. Here, systematic analysis of the response is made through extended and accurate computer simulations, and low-order approximate solutions are used for predicting bifurcations with meaning of possible precursors to chaos. Aims of the paper are: i) to obtain regions of different periodic or chaotic responses in a control parameters space of the system; ii) to characterize type of bifurcations and chaotic behaviour through different dynamic measures; iii) to relate the results of computer simulations to the stability of leading approximate solutions occurring in various regions of resonance for the system.

2.   REGIONS OF PERIODIC AND CHAOTIC RESPONSE

The regions of responses obtained for the system through a point-by-point time integration search with zero initial conditions are shown in the parametric diagram of Fig. 1. To obtain the diagram, basic steps of the control parameters Ω and P equal to 0.04 and 0.004 - and locally narrower ones - have been considered. The degree of periodicity has been declared on the basis of the Poincaré map but further dynamic measures - such as frequency response spectrum, Lyapunov exponents and fractal dimensions - have been calculated in many specific situations. Notable care has been paid in the investigation to general and algorithmic parameters playing important rôles in the calculation of system response or of specific measures.

Roughly speaking, three main regions of the diagram can be considered. Two are located in the neighbourhood of the $1/2$ ($\Omega = 2$) and $1/3$ ($\Omega = 3$) subharmonic resonances of the system, the third one covers the zones of order 3 ($\Omega = 0.33$) and 2 ($\Omega = 0.5$) superharmonic resonances and extends approximately up to primary resonance ($\Omega = 1$). While in the first two regions the response, though becoming quite complex, shows some fairly well-defined zones of periodicity with clearly established zones of chaos, in the third region, that is richer in terms of resonance frequencies, several transition zones occur, in which the periodicity of the response is strongly sensitive to small variations of control parameters and the zones of chaotic response are not so well-established. This different behaviour of the system is also concerned with the values of forcing amplitude for which chaos is observed: these are rather low and thus of practical interest in the $1/2$ - subharmonic range, much higher in the $1/3$ - subharmonic range - which is consistent with the associated weaker nonlinearity of the system response and with the findings obtained for systems with only symmetric nonlinearities -; finally in the super-harmonic range they are distributed all over the range of forcing amplitudes considered, following a certain regular pattern of parallel stripes.

Bifurcation diagrams obtained as results of detailed investigations made in the three frequency ranges at given values of the forcing amplitude are reported in Figs. 2-4 in terms of number of periods of the response and associated Lyapunov exponents. Different routes to chaos are observed. In the $1/2$ - subharmonic range there occurs a fairly clean sequence of period doubling bifurcations with decreasing frequency and instead sudden transition with increasing frequency. In the $1/3$ - subharmonic range transition to chaos from the left occurs via a basic period 6 response and responses with period multiple of 6 originating from the former, transition from the right is smoother and characterized mainly by a sequence of period doubled solutions somewhat dirtied by further period multiple than 6 responses. In the superharmonic range, at least at the forcing amplitude value considered, narrow zones of chaos exist, the transition to which seems to occur with a fairly clean period doubling sequence when decreasing the frequency, while it is preferably of sudden type when increasing the frequency.

The chaotic attractors obtained in the three zones are shown in Fig. 5. Remarkable differences occur among them. In the $1/2$ - subharmonic range chaos is quite well-established as denoted by all measures considered; in the $1/3$ - subharmonic range, the strange attractor exhibits 6 independent bundles originating from the period 6 solution around which the motion fluctuates and, correspondingly, lower values of its dimension are obtained; finally, the attractor in the superharmonic range resembles to the first one but has much thinner structure.

3.   APPROXIMATE SOLUTIONS, STABILITY ANALYSIS AND BIFURCATIONS

Approximate analytical techniques are used to predict bifurcations of the system response [2]. As simple as possible expansions are considered for the solution, with the aim of examining the predictive capability of low-order approximations:

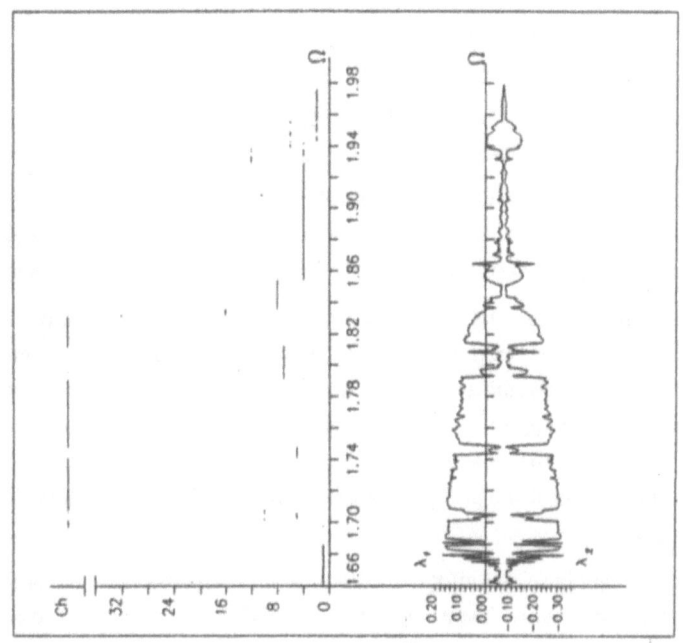

Fig. 2

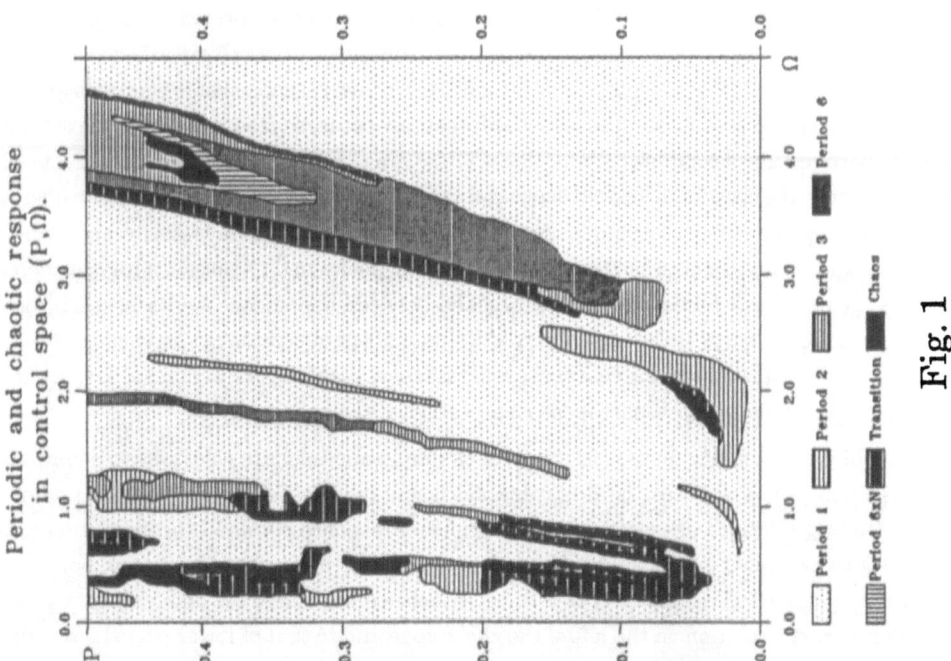

Periodic and chaotic response
in control space (P,Ω).

Fig. 1

$$q(t) = a_0 + a_1 \cos(\Omega t + \vartheta)$$

$$q(t) = a_0 + a_1 \cos(\Omega t + \vartheta) + a_{1/2} \cos(\Omega t/2 + \varphi)$$

$$q(t) = a_0 + a_1 \cos(\Omega t + \vartheta) + a_{1/3} \cos(\Omega t/3 + \varphi)$$

$$q(t) = a_0 + a_1 \cos(\Omega t + \vartheta) + a_2 \cos(2\,\Omega t + \varphi) + a_3 \cos(3\,\Omega t + \psi)$$

2 (a-d)

Period 1 solution (2a) is assumed as the reference fundamental one, period 2 (2b) and 3 (2c) solutions are assumed in the neighbourhood of the 1/2 and 1/3 subharmonic resonances respectively, improved period 1 (2d) in the range of superharmonic resonances.

By applying the harmonic balance method, four sets of algebraic nonlinear equations are derived and then solved through a numerical procedure, obtaining the resonance curves for each solution. The relevant stability analysis is made through the Floquet theory by calculating the eigenvalues of the monodromy matrix associated with the linearized variational equation [3]. The manner in which they leave the unit circle when varying a control parameter characterizes the local bifurcations of the periodic solutions. Jump bifurcation and period doubling bifurcation can occur.

Resonance curves of period 1 and period 2 solutions are plotted in Fig. 6a in terms of amplitudes of the relevant harmonics. Their stability analysis has been made through repeated computations of the eigenvalues. In the neighbourhood of the 1/2 - subharmonic resonance the period 1 solution of lower amplitude becomes unstable through p. d. bifurcations. Correspondingly, period 2 solution establishes and then it loses stability on its upper branch through p. d. bifurcations again. Use of higher subharmonic approximations would likely allow to put into evidence the possibility of further p. d. bifurcations. However, successive predictions are not necessary given the small region in which bifurcations occur, so it is preferable to compare directly with results of computer simulations. These are reported in Fig. 6b for different sets of i.c., namely the zero ones (black dots) and conditions corresponding to the considered solution via Eq. 2b (white dots). 2nd-order instability of period 2 solution leads to chaos, which is located approximately between points of vertical tangencies to the two solutions considered [2]. Fig. 5b also shows jumps of numerical results between the various solutions: their local shifts heuristically suggest the idea of a fractal-like nature of boundaries between different periodic (and/or chaotic) regions in a control parameters space depending on the i.c.; accordingly, several motions coexist mostly for some values of the parameters.

The capability of stability numerical analysis of simple approximate solutions to delimitate regions where chaos can occur is illustrated more generally in Fig. 7: the stability limits of period 1 and 2 solutions are plotted on the regions of response obtained through numerical integrations around the 1/2 - subharmonic resonance. For a correct comparison, one must remind that numerical results refer to zero i.c., which, e.g., belong to the domain of attraction of higher period 1 solution on the left of the chaos zone and to that of lower period 2 solution on its right, while the stability analysis does not account for the i.c..

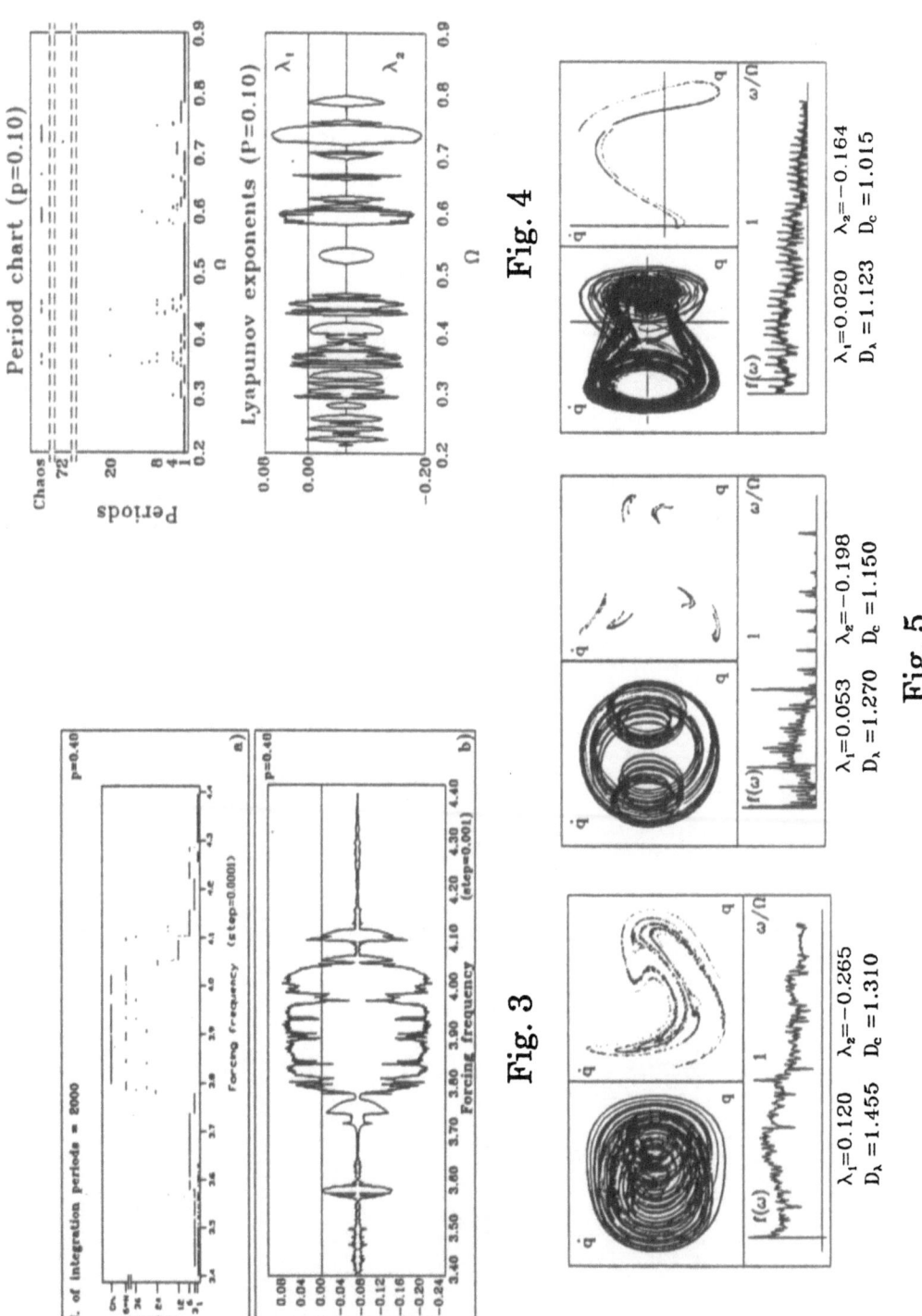

Fig. 3

Fig. 4

Fig. 5

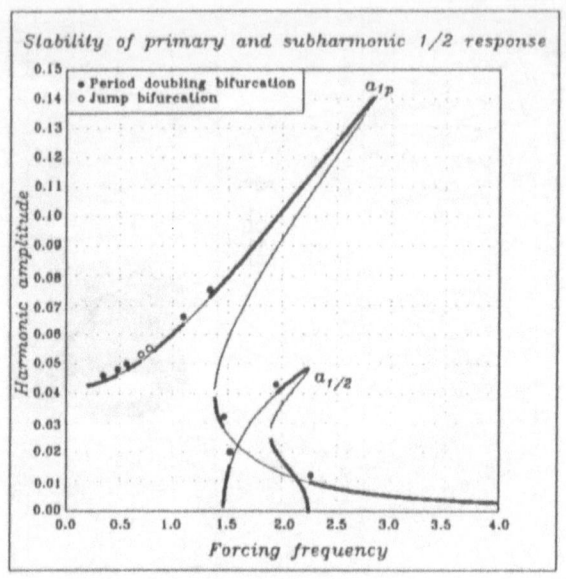

a)

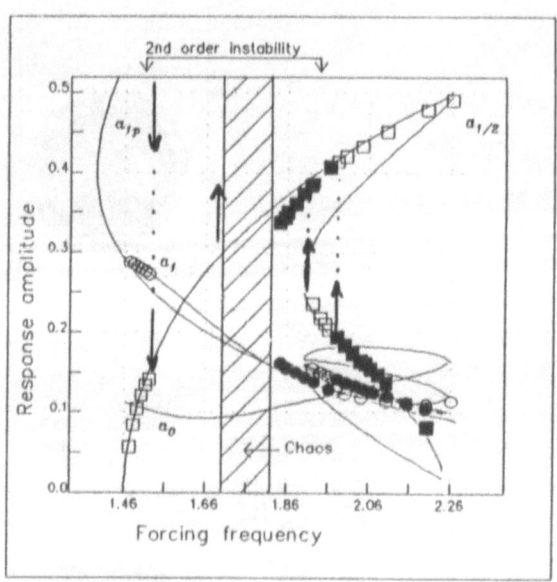

b)

Fig. 6

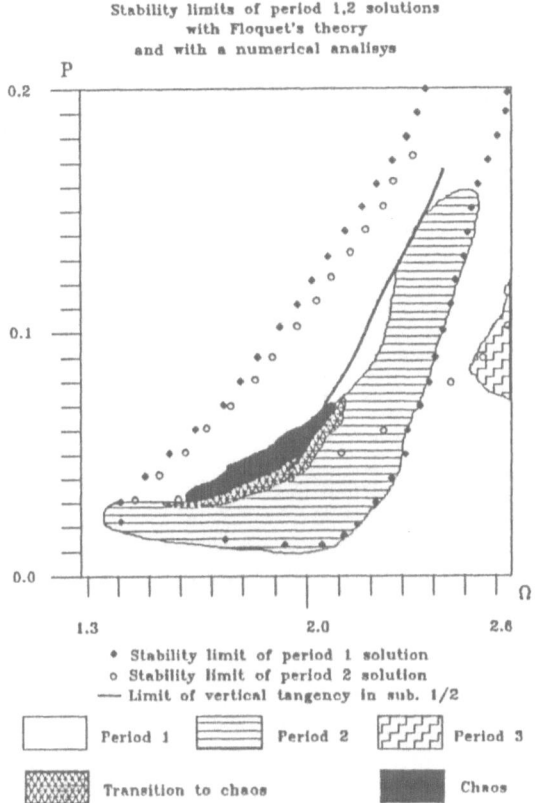

Stability limits of period 1,2 solutions
with Floquet's theory
and with a numerical analisys

- • Stability limit of period 1 solution
- o Stability limit of period 2 solution
- — Limit of vertical tangency in sub. 1/2

| Period 1 | Period 2 | Period 3 |

Transition to chaos   Chaos

Fig. 7

Similar comparisons have been made around the 1/3 - subharmonic and the superharmonic resonances with rather satisfactory results. However, it must be stressed that in certain situations the instability of the leading periodic solution does not end into chaos, which occurs only with particular sets of i.c., thus confirming the determinant rôle that they play for chaotic dynamical systems.

Acknowledgment: This work was partially supported with M.P.I. 40% Grants. The authors thank Dr. A. Salvatori for his help with the computations.

REFERENCES

1. Benedettini, F., Rega, G. (1990). 1/2 - Subharmonic Resonance and Chaotic Motions in a Model of Elastic Cable, *Proc. IUTAM Symp. Nonlinear Dynamics in Engineering Systems*, Springer-Verlag, 27-34.

2. Szemplinska-Stupnicka, W. (1987). Secondary Resonance and Approximate Models of Routes to Chaotic Motion in Non-linear Oscillators, *J. Sound Vibrat.* 113, 155-172.

3. Seydel, R. (1988). *From Equilibrium to Chaos*, Elsevier.

International Series of Numerical Mathematics, Vol. 97, © 1991 Birkhäuser Verlag Basel

Turing Structures in Anisotropic Media

J. Boissonade, V. Castets, E. Dulos, and P. De Kepper

Centre de Recherche Paul Pascal,
Université Bordeaux I, Avenue Schweitzer, F-33600 Pessac, France

When a convection-free chemical system is kept far from equilibrium by a permanent supply of fresh reactants, stationary spatial concentration patterns may spontaneously form as a result of the coupling of the sole reaction and diffusion processes. In view of its possible implication in morphogenesis, this phenomenon, first predicted by Turing in 1952 [1], has instigated numerous theoretical studies. Various well documented reviews can be found, both from the nonlinear physics [2-4] and biological [5-8] standpoints. However, their experimental observation has long been delayed in regard of practical impediments. Recently, we have reported the first experimental evidence of a Turing-type pattern [9,10]. The key to the success was the use of open spatial reactors with nonhomogeneous distribution of the feeding species, which induces a basic anisotropy in the reactive medium. Thus, all properties of Turing patterns have to be revisited in consideration of this anisotropy. Here we report numerical simulations performed with a geometry and boundary conditions similar to those of the experiments. Our results are in excellent agreement with the observations. They also raise a new class of problems, namely the formation of pattern defects present both in the computations and in the real experiments.

1 Turing structures

1.1 Spatial instabilities and Turing bifurcations

We shall only consider single phase chemical systems – excluding physical heterogeneities such as catalytic particles – free from any form of convection so that matter transport is restricted to molecular diffusion. The system is supposed to be maintained far from equilibrium by keeping constant and *uniform* the concentrations of the *"pool species"*, *i.e.* the fresh reactants and the products kinetically active through reverse reactions. These conditions, somewhat unrealistic from a practical point of view, are retained in most theoretical works for sake of simplicity. The concentrations c_i of the other species, the *"intermediate species"*, are let free to change according to the reaction rate laws and define the dynamical state of the system. They obey a system of so-called *reaction-diffusion equations*:

$$\frac{\partial c_i}{\partial t} = f_i(\ldots, c_j, \ldots) + D_i \Delta_{\mathbf{r}} c_i \tag{1}$$

where f_i accounts for the reaction rate (generally nonlinear), D_i is the diffusion coefficient of species i and $\Delta_{\mathbf{r}}$ is the Laplacian operator. Let us consider the linear stability of the uniform

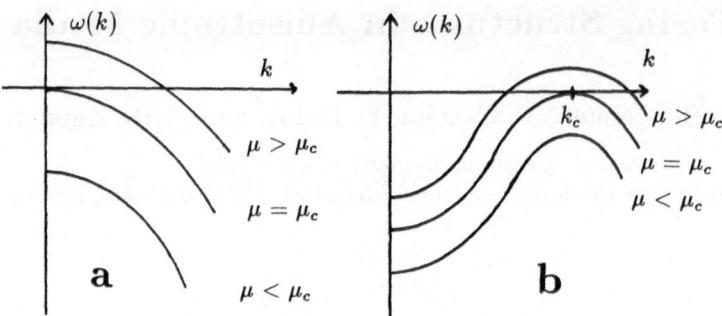

Figure 1:   *a) Homogeneous instability*   *b) Turing instability*

stationary state $\mathbf{c}_s$, solution of the homogeneous equation $\mathbf{f}_i(\mathbf{c}_s) = 0$, to a nonhomogeneous perturbation $\delta\mathbf{c}_s = \sum_\mathbf{k} a_\mathbf{k} e^{\omega \mathbf{k} t + i \mathbf{k} \mathbf{r}}$. The amplification factors $\omega_\mathbf{k}$ of the modes with wavevector $\mathbf{k}$ are the eigenvalues of the linear operator $L_{ij} = F_{ij} - D_i k^2 \delta_{ij}$, where F_{ij} is the jacobian matrix $\left\{ \dfrac{\partial f_i}{\partial c_i} \right\}_{\mathbf{c} = \mathbf{c}_s}$ of the reactive terms. Thus, for each eigenvalue, the dispersion law takes the form $\omega_\mathbf{k} = \omega(k^2, D_i, \mu)$, where μ holds for any external chemical constraint retained as a bifurcation parameter.

When all the diffusion coefficients D_i are equal to a single value D, these eigenvalues are given by $\omega_i = \omega_{0i} - Dk^2$, the ω_{0i} being the eigenvalues of F_{ij}. In this case, the first possible instability necessarily occurs at $\mathbf{k} = 0$ when ω_{0i} becomes positive and corresponds to a *homogeneous* instability without spatial organization (Fig 1a). When several diffusion coefficients are *different*, the dispersion law can exhibit a positive curvature and the first bifurcation occurs for a finite positive critical value of the wavenumber $|\mathbf{k}| = k_c$ (Fig.1b) leading to a Turing stationary spatial structure. This necessary condition is not as stringent as it is sometimes asserted, since, in the vicinity of singularity points, the differences between the diffusion coefficients can be arbitrary small [11] and can remain realistically small over a significant range of parameters [12], without precluding the emergence of the pattern. Contrary to the nonequilibrium hydrodynamic structures, such as the well-known Bénard and Couette flow patterns [13], which depend on some geometrical length of the system, this Turing critical wavenumber and the associated critical wavelength only depend on intrinsic properties of the medium (concentrations, diffusion coefficients, and kinetic constants).

1.2   Model schemes for Turing structures

Numerous model schemes able to exhibit Turing type instabilities have been proposed, most of them with only two intermediate species. All of them encompass two antagonistic non-linear processes which, as a result of the differences between diffusion coefficients of the associated species, take alternatively precedence over each other. Among these, two classes have received a special attention. Mathematical criteria to differentiate them are explicitly given in Fuji *et al* [14].

1.2.1 Activator-inhibitor models

In these schemes, the antagonistic processes are provided by an activator species which increases its own rate of production but simultaneously produces its own inhibitor. When the inhibitor diffuses faster than the activator, a Turing pattern may form. A good example is provided by the Gierer and Meinhart model [15], proposed for the interpretation of biological morphogenesis. The properties of these models are extensively discussed in ref. 5 and 8.

1.2.2 Activator-substrate depleted models

In these models, an activator is still present but the antagonistic process is the depletion of a highly diffusible substrate [5]. Some of them include an autocatalytic trimolecular step of the type $2X + Y \rightarrow 3X$ among which the best known is the *"Brussellator"*, introduced by Prigogine and Lefever [16] (for reviews see ref. 2 and 6). Another convenient trimolecular scheme of the same type, which, for computational facilities, we shall consider in the rest of this paper, has been proposed by Schnackenberg [17] and extensively studied by Murray [8,18]. The reaction steps are:

$$A \xrightarrow{k_1} X$$
$$X \xrightarrow{k_2} Products$$
$$2X + Y \xrightarrow{k_3} 3X$$
$$B \xrightarrow{k_4} Y$$

where A and B are the pool species and X and Y the intermediate species. This scheme is identical to the "Brussellator" except that the last step of the latter is $B + X \rightarrow Y$. Let L be a typical length scale and, using the same character for the species and their concentration, we define the following dimensionless variables:

$$t^* = \frac{D_X t}{L^2}, \quad \mathbf{r}^* = \frac{\mathbf{r}}{L}, \quad d = \frac{D_Y}{D_X}, \quad \gamma = L^2 k_2 / D_X,$$

$$a = \frac{k_1}{k_2}\left(\frac{k_3}{k_2}\right)^{1/2} A, \quad b = \frac{k_4}{k_2}\left(\frac{k_3}{k_2}\right)^{1/2} B, \quad u = \left(\frac{k_3}{k_2}\right)^{1/2} X, \quad v = \left(\frac{k_3}{k_2}\right)^{1/2} Y,$$

From the mass action law, it comes

$$\frac{\partial u}{\partial t} = \gamma(a - u + u^2 v) + \Delta_{\mathbf{r}} u \tag{2}$$

$$\frac{\partial v}{\partial t} = \gamma(b - u^2 v) + d\Delta_{\mathbf{r}} v \tag{3}$$

where the asterisks were dropped for convenience. It results from the linear stability analysis that a Turing bifurcation occurs in the parameter regions where the following conditions are simultaneously fulfilled [8,18], defining the so-called *"Turing space"*:

$$0 < b - a < (a+b)^3, \quad d(b-a) > (a+b)^3, \quad [d(b-a) - (a+b)^3]^2 > 4d(a+b)^4 \tag{4}$$

with a critical wavenumber $k_c^2 = \sqrt{\gamma} f(a, b, d)$.

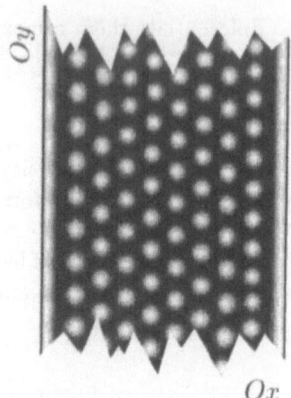

Figure 2: *Turing structures in the Schnack-enberg model (detail): pattern of u*
System size:1×2.5
$d = 20, \gamma = 10000$
$a = 0.15, b = 1$
On boundaries $x = 0$ et $x = 1$:
$u(0) = u(1) = 1.15$
$v(0) = v(1) = 0.756$

1.3   Turing patterns in multidimensional systems

While most studies are limited to 1-D systems, a few authors have considered mode selection and patterning in higher dimensionalities where the wavevector degeneracy close to the bifurcation point can lead to different solutions. In two dimensional systems, Pismen predicts that, among the regular patterns which tesselate the plane (bands, hexagons, diamonds), the structures will more likely develop according to the hexagonal symmetry [19]. We have performed a numerical simulation of the Schnackenberg model with parameters corresponding to a Turing instability, with periodic boundary conditions in direction Oy and fixing the concentrations u and v to the stationary state values at $x = 0$ and $x = 1$ (Dirichlet conditions). Starting from a randomly distributed state, the final state actually exhibits a stationary structure with hexagonal symmetry (Fig.2).

In three-dimensional systems, Walgraef *et al.* have developed a Ginzburg-Landau theory including fluctuations and have shown that the most stable patterns, thus the most likely selected, should be successively body centered cubic (b.c.c.), triangular (hexagonal prisms), and roll-type (bands) when the bifurcation parameter increases [20] but no computational results are yet available.

2   Turing structures with feed-induced anisotropy

2.1   The boundary-fed reactor

Keeping a uniform pool of chemicals, preserving molecular diffusion, and avoiding convection at the same time is an unpracticable task. The following arrangement was proposed in order to make the experimental observation of Turing structures workable [21,22]. The reaction is performed inside a long thin strip of an inert gel in order to avoid any convection phenomena.

We shall assume that the two opposite edges are parallel to the Oy axis. This "reactor" is fed by diffusion of the pool reactants from the two opposite edges, maintained in contact with well-mixed and permanently replenished reservoirs. Thus the concentrations of all species are kept constant on these boundaries (Dirichlet boundary conditions). Since there is a continuous change of the input chemical concentrations from one edge to the other, there is a continuous change of the actual parameters along the transverse direction Ox, so that Turing structures can only develop in a ribbon-like region parallel to Oy, where these parameters belong to the Turing space. Localized structures had previously been computed in 1-D systems with Dirichlet conditions at both ends but, the introduction of a second direction Oy, along which there is no imposed gradient, allows symmetry breaking to be more explicit [23,24]. When this region is strongly localized in a zone where concentrations change rapidly, the instability can be described in terms of a self-organizing front (quasi 1-D structure) [12,21,22]. When the gradients in direction Ox are smoother, the problem is better described in terms of a 2-D structure developing in an anisotropic medium [25], where the anisotropy results from these gradients. This is the point of view retained in the rest of this paper.

Note that similar boundary-fed reactors with annular geometry have also been success-fully used to study sustained *nonstationary* localized structures , namely systems of traveling waves in the B.Z. reaction [26,27].

2.2   Numerical simulations of Turing patterns in the boundary-fed reactor

Several types of patterns should develop according to the ratio of the width l_T of the Turing region to the wavelength λ of the pattern. For $l_T <\sim \lambda$ one expects a quasi 1-D pattern, made of a single row of concentration peaks, parallel to the edges. When the ratio becomes larger and larger, new rows should appear, building progressively a genuine 2-D structure. We have checked these qualitative predictions in a series of numerical simulations with the Schnackenberg model.

The computations were performed by finite differences on a 100×250 rectangular grid with Dirichlet boundary conditions along the edges at $x = 0$ and $x = 1$, and periodic boundary conditions in direction Oy. This arrangement corresponds to an annular strip of gel where one neglects the curvature. aThe concentrations of the pool species were fixed to $a \neq 0$; $b = 0$ at $x = 0$ and $a = 0$; $b \neq 0$ at $x = 1$, whereas the concentrations u and v of the intermediate species were fixed to zero on these boundaries in order to account for the fast renewal of the reactants along the edges. The pool species are assumed to be in large excess so that they are distributed according to a linear gradient in the direction Ox. In order to allow for the development of the instability in the direction Oy, the system was initialized in a non uniform state obtained by adding to a state invariant in the direction Oy either a localized perturbation, or a distributed random noise. The computations were carried on until the system reaches a stationary state, or a quasi-stationary state when the relaxation times become prohibitively long. Several representative distributions of u, corresponding to different sets of pool species concentrations are displayed on the figure 3 with 32 grey equidistant intensity levels (full scale).

These results are in excellent agreement with the predictions. The ratio l_T/λ increases

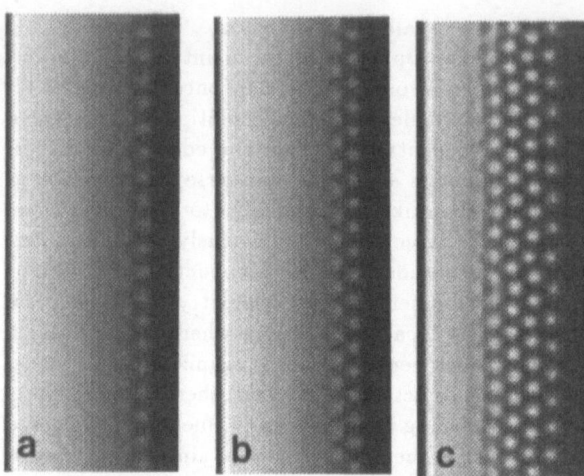

Figure 3: *Turing structures in anisotropic media (Schnackenberg model)*

System size$= 1 \times 2.5$,
$d = 20, \gamma = 10000$
$u(0) = u(1) = v(0) = v(1) = 0$

3a : $a(0) = 0.3, \ b(1) = 3.9$
3b : $a(0) = 0.3, \ b(1) = 3.5$
3c : $a(0) = 0.3, \ b(1) = 2.4$

from Fig. 3a to Fig. 3c. In Fig. 3a the pattern is actually a quasi 1-D single row of spots, whereas in Fig 3c, in spite of the anisotropy, the hexagonal symmetry of the 2-D uniform pattern is recovered, apart from a small residual modulation.

3   Experimental data

In order to provide a basis for comparison, we briefly report our experimental results obtained with a gel reactor fed by diffusion from the boundaries. They constitute the first experimental evidence of a Turing type structure. More extensive descriptions of these experiments can be found elsewhere [9,10]. Experiments have been carried on successfully in rectangular [9] and annular [10] reactors.

3.1   The reactor

The structures presented here were obtained in an annular, 1mm thick, 3mm width, film of an inert polyacrylamide gel of radius 2cm, thermostated by a water jacket. The width being much smaller than the diameter, the curvature can be neglected and the length is large in regard of the width. The two pool species solutions – separately nonree – are regenerated in the reservoirs by permanent flows of fresh chemical. We have used the chlorite-iodide-malonic acid reaction [28], a reaction already known to exhibit interesting oscillating and spatiotemporal properties (for a review, see ref. 29). The spatial distribution of the local chemical state is revealed by a color indicator, sensitive to the intermediate species I_3^-.

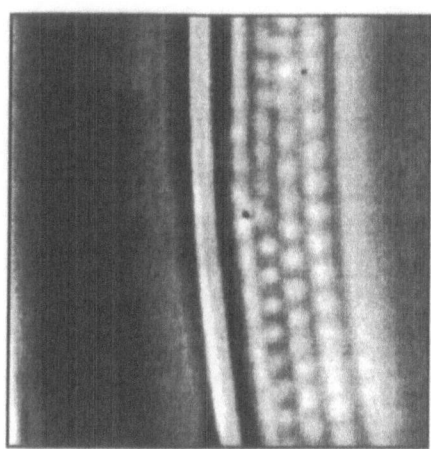

Figure 4: *Turing patterns in an annular reactor ($R = 2cm$): central portion of the gel strip. The slight curvature of the structure is due to the circular geometry*

Boundary compositions:
Left edge: $[CH_2(COOH)_2]$= 8.3×10^{-3} mole/l,
$\qquad\qquad [H_2SO_4] = 10^{-2}$ mole/l
Right edge: $[NaClO_2^-] = 2.4 \times 10^{-2}$ mole/l
Both edges: $[NaOH]$=3×10^{-3} mole/l,
$\qquad\qquad [Na_2SO_4] = 3 \times 10^{-3}$ mole/l,
$\qquad\qquad [I^-] = 2.66 \times 10^{-3}$ mole/l
Temperature: $7°C$

3.2   Experimental structures

Turing type patterns are found to settle over a significant range of parameters. After a while, bands, alternately dark and clear, form parallel to the edges, then split up into spots, leading ultimately to a stationary localized structure with hexagonal symmetry. An example of the final hexagonal structure is shown on Fig. 4 where contrasts have been enhanced by image processing. In the well organized pattern, localized in the median region, the clear spots correspond to oxidized iodine states. The wavelength $\lambda \sim 0.18$mm does not change much from one experiment to another. This value is much lower than any geometrical parameter and is independent of the reactor geometry – rectangular or annular. Thus, this wavelength is more likely intrinsic as expected for a Turing structure.

There is a striking similarity between the structure of Fig. 4 and the results of the simulations in section 3. Other patterns, which exhibit only one or two visible rows of spots, as in the computations of Fig. 3a and 3b, have also been observed. Nevertheless, according to the wavelength value – five times smaller than the gel thickness – the structure is clearly tridimensional, whereas our computations are limited to two dimensional systems. So far, we cannot distinguish between a 3-D triangular pattern with hexagonal symmetry or a body centered cubic which both fit our top-view observations (see section 2).

4   Structural defects

Most of the structures observed in the experiments present structural defects as shown on the example of Fig. 5. A straightforward explanation would lie in experimental imperfections, like the small inhomogeneities of the gel. Nevertheless, there are more fundamental intrinsic sources of geometrical defects, as confirmed by our numerical simulations.

The first type of defects is illustrated on Fig. 6 where the simulations have been performed with uniform control parameters. These topological defects result from a memory of the initial conditions. For instance, on Fig. 6b, the defect preserves the symmetry of an initial

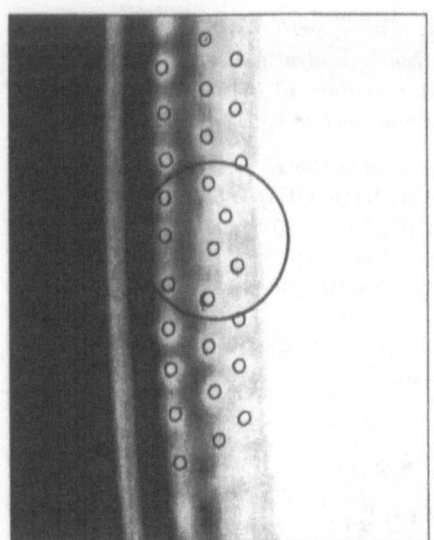

Figure 5: *Structural defect in an experimental Turing structure. Circular marks have been added on intensity maxima to make easier the visualization of the structure.*

symmetric perturbation. In an unbounded system – although the process can last for an infinite time – the pattern can progressively reorganize into a regular structure, but, in a confined system, the defects can remain "trapped" indefinitely.

Another type of defect – more fundamental in our problem – results from the anisotropy of the system. The progressive change of the input species concentrations along Ox induces a progressive change of wavelength. If the pattern extends over a rather large area this creates a frustration in the successive rows. When this frustration becomes too strong the pattern reorganize locally around a defect. This is the case in Fig. 7 when the pattern has been obtained by a small randomly distributed perturbation of a uniform state , and where the change of wavelength is clearly visible.

Finally, on Fig. 8 we show that the influence of defects associated with frustration either can persist over lengthes far beyond the wavelength, or be localized. In this example these two types of defects are present. We were not able to determine whether this structure would ultimately relax to a more regular pattern or not, but in the former case the required time would be so long that the regular structure could not be attained from a practical point of view.

The problem of topological defects has been extensively adressed in the field of hydrodynamical instabilities (Bénard, liquid crystals,...) [30]. In chemical systems, this problem has been well discussed in the case of nonstationary systems, such as excitable media and phase waves [31,32] , but the problem of defects in chemical stationary patterns has not yet received a special attention.

5   Conclusion

We have shown both in numerical simulations and experimentally that it is possible to produce Turing-type structures by introducing the fresh reactants by pure diffusive processes from the edges. This process induces a basic gradient of pool species, but this does not

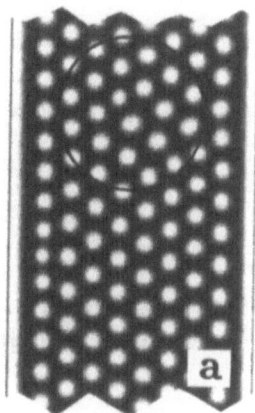
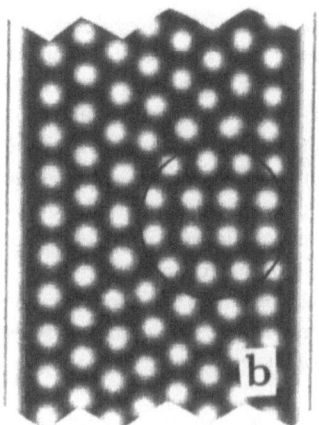

Figure 6: *Structural defects in computed Turing structures. (Same conditions as in Fig 2).*

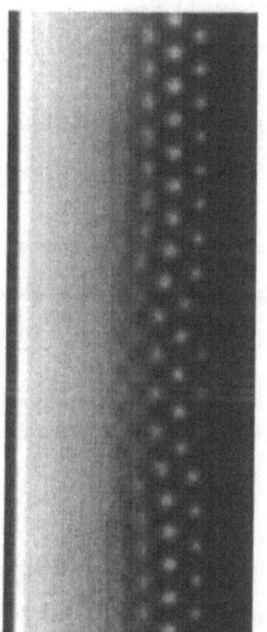

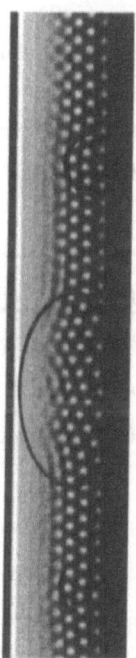

Figure 7: *Structural defect induced by frustration.*
Same parameters as Fig. 3 except $b(1) = 3$.

Figure 8: *Structural defects*
Same parameters as in Fig. 3c except that the system size is 1×5. Localized defects:circles; long range defect: ellipse

preclude an unambiguous evidence of a symmetry breaking in the direction orthogonal to the gradient. Nevertheless the anisotropy associated to this gradient has to be seriously considered. First, it leads to a localization of the structure which, apart from small deformations, preserves the symmetry of the pattern (hexagonal in the 2-D simulations). Moreover, due to the small wavelength variations associated with this anisotropy, structural defects can result from frustration effects. Since this is the sole example of an experimental Turing structure, more attention should be paid to the development of the theory of these structures in *anisotropic* media, rather than in the more commonly assumed system of uniformly distributed pool species.

Aknowledgements: We are grateful to A. Arneodo, P. Borckmans and G. Dewel for profitable discussions. This work has been supported by the Venture Research Unit of the British Petroleum Company.

References

[1] A.M. Turing, *Philos. Trans. R. Soc. London*, **B 327**, 37 (1952)

[2] G. Nicolis and I. Prigogine, *"Self-organization in Nonequilibrium Chemical Systems"* (Wiley, New York, 1977)

[3] H. Haken, *"Synergetics, an Introduction"*, (Springer-Verlag, 1977)

[4] R.J.Field and M.Burger Edts, *"Oscillations and Traveling Waves in Chemical systems"* (Wiley, N.Y., 1985)

[5] H.Meinhardt, *"Models of Biological Patterns Formation"*, (Academic Press, N.Y., 1982)

[6] A. Babloyantz, *"Molecules, Dynamics and Life"* (Wiley, N.Y., 1986)

[7] L.G. Harrison, *J. theor. Biol.*, **185**, 369 (1987)

[8] J.D. Murray, *"Mathematical Biology"* (Springer-Verlag, 1989)

[9] V. Castets, E. Dulos, J. Boissonade and P. De Kepper, *Phys. Rev. Lett.*, **64**,2953 (1990).

[10] P. De Kepper, V. Castets, E. Dulos, and J. Boissonade, *Physica D*, to appear.

[11] J.E. Pearson and W. Horsthemke, *J. Chem. Phys.*, **90**, 1588 (1989)

[12] A. Arneodo, J. Elezgaray, J.E. Pearson and T. Russo, *Physica D*, in press

[13] S. Chandrasekhar, *"Hydrodynamic and Hydromagnetic Stability"*, (Oxford Univ. Press, 1961); H.L. Swinney and J.P. Gollub Edts, *"Hydrodynamic Instabilities and the Transition to Turbulence"*, (Springer, 1981)

[14] H. Fuji, M. Mimura, and Y. Nishiura, *Physica D*, **5**,1 (1982)

[15] A. Gierer and H. Meinhardt, *Kybernetik*, **12**, 30 (1972)

[16] I. Prigogine and R. Lefever, *J. Chem. Phys.*, **48**, 1695 (1968)

[17] J. Schnackenberg, *J. theor. Biol.*, **81**,389 (1979)

[18] J.D. Murray, *J. theor. Biol.*, **98**, 143 (1982)

[19] L.M. Pismen, *J. Chem. Phys.*, **72**, 1900 (1980)

[20] D. Walgraef, G. Dewel, and P. Borckmans, *Phys. Rev.A*, **21**, 397 (1980); D. Walgraef, G. Dewel, and P. Borckmans, *Adv. Chem. Phys.*, **49**, 311 (1982); D. Walgraef, *"Structures spatiales loin de l'équilibre"* (Masson, Paris, 1988)

[21] J. Boissonade, *J. Physique (France)*, **49**, 541 (1988);

[22] J. Boissonade, M. Boukalouch and P. De Kepper, in *"Spatial inhomogeneities and transient behaviour in chemical kinetics"*, Ed. P. Gray, G. Nicolis, F. Baras, P. Borckmans and S.K. Scott, p. 433 (Manchester University Press, 1990)

[23] M. Herschkowitz-Kaufman and G. Nicolis, *J. Chem. Phys.*, **56**,1890 (1972); M. Herschkowitz-Kaufman, *Bull. Math. Biol.* **37**,585 (1975)

[24] L.L. Bonilla and M.C. Velarde, *J. Math. Phys.*, **21**, 2586 (1980)

[25] G. Dewel, D. Walgraef, and P. Borkmans, *J. Chim. Physique (Paris)*, **84**, 1335 (1987)

[26] Z. Noszticzius, W. Horsthemke, W.D. McCormick, H.L. Swinney, and W.Y. Tam, *Nature*, **329**,619 (1987); Z. Noszticzius, W. Horsthemke, W.D. McCormick, and H.L. Swinney in *"Spatial inhomogeneities and transient behaviour in chemical kinetics"*, Ed. P. Gray, G. Nicolis, F. Baras, P. Borckmans and S.K. Scott, p. 429 (Manchester University Press, 1990); N. Kreisberg, W.D. McCormick, and H.L. Swinney, *J. Chem. Phys.* **91**, 6532 (1989)

[27] E. Dulos, J. Boissonade and P. De Kepper, *in "Nonlinear Wave Processes in Excitable Media"*, Edt A.V. Holden, M. Markus, H.G. Othmer, Pergamon press, in press.

[28] P. De Kepper, I.R. Epstein, K. Kustin and M. Orbán, *J. Phys. Chem.* **86**, 170 (1982); C.E. Dateo, M. Orbán, P. De Kepper and I. R. Epstein, *J. Am. Chem. Soc.*, **104**, 504 (1982); M. Orbán, P. De Kepper, I.R. Epstein,and K. Kustin, *Nature*, **292**, 816 (1981)

[29] P. De Kepper, J. Boissonade, and I. Epstein, *J. Phys. Chem.* , **94**, 6525 (1990)

[30] P. Manneville, *"Dissipative Structures and Weak Turbulence"*, (Academic Press, 1990)

[31] Y. Kuramoto, *"Chemical Oscillations, waves, and turbulence"*, (Springer, 1984)

[32] V.S. Zykov *"Simulation of wave processes in excitable media"*, (Manchester University Press, 1987)

International Series of Numerical Mathematics, Vol. 97, © 1991 Birkhäuser Verlag Basel

REGULAR AND CHAOTIC PATTERNS OF RAYLEIGH-BÉNARD CONVECTION

by F.H. Busse and M. Sieber
Institute of Physics, University of Bayreuth, 858 Bayreuth, FRG

1. INTRODUCTION

Regular and irregular patterns in nature are often observed in close association. Clouds are cited as typical chaotic structures on the one hand, but can also be observed as highly regular bands with well defined wavelength. Satellite pictures sometimes reveal regular hexagonal cells of mesoscale convection with diameters up to one hundred km, while most of the other large scale cloud structures can be characterized only by their fractal dimension. A similar coexistence of regular and chaotic features can be observed in the atmosphere of Jupiter. Well defined anticyclonic eddies with approximately constant separation are observed at certain latitudes while other latitudes exhibit cloudbands in seemingly chaotic motions (Smith et al., 1979).

The duality between regular and random phenomena is also known from laboratory studies of various types of hydrodynamic instability. On the one hand transitions to steady regular secondary motions are observed as for example in the form of Taylor vortices in the case of differentially rotating cylinders or as Bénard cells in a fluid layer heated from below. In other cases such as pipe or channel flow the secondary motion is characterized by a chaotic structure in space and in time. Typically the latter cases correspond to inverse bifurcations, i.e. the bifurcation point predicted by linear theory is delayed and the bifurcating solutions can not be physically realized because they are unstable. Rayleigh-Bénard convection and Taylor vortices are the prime examples for forward (or supercritical) bifurcations in hydrodynamic stability theory and their finite amplitude properties have thus been analyzed in detail and compared with experimental observations.

The presence of an inverted bifurcation is not the only reason for a transition to chaotic motion from the basic state of a system. Nor do subcritical bifurcations necessarily lead to turbulent flow. As many examples of convection flows indicate, regular patterns can be realized when the bifurcating branch is subcritical and thus unstable. The only requirement seems to be that the secondary solution reverts to a forward dependence on the control parameter not too far from the original point of bifurcation. There is another way in which the duality of regular and chotic structures can be studied. Even at the first forward bifurcation slight changes in the parameters of a system may lead to a replacement of stable steady patterns by a dynamical state that is chaotic in space as well as in time. Rayleigh-Bénard convection offers some interesting examples for this behaviour as we like to point out in the following.

While problems of hydrodynamic instability typically lead to a large manifold of solution branches of the basic nonlinear equations as subsequent bifurcations occur, Rayleigh-Bénard convection is distinguished in that an infinite number of solutions bifurcate right at the onset of instability if the limit of a horizontally extended layer is considered. Thus some problems associated with the interaction of a large number of bifurcating solutions can be investigated in a simpler way than in other cases of hydrodynamic instability. The analysis outlined in this paper is based on an expansion of the nonlinear basic equations in powers of the amplitude of convection. The solvability conditions in the cubic order yield an infinite set of nonlinear equations for the evolution in time of the amplitudes of the bifurcating modes. In many cases these equations can be written in gradient form in which case the existence of at least one stable steady state is guaranteed. But the introduction of additional physical effects can destroy the gradient character and lead to the instability of the single previously stable state. It is in this situation that phase turbulence is realized.

2. THE EXISTENCE OF STABLE STEADY STATES OF RAYLEIGH-BÉNARD CONVECTION

We consider the usual configuration of a horizontally extended Rayleigh-Bénard layer (see, for example, Drazin and Reid, 1981; or Busse, 1978, 1989). Using the height d of the layer as length scale, d^2/ν as time scale, where ν is the kinematic viscosity of the fluid, and $(T_2-T_1)/R$ as scale of the temperature we can write the basic equations in dimensionless form. T_1 and T_2 are the temperatures of the upper and lower boundaries and R is the Rayleigh number. Since the velocity $\underset{\sim}{v}$ satisfies the equation of continuity, $\nabla \cdot \underset{\sim}{v}=0$, it is

convenient to introduce the general representation for a solenoidal vector
field

$$\underset{\sim}{v} = \nabla \times (\nabla \times \underset{\sim}{\lambda}\varphi) + \nabla \times \underset{\sim}{\lambda}\psi \equiv \underset{\sim}{\delta}\varphi + \underset{\sim}{\varepsilon}\psi \qquad (1)$$

where $\underset{\sim}{\lambda}$ denotes the vertical unit vector. In order to eliminate any
arbitrariness we shall require that the average of φ and ψ over horizontal
planes vanishes. The equations for the scalar functions φ, ψ and for the
deviation ϑ of temperature from the distribution in the basic static state of
the problem can be written in the form

$$\left(\nabla^2 - \frac{\partial}{\partial t}\right)\nabla^2\Delta_2\varphi - \Delta_2\vartheta = \underset{\sim}{\delta}\cdot[(\underset{\sim}{\delta}\varphi + \underset{\sim}{\varepsilon}\psi)\cdot\nabla(\underset{\sim}{\delta}\varphi + \underset{\sim}{\varepsilon}\psi)] \qquad (2a)$$

$$\left(\nabla^2 - \frac{\partial}{\partial t}\right)\Delta_2\psi = \underset{\sim}{\varepsilon}\cdot[(\underset{\sim}{\delta}\varphi + \underset{\sim}{\varepsilon}\psi)\cdot\nabla(\underset{\sim}{\delta}\varphi + \underset{\sim}{\varepsilon}\psi)] \qquad (2b)$$

$$\left(\nabla^2 - P\frac{\partial}{\partial t}\right)\vartheta - R\Delta_2\varphi = P(\underset{\sim}{\delta}\varphi + \underset{\sim}{\varepsilon}\psi)\cdot\nabla\vartheta \qquad (2c)$$

where Δ_2 denotes the horizontal Laplacian, $\Delta_2 \equiv \nabla^2 - (\underset{\sim}{\lambda}\cdot\nabla)^2$, and the Rayleigh and
Prandtl numbers are defined by

$$R = \frac{\gamma(T_2 - T_1)gd^3}{\nu\kappa}, \qquad\qquad P = \frac{\nu}{\kappa}.$$

Here γ, g, κ denote the coefficient of thermal expansion, the acceleration of
gravity, and the thermal diffusivity, respectively.

At the critical Rayleigh number R_c the right hand sides of equation (2) and
the time derivatives can be dropped and the solution can be written in the
form

$$\varphi = f(z) \sum_{n=-N}^{N} C_n \exp\{i\underset{\sim}{k}_n\cdot\underset{\sim}{r}\} \qquad \text{with } |\underset{\sim}{k}_n| = \alpha, \qquad \underset{\sim}{k}_n\cdot\underset{\sim}{\lambda} = 0 \qquad (3a)$$

$$\psi \equiv 0, \qquad \left(\frac{\partial^2}{\partial z^2} - \alpha^2\right)\vartheta = -R_c\alpha^2\varphi \qquad (3b)$$

where the summation limit N may tend to infinity and where the horizontal

directions of the vectors $\underset{\sim}{k}_n$ are arbitrary because of the horizontal isotropy
of the problem. In order to ensure a real expression (3a) the convention

$$C_{-n} = C_n^+ , \qquad \underset{\sim}{k}_{-n} = - \underset{\sim}{k}_n \tag{3c}$$

has been assumed. We also have introduced a Cartesian system of coordinates
with the z-coordinate in the vertical direction and the origin on the midplane
of the layer.

In order to determine the amplitudes C_n for a given distribution of vectors $\underset{\sim}{k}_n$
the nonlinear terms on the right hand side of equation (2) must be taken into
account. When terms up to cubic order are considered and when a weak time
dependence of the amplitudes C_n is allowed for, the following system of
equations is obtained (Schlüter et al., 1965; Busse, 1967)

$$M \frac{d}{dt}C_n = \{R-R_c - \sum_{m=1}^{N} [A(\underset{\sim}{k}_n \cdot \underset{\sim}{k}_m) + B(\underset{\sim}{k}_n \cdot \underset{\sim}{k}_m)\underset{\sim}{\lambda} \cdot \underset{\sim}{k}_n \times \underset{\sim}{k}_m \Omega]|C_m|^2\}C_n \qquad \text{for } -N \leq n \leq N \tag{4}$$

where M is a constant and the functions A and B depend only on the inner
product $\underset{\sim}{k}_n \cdot \underset{\sim}{k}_m$. We have extended the problem by admitting the possibility of a
rotation of the layer about a vertical axis. In this case the terms $-\Omega\underset{\sim}{\lambda} \cdot \nabla\Delta_2\psi$
and $\Omega\underset{\sim}{\lambda} \cdot \nabla\Delta_2\varphi$ must be added on the left hand sides of equation (2a), (2b),
respectively, and ψ no longer vanishes in solution (3). The dimensionless
rotation parameter Ω is related to the angular velocity Ω_D by

$$\Omega = \Omega_D d^2/\nu$$

Using suitable values of d and ν the centrifugal force can be kept small in
comparison with gravity over an extended region of the layer and its
horizontal isotropy can thus be preserved approximately. The introduction of
the parameter Ω serves to make an important point. For vanishing Ω the right
hand side of (4) depends only on the inner product between two $\underset{\sim}{k}$-vector and
can be written as the gradient $-\partial F/\partial C_n^+$ of the function

$$F = -\tfrac{1}{2}(R-R_c)K \sum_{m=-N}^{N} |c_m|^2 + \tfrac{1}{4} \sum_{m,n} A(\underset{\sim}{k}_n \cdot \underset{\sim}{k}_m)|C_m|^2|C_n|^2 \tag{5}$$

Because this function serves as a Lyapunov functional,

$$M \frac{d}{dt} F(C_1(t), \ldots, C_N(t)) = - \sum_n \left| \frac{\partial}{\partial C_n} F \right|^2 \tag{6}$$

the gradient property guarantees the existence of at least one stable steady solution of the form (3). Examples for those solutions are two-dimensional convection rolls in the presence of rigid, well conducting boundaries (Schlüter et al., 1965) or square cells in the case of nearly insulating boundaries (Busse and Riahi, 1980). The gradient property continues to hold when temperature dependent material properties of the fluid are taken into account in equation (2). The resulting asymmetry of the layer about the midplane tends to favor hexagonal cells which appear in two different forms depending on the sign of the asymmetry. The four patterns that are found to be physically realizable are shown in figure 1. They correspond to regular distributions of vectors k_n with N=1,2,3. Incidentally these cases are also the only ones which yield periodic patterns in the horizontal plane.

In the case of a finite rotation rate the gradient property of the system (4) is lost and the situation that all steady solutions of equations (4) become unstable does indeed occur (Küppers and Lortz, 1969). The resulting nonlinear evolution in time has been investigated with a restricted set of modes by Busse and Clever (1979) and by Busse (1984). Most striking, however, are the laboratory observations of the convection pattern with its chaotic spatial and temporal features (Busse and Heikes, 1980; Heikes and Busse, 1980) which are best visualized by the experimental movie. We do not consider here this case in more detail and instead turn to another case of phase turbulence which occurs in the presence of stress-free boundaries.

3. PHASE TURBULENCE IN THE PRESENCE OF STRESS-FREE BOUNDARIES

When Lord Rayleigh (1916) first analyzed the Bénard problem of convection in a layer heated from below, he introduced the assumption of stress-free upper and lower boundaries in order to derive a simple sine as solution for the function f(z) in expression (3a). Theoreticians have favored these boundary conditions ever since for their mathematical convenience. It is ironic that these simple

boundary conditions lead to the complex phenomenon of phase turbulence at the
onset of convection in a non-rotating layer. The cause for this complication
lies in the fact that stress-free boundaries permit large scale flows of the
form

$$\underset{\sim}{u} = -\underset{\sim}{\lambda}\times\nabla\bar{\psi}(x,y) \tag{7}$$

which possess vanishing viscous dissipation in the limit $\Delta_2\bar{\psi} \to 0$. Thus in
addition to the modes included in the representation (3c) we are forced to
admit large scale flows of the form (7). The resulting additional terms in the
system (4) of equation are responsible for the loss of the gradient property
(6) and for the appearance of additional instabilities. Busse and Bolton
(1984) have shown that a skewed varicose and an oscillatory skewed varicose
instability of rolls occur with positive growthrates in the regime

$$R-R_c > \alpha(\alpha_c-\alpha)\ 108\pi^2/7 \quad \text{for skewed varicose instability} \tag{8a}$$

$$R-R_c < \alpha(\alpha_c-\alpha)\ h(P) \quad \text{for oscillatory skewed varicose instability} \tag{8b}$$

where $R_c=27\pi^4/4$, $\alpha_c=\pi/\sqrt{2}$ are the well known critical parameters of the problem
and where $h(P)$ is an increasing function of the Prandtl number P which has
been given explicitly in the above mentioned paper. Earlier results on this
problem obtained by Zippelius and Siggia (1983) are not correct. As a
consequence of the two instabilities all steady convection solutions of the
form (3a) are unstable when the right hand side of inequality (8b) exceeds
that of inequality (8a). This situation occurs for P<0.543.

As in the case of a rotating convection layer phase-turbulent convection
occurs when the only stable steady solution in the form of rolls succumbs to
instability. The name "phase turbulence" derives from the property that the
absolute value of the wavevectors remain fixed at the critical value α_c of the
wavenumber or, more exactly, within a small neighborhood of α_c. The directions
of the $\underset{\sim}{k}$-vectors and the phases of the complex coefficients $C_n(t)$ are free to
vary, however. In order to exhibit phase turbulent convection we must rely on
the numerical integration of equations (4) with the additional terms added
that have been mentioned above. It would be nice to demonstrate this type of
phase turbulence experimentally as has been done in the case of the rotating

layer. But this possibility appears to be remote even though it is possible in principle to generate thermal convection in the presence of approximately stress-free boundaries (Goldstein and Graham, 1969).

We start with the small amplitude solutions

$$\varphi = \sin\pi(z+\tfrac{1}{2}) \sum_{n=-N}^{N} C_n(t)\exp\{ik_n \cdot r\}+\sin2\pi(z+\tfrac{1}{2}) \sum_{n,m} C_n(t)C_m(t)A_{nm}\exp\{i(k_n+k_m)\cdot r\} \tag{9a}$$

$$\vartheta = \sin\pi(z+\tfrac{1}{2}) \sum_{n=-N}^{N} \frac{C_n(t)}{\pi^2+|k_n|^2} \exp\{ik_n \cdot r\}+\sin2\pi(z+\tfrac{1}{2}) \sum_{n,m} C_n(t)C_m(t)B_{nm}$$
$$\exp\{i(k_n+k_m)\cdot r\} \tag{9b}$$

$$\psi = \sum_{n,m} G_{nm}(t) \exp\{i(k_n+k_m)\cdot r\} \tag{9c}$$

where the constants A_{nm}, B_{nm} depend on the inner products $k_n \cdot k_m$ and where only the z-independent component of ψ generated according to equation (2b) has been taken into account. The convention (3c) applies for the representation (9). In contrast to case of phase turbulence in a rotating layer it is actually necessary to take into account a finite bandwidth of values α since it can be shown that the terms proportional to C_nC_m on the right head side of equation (2.9) vanish if all vectors k_n have the same length (Schlüter et al., 1965).

The equations for the amplitudes $C_n(t)$, $G_{nm}(t)$ can be written in the form

$$M_n \frac{d}{dt} C_n = (R-R_c)C_n - \sum_{m,p,r} \delta(k_m+k_p+k_r-k_p) [d_{mprn}C_mC_pC_r+s_{mprn}G_{mp}C_r] \tag{10a}$$

$$\frac{d}{dt} G_{nm} = -|k_n+k_m|^2G_{nm} + |k_n+k_m|^{-2}q_{nm}C_nC_m \tag{10b}$$

where R_n is defined by $R_n \equiv (\pi^2+|k_n|^2)^3|k_n|^{-2}$

Equations (10a) differ from equations (4) not only through the addition of the terms involving the coefficients $G_{nm}(t)$ but also through the admission of vectors k_n with slightly different length. The δ-function can no longer be evaluated easily and the constants M_n, d_{mprn} replace the simpler expressions M and $A(\underset{\sim}{k}_m \cdot \underset{\sim}{k}_n)$ of equations (4). In fact, it would be desirable to replace the discrete vectors $\underset{\sim}{k}_n$ by a continuum and to consider partial differential equations akin to the Ginzburg–Landau type equations derived by Newell and Whitehead (1969) and by Segel (1969) for convection in the form of nearly two-dimensional rolls. From the point of view of numerical analysis, however, equations (10) are well suited, and there just remains the task of introducing appropriate selection rules.

To obtain a finite system of equations (10a) we impose periodic boundary conditions in the horizontal plane, i.e. admissable vectors $\underset{\sim}{k}$ are selected from the manifold

$$\underset{\sim}{k} = \left(\frac{2l\pi}{a} \, , \, \frac{2m\pi}{a} \right) , \qquad -\infty < l, m < \infty \qquad (11)$$

where a is the aspect ratio. Since only vectors $\underset{\sim}{k}_n$ with a length near the critical wavenumber α_c are of interest, we further restrict the set of vectors $\underset{\sim}{k}_n$ to those satisfying the condition

$$R_n < R_L \qquad (12)$$

where R_L is a parameter of the analysis. For example, using a=8, R_L=800 we find 40 vectors $\underset{\sim}{k}_n$ satisfying relationships (11), (12). This count includes their negative counterparts such that 20 complex equations (10a) must be solved since the other 20 equations represent the complex conjugates. The much larger number of equations (10b) could be reduced, in principle, since only amplitudes G_{nm} corresponding to small values of $|\underset{\sim}{k}_n + \underset{\sim}{k}_m|$ make significant contributions. But such a restriction has not been used in order to avoid the introduction of still another parameter. The conditions a=8, R_L=800, seem to produce about the right number of modes which seem to be necessary to obtain the phenomenon of turbulence. Chaotic solutions have been obtained also with 16 instead of 20 equations, but not yet with a lesser number.

Even when R_L and a are fixed the system of equations still depends on R and P and thus extensive computations must be performed in order to survey the parameter space. The numerical integrations in time were performed originally with a Runge–Kutta method, but this method turned out to be insufficiently reliable and later computations have been carried out with the Adams–Bashforth scheme. By decreasing the time step we have tested the accuracy of the solutions and found excellent convergence except for small expansions in the time scale.

A typical set of solutions is shown in figure 2, which demonstrate the rich dynamical behavior that is produced by the system. At low values of the Rayleigh number R only a few modes are excited since the decay rates are proportional to R_n-R. The time scale for the growth of instabilities is of the order $(R-R_c)^{-1}$ and thus rather long. In the case R=661 the roll solution becomes unstable through the skewed varicose instability and the convection heat transport decreases to a very low level. As the roll solution grows again to its equilibrium amplitude the process repeats itself in the form of a relaxation oscillation. At slightly higher Rayleigh numbers an aperiodic time dependence is realized as a growing number of modes participates actively in the dynamics of the system. As the Rayleigh number is further increased the time dependence of the convection heat transport becomes more chaotic in general. But there are also ranges of the Rayleigh number in which the system settles into a stable time periodic state as in the cases R=680 and R=710 of figure 1. Ultimately, as the Rayleigh number reaches a value of the order R_L the system is able to reach a stable steady state as seen in the uppermost graph of figure 1a. This steady state corresponds to convection rolls with a vector $\underset{\sim}{k}_n$ which has the lowest length permitted by inequality (12). There thus does not exist a sufficient variety of vectors $\underset{\sim}{k}_m$ which permit the growth of one of the skewed varicose instabilities. In other words, the finite aspect ratio of the convection box stabilizes the large wavelength convection roll.

4. CONCLUDING REMARKS

Because of the large number of parameters involved in the numerical integration of equations (19) it is not yet possible to make definitive statements about general properties of the solutions. The finite aspect ratio a of the problem permits a clear separation of regions of time periodic and of chaotic behavior of the system as a function of the Rayleigh number. As a and the number of participating modes increase, stable periodic solution will become rarer and disappear ultimately. The emergence of a stable steady roll solution at high values of R also seems to be a property restricted to low aspect ratios. Apart from these limitations the system (10) of equations can be useful to probe the transition from the dynamics of a few interacting modes to a large number approaching a continuum.

The onset of turbulence observed in a convection layer with rigid boundaries appears to be similar to the phase turbulence studied in this paper, at least at moderate Prandtl numbers where the skewed varicose instability is primarily responsible in limiting the stability of convection rolls with the critical wavenumber. There is no evidence for the existence of an oscillatory skewed varicose in the presence of no-slip boundaries; but the ordinary skewed varicose instability occurs in a similar fashion as in the stressfree case except that a finite excess of the Rayleigh number over its critical value is needed. In their experiments on convection in water with the Prandtl number 2.5 Heutmaker and Gollub (1987) find onset of aperiodic convection near the skewed varicose instability boundary in agreement with earlier observations by Gollub et al. (1982) and Pocheau et al. (1985). The experimental observations are complicated by the fact that another mechanism of instability called focus instability (Newell et al., 1990) which occurs near the sidewalls of the experimental convection layers and which dominates at low Prandtl numbers also introduces an aperiodic time dependence. We refer to the detailed discussion in the latter paper of the experimental evidence in the light of the phase diffusion theory.

The support of the Volkswagenstiftung for some of the research reported in this paper is gratefully acknowledged.

5. REFERENCES

Busse, F.H., The stability of finite amplitude cellular convection and its relation to an extremum principle, J. Fluid Mech. 30, 625-649, 1967.

Busse, F.H., Nonlinear properties of convection, Rep. Progress in Physics 41, 1929-1967, 1978.

Busse, F.H., Transition to turbulence via the statistical limit cycle route, in "Turbulence and Chaotic Phenomena in Fluids", ed. by T. Tatsumi, Elsevier, 1984.

Busse, F.H., Fundamentals of thermal convection, pp. 23-95, in "Mantle Convection", W. Peltier, ed., Gordon and Breach, Publ. 1989.

Busse, F.H., and R.M. Clever, Nonstationary convection in a rotating system, pp. 376-385 in "Recent Developments in Theoretical and Experimental Fluid Mechanics" (U. Müller, K.G. Roesner, B. Schmidt, eds.), Springer, 1979a.

Busse, F.H., and K.E. Heikes, Convection in a rotating layer: A simple case of turbulence, SCIENCE 208, 173-175, 1980.

Busse, F.H., and E.W. Bolton, Instabilities of convection rolls with stress-free boundaries near threshold, J. Fluid Mech., 146, 115-125, 1984.

Busse, F.H., and N. Riahi, Nonlinear convection in a layer with nearly insulating boundaries, J. Fluid Mech. 96, 243-256, 1980.

Drazin, P., and W. Reid, Hydrodynamic stability, Cambridge University Press, 1981.

Gollub, J.P., A.R. McCarriar, and I.F. Steinman, Convection pattern evolution and secondary instabilities, J. Fluid Mech. 125, 259-281, 1982.

Goldstein, R.J., and D.J. Graham, Stability of a horizontal fluid layer with zero shear boundaries, Phys. Fluids 12, 1133-1137, 1969.

Heikes, K.E., and F.H. Busse, Weakly nonlinear turbulence in a rotating convection layer, Annals N.Y. Academy of Sciences 357, 28-36, 1980.

Heutmaker, M.S., and J.P. Gollub, Wave-vector field of convection flow patterns, Phys.Rev. A35, 242-260, 1987.

Küppers, G., and D. Lortz, Transition from laminar convection to thermal turbulence in a rotating fluid layer, J. Fluid Mech. 35, 609-620, 1969.

Newell, A.C., T. Passot, and M. Souli, The phase diffusion and mean drift equations for convection at finite Rayleigh numbers in large containers I., J. Fluid Mech., in press, 1990.

Newell, A.C., and J.A. Whitehead, Finite bandwidth, finite amplitude convection, J. Fluid Mech. 38, 279-303, 1969.

Pocheau, A., V. Croquette, and P. LeGal, Turbulence in a cylindrical container of Argon near threshold of convection, Phys. Rev. Lett. 55, 1094-1097, 1985

Rayleigh, Lord, On convection currents in a horizontal layer of fluid, when the higher temperature is on the under side, Phil. Mag. 32, 529-46, 1916.

Schlüter, A., D. Lortz, and F.H. Busse, On the stability of steady finite amplitude convection, J. Fluid Mech. 23, 129-144, 1965.

Segel, L.A., Distant side-walls cause slow amplitude modulation of cellular convection, J. Fluid Mech. 38, 203-224, 1969.

Smith, B.A., et al., The galilean satellites and Jupiter: Voyager 2 imaging science results, SCIENCE <u>206</u>, 927-950, 1979.

Zippelius, A., and E.D. Siggia, Stability of finite-amplitude convection, Phys. Fluids <u>26</u>, 2905-2915, 1983.

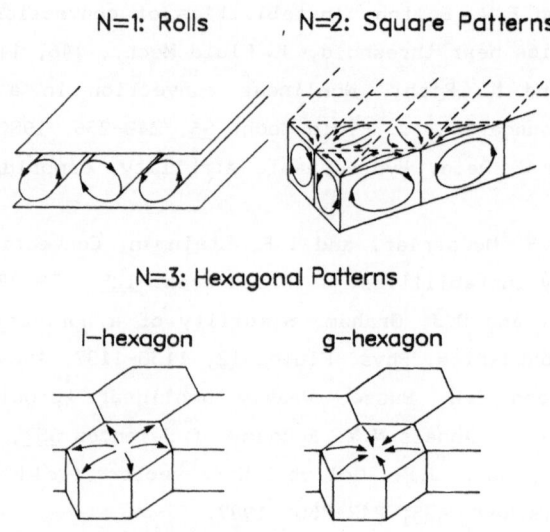

Figure 1:   Patterns of steady convection in a layer heated from below. N denotes the number of terms necessary for the description of the pattern in expression (3a).

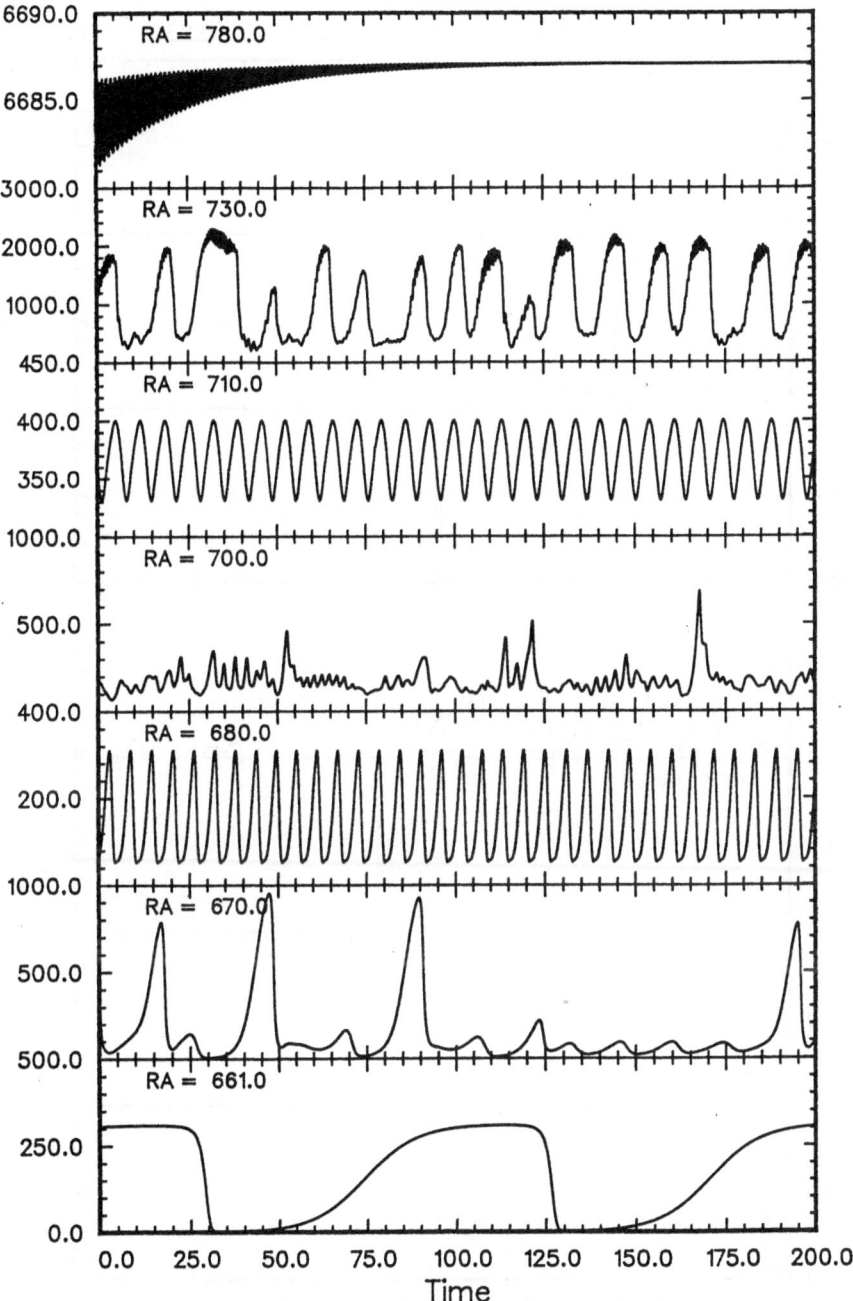

Figure 2a: Seven examples of the dependence on time of the convective heat
             transport of solutions of equation (10) for P=0.15, R_L=800, a=8. A
             timestep of Δt=0.02 has been used. RA gives the Rayleigh number
             referred to by R in the text.

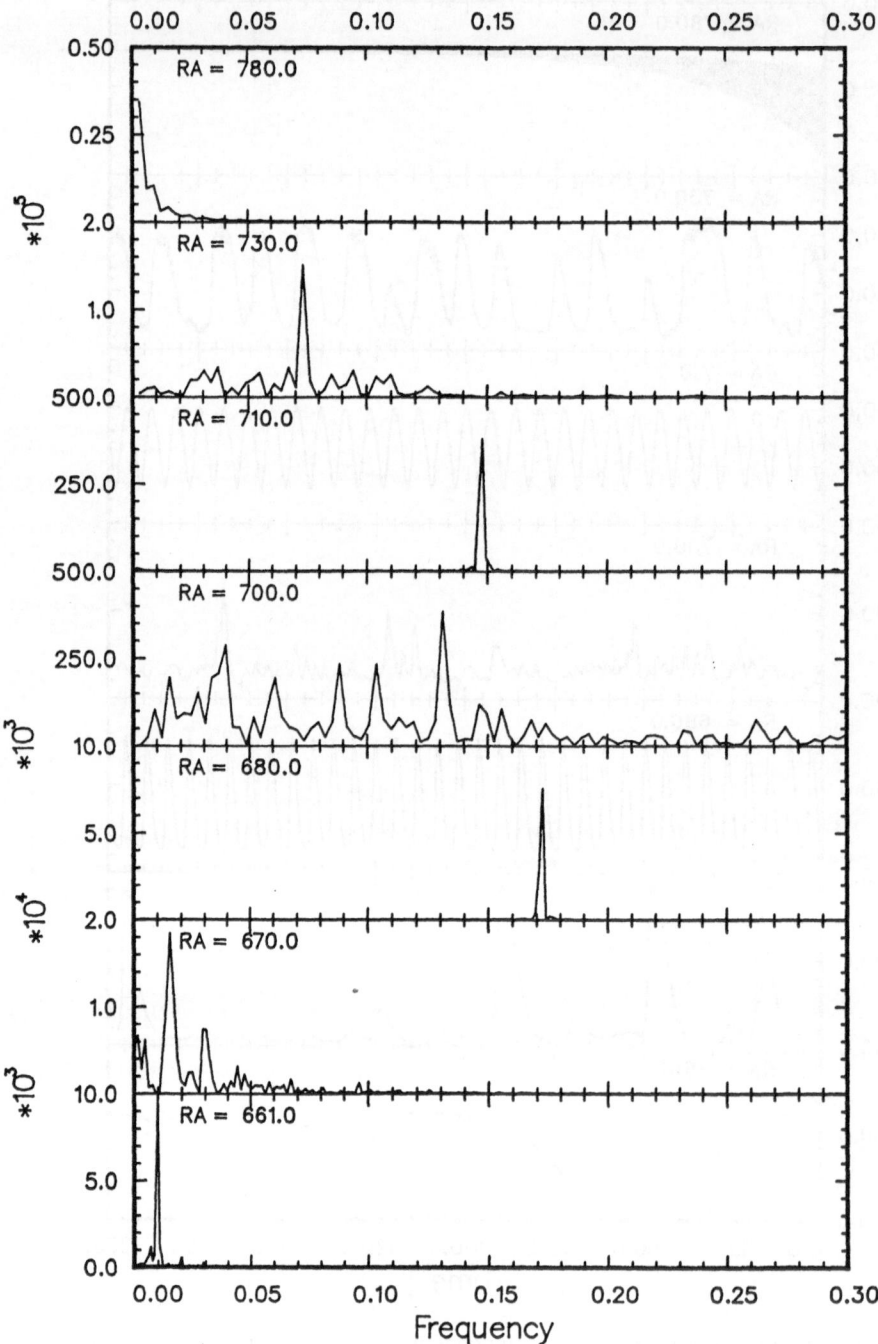

Figure 2b: Power spectra corresponding to the runs of figure 2a.

International Series of Numerical Mathematics, Vol. 97, © 1991 Birkhäuser Verlag Basel

Bifurcations in slowly rotating systems with spherical geometry

G. Dangelmayr and C. Geiger

Institut für Informationsverarbeitung, Universität Tübingen

Köstlinstr. 6, D-7400 Tübingen, FR Germany

Abstract

The effect of a slow rotation on a stationary bifurcation in an $O(3)$-equivariant system is considered. The Coriolis force induces a perturbed normal form, defined in the space of spherical harmonics of order l, in which the full $O(3)$-symmetry is broken down to $SO(2)$. We discuss the case $l = 1$ in some detail and report briefly about corresponding results for $l = 2$.

1. Introduction

In this paper we are concerned with bifurcations occuring in spherically symmetric systems which rotate slowly about a preferred axis with a constant angular velocity Ω. A prototype for this kind of problem is provided by Benard convection in a rotating spherical shell [1], e.g., the earth. The resting system ($\Omega = 0$) possesses a full $O(3)$-symmetry so that generic bifurcations can be described by normal forms that are equivariant under a representation of $O(3)$ in V_l, the space of spherical harmonics of order l. When the rotation is switched on ($\Omega \neq 0$) the $O(3)$-symmetry is broken down to $SO(2)$. Our approach to this problem is to consider Ω as a small perturbation parameter so that the symmetry breaking is weak. Related problems have been discussed by Golubitsky and Schaeffer [6] and by Dangelmayr and Knobloch [4] who analyzed the generic effects which occur when $O(2)$ is broken to $Z(2)$ in generic $O(2)$-equivariant stationary and Hopf bifurcations, respectively.

In section 2 we introduce an abstract evolution equation describing general spherically symmetric systems rotating with constant rate Ω. This equation is adopted from Sattingers [9] approach to pattern formation in a plane. In our case the presence of the rotation induces a "perturbed normal form" that describes the bifurcation behaviour close to a stationary bifurcation of the resting system. The leading term of the linear part of the perturbation possesses a well-defined structure by virtue of the special form of the Coriolis force. We have analyzed the perturbed normal forms for $l = 1$ and $l = 2$, but confine ourselves in this paper (section 3) to a discussion of the case $l = 1$. The results for the case $l = 2$ are described only briefly for reasons of space.

2. Perturbed stationary bifurcations in slowly rotating spherical systems

Following [9] we first introduce an abstract formulation for rotating spherically symmetric systems. Let H_0 be a Hilbert space with inner product $(.,.)_0$ and let H be the Hilbert space of functions $u : S^2 \to H_0$, $\xi \to u(\xi)$ ($\xi \in S^2$) for which $(u(\xi), u(\xi))_0$ is integrable. We consider an evolution equation in H of the form

$$\dot{u} = F(u, \lambda) + \Omega K \frac{\partial u}{\partial \phi}, \tag{1}$$

where $F : H \times \mathbf{R} \to H$ is a nonlinear mapping, $F(0, \lambda) = 0$, $\lambda \in \mathbf{R}$ is the bifurcation parameter and $K : H \to H$ is linear. In writing (2) we have introduced polar angles $\xi = (\theta, \phi)$ on S^2 with the polar axis coinciding with the axis of rotation. The effect of the rotation is then incorporated by the second term $\Omega K \partial u/\partial \phi$ where Ω is considered to be a small quantity. In general this form of the Coriolis force is valid provided the velocities of the physical system are described by stream functions [1]. The spherical symmetry of (1) is taken into account by assuming that F and K are equivariant under a representation T_γ of $O(3)$ in H, i.e., $F(T_\gamma u, \lambda) = T_\gamma F(u, \lambda)$, $KT_\gamma = T_\gamma K$ for all othogonal (3×3)-matrices $\gamma \in O(3)$. A general action would be of the form $T_\gamma u(\xi) = T_\gamma^0 u(\gamma \xi)$ with an action T_γ^0 in H_0. To avoid unnecessary complications we assume that T_γ^0 acts trivially. Physically this means that $u(\xi)$ represents a collection of scalar field variables.

Let $V_l = \{Y_{lm}(\theta, \phi) | -l \leq m \leq l\}$ be the space of spherical harmonics of order l. If $v \in H_0$ and $A : H \to H$ is linear and equivariant, the operation $A(vY_{lm})$ can be written in the form $(A_l v)Y_{lm}$ with $A_l : H_0 \to H_0$. We assume that for $\Omega = 0$ (1) undergoes a stationary bifurcation at $\lambda = 0$, i.e., $L := d_u F(0, 0)$ possesses a semi-simple zero eigenvalue whereas all remaining eigenvalues have negative real parts. Typically then L_l has an one dimensional kernel spanned by a $v_0 \in H_0$ for some specific l and all other $L_{l'}$ ($l' \neq l$) are invertible. Thus the center eigenspace is $(2l + 1)$-dimensional and spanned by $\{v_0 Y_{lm} | -l \leq m \leq l\}$. If a center manifold reduction is pursued for (1) we can describe the flow on the center manifold by a $(2l + 1)$-dimensional system of o.d.e.'s. For $\Omega = 0$ this system is equivariant under the standard representation of $O(3)$ in the space V_l. When the perturbation is switched on ($\Omega \neq 0, |\Omega| \ll 1$) the $O(3)$ symmetry is broken down to $SO(2)$, but the symmetry breaking terms are small. In general, if the center eigenspace is parametrized in the standard form

$$u = v_0 \{xY_{l0} + \sum_{m=1}^{l} (z_m Y_{lm} + (-1)^m \bar{z}_m Y_{l,-m}\}, \tag{2}$$

where $x \in \mathbf{R}$, $z_m \in \mathbf{C}$ ($1 \leq m \leq l$), the resulting vector field for $z \equiv (x, z_1, ..., z_l)$ has the form

$$\dot{z} = f_s(z, \bar{z}, \lambda) + f_a(z, \bar{z}, \lambda, \Omega) \equiv f(z, \bar{z}, \lambda, \Omega), \tag{3}$$

with $f_a = 0$ for $\Omega = 0$. Here, f_s is the equivariant part of f that represents the "unperturbed" system, whereas f_a contains the symmetry breaking perturbation that vanishes for $\Omega = 0$. f_a is equivariant only under the remaining $SO(2)$-symmetry which acts here according to $x \to x$, $z_m \to e^{im\alpha}z_m$ ($0 \leq \alpha \leq 2\pi$). Since f_a results from the special form $\Omega K \partial u / \partial \phi$ of the Coriolis force the symmetry breaking term f_a does not contain arbitrary $SO(2)$-symmetric perturbations. Specifically, the leading terms of f_a are given by $f_{a,0} = O(\omega^2)$ and $f_{a,m} \approx m i \omega z_m + ...$, where $\omega = \Omega(K_l v_0, v_0^*)_0$, with v_0^* spanning the kernel of the adjoint L^* of L. We have calculated the relevant terms of f_s and f_a on the basis of general equivariant operators F and K, but omit the details here. In the next section we focus onto the case $l = 1$ and comment shortly on the case $l = 2$.

3. Perturbed bifurcation equations

For $l = 1$ the vector field for (x, z) ($z \equiv z_1$) takes the form

$$\dot{x} = [\lambda + q\omega^2 + q_1\omega^4 - a(x^2 + 2|z|^2)]x + \omega^2(a_{00}x^2 + a_{01}|z|^2)x + ... \tag{4a}$$

$$\dot{z} = [\lambda + i\omega + i\omega\lambda s + p\omega^2 + p_1\omega^4 - a(x^2 + 2|z|^2)]z$$
$$+ [(i\omega b_2 + \omega^2 a_{10})x^2 + (i\omega b_2 + \omega^2 a_{11})|z|^2]z + ..., \tag{4b}$$

where the dots denote higher order terms and the coefficients q, q_1, a, p etc are real. For $\omega = 0$ we recover the common $O(3)$-equivariant vector field for the $(l = 1)$-representation. If $a \neq 0$ the relevant scaling regime is $\omega \sim \lambda \sim |z|^2 \sim x^2$ so that the dominant terms consist of the $O(3)$-equivariant part together with the rotation part $i\omega z$. Assuming that $a > 0$, the solutions of (4) tend towards the origin if $\lambda < 0$. Thus we consider only the case $\lambda > 0$ for which we introduce scaled variables $\omega = \lambda k$, $(x, z) = \sqrt{\lambda}(X, Z)$. For simplicity we set $a = 1$ which may be achieved by a further proper scaling. In terms of polar coordinates $Z = Re^{i\phi}$, (4) becomes

$$\dot{X}/\lambda^{\frac{3}{2}} = \{1 - (X^2 + 2R^2)\}X + qk\lambda X + O(\lambda^2)$$

$$\dot{R}/\lambda^{\frac{3}{2}} = \{1 - (X^2 + 2R^2)\}R + pk\lambda R + O(\lambda^2) \tag{5}$$

$$\dot{\phi}/\lambda = k + k\lambda\{s + b_1 X^2 + b_2 R^2\} + O(\lambda^2).$$

For $\lambda = 0$ the right hand sides of the (X, R)-system possess a 1-parameter family of zeros given by $(X_0(\alpha), R_0(\alpha)) = (\cos \alpha, \frac{1}{\sqrt{2}} \sin \alpha)$ ($0 \leq \alpha \leq \pi$). They correspond to the familiar axisymmetric solutions for the non-rotating ($O(3)$-equivariant) system. A member of this family can be extended to a stationary solution for $\lambda \neq 0$ if the vector $(qX_0(\alpha), pR_0(\alpha)) \in \mathbf{R}^2$ is in the image of the linearization around $(X_0(\alpha), R_0(\alpha))$ of the vector field $[\lambda^{3/2}(\dot{X}, \dot{R})]|_{\lambda=0}$, which possesses a zero eigenvalue for all α. This leads to the solvability condition $(p - q)\sin 2\alpha = 0$. Consequently, if $p \neq q$ there exist two solution branches of the form

$$R = 0, \quad X = \pm 1 + O(\lambda) \qquad\qquad (\phi = 0, \pi) \qquad (6a)$$

$$X = 0, \quad R = \frac{1}{\sqrt{2}} + O(\lambda) \qquad\qquad (\phi = \tfrac{\pi}{2}). \qquad (6b)$$

The eigenvalues of the solutions (6) in the directions of the non-zero coordinates are negative $(-2+O(\lambda))$ whereas the other eigenvalues are $(p-q)k\lambda + O(\lambda^2)$ for (6a) and $-(p-q)k\lambda + O(\lambda^2)$ for (6b). It follows that if $p \neq q$ one of the solutions (6) is a saddle and the other is a stable node. In this case there are no other fixed points in the (X, R)-plane that bifurcate from the symmetric family. For the full 3-dimensional system (6a) corresponds to stationary solutions and (6b) induces a periodic orbit in the (X, Z)-space. Both solutions lead to axisymmetric patterns of the original evolution equation, one being stationary and the other being slowly rotating. Non-axisymmetric solutions may occur if $p = q$. In this case we have to go to second order in λ. Then, putting $p = q = 0$ for simplicity, the solvability condition becomes $(A - B\cos^2 \alpha)\sin \alpha = 0$, where $A = p_1 - q_1 + \frac{1}{2}(a_{11} - a_{01})$ and $B = \frac{1}{2}(a_{11} - a_{01}) - a_{10} + a_{00}$. Again we find the stationary and rotating axisymmetric solutions (6) via $\sin 2\alpha = 0$. In addition, if $0 < A/B < 1$, there exists a further solution (X_1, R_1) that also corresponds to a periodic orbit in the (X, Z)-system and thus to a rotating pattern for the original evolution equation. This pattern is generically non-axisymmetric at order $\lambda^{\frac{5}{2}}$. A comparsion of (5) with the local normal forms for multiple bifurcations of codimension two [7] shows that if (X_1, R_1) exists, then it is stable whereas the fixed points (6) are saddles. In Figure 1 the phase portraits in the (X, R)-plane for the three cases discussed so far are shown. We note that the system (5) is a degenerate version (in virtue of the dominant $O(3)$-equivariant parts) of a system discussed by Holmes [8] in the context of generic interactions between a stationary and a Hopf bifurcation with Z_2-symmetry.

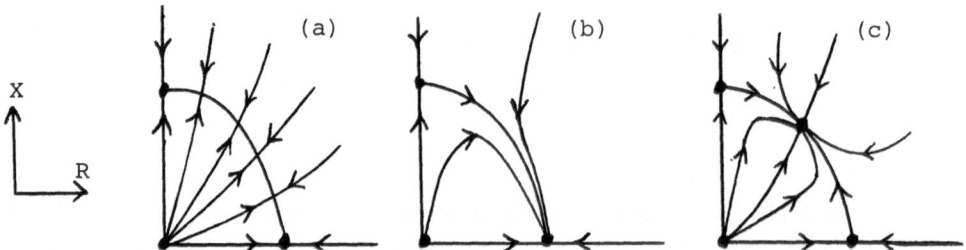

Figure 1. Phase portraits in the (X, R)-plane. (a) limiting case for $\lambda \to 0$ ($O(3)$-equivariant case) showing the family of orbitally stable fixed points. (b) $p > q$, (c) $p = q$ with a stable non-axisymmetric fixed point.

We briefly comment on the case $l = 2$, leaving further details for another occasion. In this case there exists a non-vanishing quadratic $O(3)$-equivariant term so that generically the relevant scaling is $\omega \sim \lambda \sim x \sim |z_m|$. This implies that the unperturped system possesses one parameter families of axisymmetric stationary solutions [5] for both signs of λ. For the scaled system one member of this family persists also as an axisymmetric solution for $\lambda \neq 0$ and is either rotating or non-rotating, depending on the sign of the equivariant quadratic term. These solutions were also obtained by Chossat [2,3], who analyzed their stability for the Benard convection problem without using the concept of perturbed normal forms. When the relevant perturbation coefficients are in a certain range there exists a further rotating, but non-axisymmetric solution, which, to our knowledge, was not obtained so far. The important fact with the generic ($l = 2$)-case is that all solutions which can be extended from the basic family for $\lambda = 0$ to $\lambda \neq 0$ are unstable because the family itself is unstable. A more interesting situation occurs when the quadratic terms of the unperturbed system are small, i.e, we perturb an $O(3)$-equivariant bifurcation of codimension two. In this case the unperturbed system already possesses a two parameter family of non-axisymmetric solutions [5] (stable or unstable depending on the coefficients) and also stable and unstable one parameter families of axisymmetric solutions. Perturbing this system by a slow rotation induces a rich variety of behaviours which will be discussed elsewhere.

References

1. F.H. Busse, Astrophys. J. **159**, 629 (1970)
2. P. Chossat, These de 3^l cycle, Universite de Nice (1977)
3. P. Chossat, SIAM J. Appl. Math. **37**, 624 (1979)
4. G. Dangelmayr and E. Knobloch, Nonlinearity (to appear)
5. M. Golubitsky, I. Stewart and D.Schaeffer, *Singularities and Groups in Bifurcation Theorie, Vol II*, Springer 1988.
6. M. Golubitsky and D. Schaeffer, Proc. Symp. Pure Math. **40**, Part I, 499 (1983)
7. J. Guggenheimer and P. Holmes, *Nonlinear Oscillations, Dynamical Systems and Bifurcation of Vector Fields*, Springer 1983
8. P. Holmes, in: *Nonlinear Dynamics*, R.H.G. Helleman (ed.), Ann. New York Acad. Sci., Vol. 357, 473 (1980)
9. D.H. Sattinger, *Group Theoretic Methods in Bifurcation Theorie* , Springer 1979

International Series of Numerical Mathematics, Vol. 97, © 1991 Birkhäuser Verlag Basel

DOMOKOS, Gábor[1]

AN ELASTIC MODEL WITH CONTINUOUS SPECTRUM

1. Introduction

Elastic structures can be divided into two classes. In the case of discrete models the number of critical loads is finite, in the case of continuous structures this number is countably infinite. However, both classes have a common feature: their spectrum is discrete, i.e. their critical loads are distinct. In conclusion we can say, that elastic structures have typically discrete spectrum. Moreover it is worth mentioning, that the number of critical loads and the degree of freedom of the structure are typically equal.

The aim of the present paper is to demonstrate an elastic model with continuous spectrum, i.e. with a critical finite interval of the load parameter.

2. The model

The model to be treated is a symmetric three-hinged structure subjected to the conservative, quasi-static vertical load Λ. (Fig.1.) Similar models have been treated in [1] ,[2] and [4], but the spring stiffness c of the torsional spring is regarded as constant. In addition [4] deliveres the finite element analysis of the compressible, continuous beam. In [3] the global stability analysis of the aforementioned structure is delivered. The unstrained length of both bars is denoted by L, the axial stiffness a of both bars is unit. The potential energy of the structure will be written in terms of the state variables x (denoting the shortening of each bar) and α (denoting the angle between the bars and the vertical direction). In addition the parameter x_0 is introduced, denoting the value of x at the last vertical position of the structure. The spring constant c in the middle torsional spring depends on x_0, i.e. the system has a certain kind of „memory“, since the stiffness depends on a previous state of the structure. The form of the function $c(x_0)$ is to be specified later. If the structure is in the vertical (trivial) position ($\alpha = 0$), then the „last vertical position“ is identic with the actual one, therefore in this case $x = x_0$ and x_0 has to be treated as a variable. In the case of $\alpha \neq 0$ typically $x \neq x_0$ and x_0 has to be treated as a constant. The potential energy can be expressed as follows:

$$V = x^2 + 2\alpha^2 c(x_0) + 2\Lambda(L - x)\cos\alpha \qquad (1)$$

[1] Tech. Univ. of Budapest, Dept. for Strength of Materials, H-1521 Budapest, Hungary

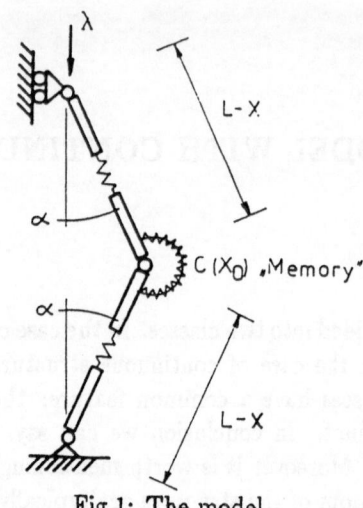

Fig.1: The model

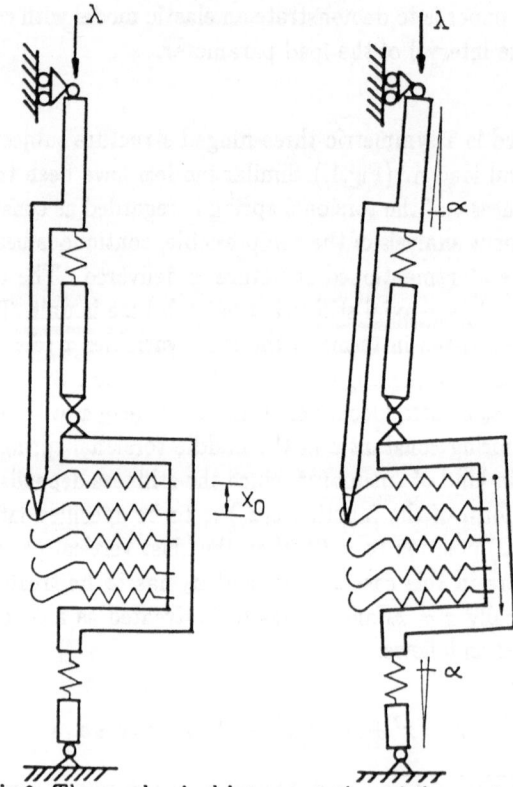

Fig.2: The mechanical interpretation of the model

The ambivalent behaviour of x_0 just described results in certain difficulties when calculating the partial derivatives of V at $\alpha = 0$. To overcome this difficulties we separate three possible cases:

Case 1. If we differentiate with respect to α, then the term $c(x_0)$ is to be treated as constant.

Case 2. If we differentiate with respect to x and no derivation with respect to α occurred previously („pure x " derivatives), then the term $c(x_0)$ is to be treated as a function of x, i.e. $c(x_0) = c(x)$. However in this case $\alpha = 0$, i.e. the derivatives of $c(x)$ are always multiplied by zero.

Case 3. If we differentiate with respect to x and the expression was already differntiated with respect to α previously („mixed " derivatives), then the term $c(x_0)$ is to be treated as constant, since in this derivatives the value of α infinitesimally differs from zero.

Since the above three cases cover all possibilities,we conclude, that when executing the differentation the term $c(x_0)$ can be treated *formally* as constant. Partial derivatives will be denoted by lower indices. The equations

$$
\begin{aligned}
V_x &= 2x - 2\Lambda \cos \alpha &= 0 \\
V_\alpha &= 4\alpha c(x_0) - 2\Lambda(L - x)\sin \alpha &= 0
\end{aligned}
\tag{2}
$$

are satisfied by the trivial solution $\alpha = 0; \Lambda = x$. As stated before, $x = x_0$ holds here, as well. In accordance with the aim formulated in the introduction we require, that the state of equilibrium should be critical for all values of Λ along this trivial path, i.e. the determinant of the Hessian matrix H^0 should vanish:

$$
\det H^0 = 4(2c(x_0) - x_0(L - x_0)) = 0
\tag{3}
$$

yielding

$$
c(x_0) = \frac{1}{2}(Lx_0 - x_0^2).
\tag{4}
$$

By substituting (4) into (1) we arrive at the energy function of the wanted system:

$$
V = x^2 + 2\alpha^2(Lx_0 - x_0^2) + 2\Lambda(L - x)\cos \alpha.
\tag{5}
$$

3. The interpretation of the model

The model presented in the previous section might have seemed somewhat „academic" to the reader. Now we are going to demonstrate, how this type of structure with memory can be interpreted mechanically. Fig.2. shows a system which displays exactly the behaviour of the model described in section 2. By applying the vertical compressive force Λ the bars will shorten by $x = x_0$ and the needle attached to the upper bar will be driven by the same distance into the hooks of the horizontal springs attached to the lower bar. If the system buckles, the torsional rigidity supplied by the horizontal spring system „remembers" its last vertical position. The discrete horizontal springs can be generalized as a continuous („Winkler") elastic support with spring constant $k(x_0)$. The latter function can be found by equating the resultant moment of the continuous spring system on the interval $[0, x_0]$ with the spring constant function $c(x_0)$ calculated in (4):

$$\frac{1}{2}(Lx_0 - x_0^2) = 2 \int_0^{x_0} x_0^2 k(x_0) \mathrm{d}x_0 \tag{6}$$

yielding

$$k(x_0) = \frac{1}{4x_0}\left(\frac{L}{x_0} - 1\right). \tag{7}$$

4. The initial post-buckling behaviour

The initial post-buckling behaviour of our model will be studied along the trivial equilibrium path by the means of Elementary Catastrophe Theory (see [1]). The 6-jet of (5) can be expressed as

$$j^6 V = x^2 + x_0 \alpha^2 x + \frac{x_0}{12}(L - x_0)\alpha^4 - \frac{x_0}{12}x\alpha^4 - \frac{x_0}{360}(L - x_0)\alpha^6. \tag{8}$$

To separate the active and the passive part the smooth transformation

$$\begin{aligned} w_1 &= x + \frac{1}{2}x_0\alpha^2 - \frac{1}{24}x_0\alpha^4 \\ w_2 &= \alpha \end{aligned} \tag{9}$$

will be applied yielding

$$j^6 V = w_1^2 + \frac{x_0}{12}(L - 4x_0)w_2^4 + \frac{x_0}{360}(16x_0 - L)w_2^6. \tag{10}$$

By excluding the degenerate case $x_0 = 0$ from the mechanical point of view the analysis of the coefficients of the powers of w_2 yields the results summarized in table. The equlibrium path is illustrated in Fig.3.

Range	Type of catastrophe
$0 < x_0 < \frac{L}{4}$	Standard cusp
$x_0 = \frac{L}{4}$	Standard butterfly
$\frac{L}{4} < x_0 \leq L$	Dual cusp

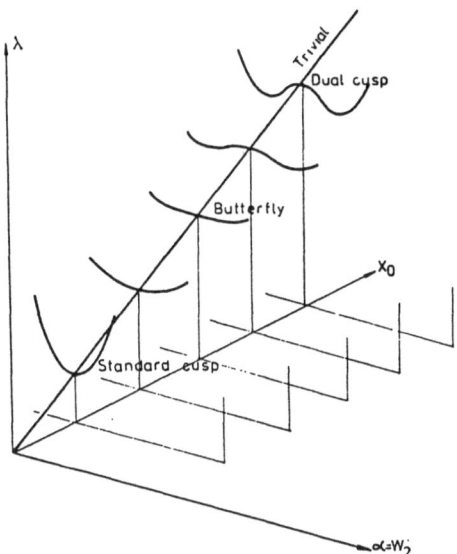

Fig.3: The equilibrium path

5. Concluding remarks

As mentioned in the introduction, the number of critical loads is typically equal to the mechanical degree of freedom of the system. Following this line of thought our model has a degree of freedom equivalent to the continuum, which is superior to the degree of freedom of a continuous beam. The explanation to this is the fact, that in the beam the infinitesimal „line elements" have a sequential causal order,i.e. the beam is continuous, whereas in our model the infinitesimal springs are arranged in a parallel way and the function describing their length can be discontinuous. By the perturbation of the spring constant function $k(x_0)$ any finite range of the trivial equilibrium path of a continuous beam may be simulated. In this respect our model can be regarded as a degenerate case of the „usual" mechanical continuum.

6. References

[1] POSTON, T. & STEWART, I.N. (1978):*Catastrophe theory and its applications* Pitman, London.

[2] GOLUBITSKY, M. & SCHAEFFER, D. (1979):*A theory for imperfect bifurcation via singularity theory* Comm. Pure and Appl. Math. **32**, 21-98.

[3] GÁSPÁR ,Zs. & DOMOKOS ,G. (1990): *Global description of a simple mechanical model* Proc. Stab. Steel Struct.,Vol 1., pp.69-76. 1990, Budapest, ed.: M. Iványi

[4] GURAN, A.: *On the stability of an elastic, imerfect column including axial compressibility* ZAMM (in press)

International Series of Numerical Mathematics, Vol. 97, © 1991 Birkhäuser Verlag Basel

Mechanistic Requirements for Chemical Oscillations

M. Eiswirth[1], *A. Freund*[2] and *J. Ross*[3]

[1] Fritz-Haber-Institut der Max-Planck-Gesellschaft, Faradayweg 4-6,
W-1000 Berlin 33, FRG

[2] Institut für Physikalische Chemie, Universität Würzburg, Marcusstr.9-11,
W-8700 Würzburg, FRG

[3] Dept. of Chemistry, Stanford University, CA 94305, USA

1. INTRODUCTION

Oscillating chemical reactions have been the subject of many studies in the last decades. The importance of both an autocatalysis and a negative feedback loop has been discussed by several authors /1-4/. The goal of the present work is to find the minimal mechanistic requirements sufficient to produce oscillations, i.e. how the postive and negative feedbacks have to be realized in a mechanism in order to allow for periodic solutions.

As a first step the species are classified into essential and nonessential ones for the occurrence of oscillations and the further considerations are limited to the mechanistic structures of the essential species. Subseqently theorems from Clarke's stoichiometric network analysis /5-7/ are used to obtain mechanistic features which allow an instability (i.e. the occurrence of unstable fixed points). We restrict ourselves to mechanism with exactly one source of instability, to which additional species and reactions are added in such a way that they give rise to oscillations. There are several qualitatively different ways how an oscillating mechanism can be constructed with this procedure, which thus leads to the distinction of oscillator categories based on mechanistic criteria.

2. MECHANISTIC FEATURES ALLOWING FOR INSTABILITY

Chemical species in an oscillatory sytem are called nonessential, if, on being held constant, oscillations in the other species continue, whereas oscillations in all species cease, if the concentration of an essential one is fixed. A detailed description of the possible roles of nonessential species and operational means to identify them have been given in /8,9/. In the following only essential species are taken into account.

In stoichiometric network analysis SNA /5-7/ a network (≡mechanism) is considered to consist of a set of reactions with given stoichiometries and rate laws (kinetic exponents), but with arbitrary nonnegative rate constants. A network is stable, if no unstable fixed points occur for any choice of the parameters, otherwise it is unstable. Using SNA, the set of all stationary states (referred to as currents) can be parametrized as linear combinations with nonnegative coefficients of so-called extreme currents. These are irreducible subsets of the orignal set of reactions for which the stationary state condition can still be fulfilled.

As shown by Clarke /5/, a current cannot be unstable, unless it contains an autocatalytic process (called irreversible current cycle). In order to exhibit a nontrivial stationary state, a current cycle must have an exit reaction by which an autocatalytic species leaves the cycle. A current cycle is denoted as strong, critical or weak, if the order of the cycle reaction is higher than, equal to or lower than the order of the exit process. Weak current cycles are always stable, strong ones are always unstable, while critical cycles can give rise to an instability depending on additional features. This leads to a distinction of two types of oscillators, namely category 1, if the instability is based on a critical, and 2, if it is caused by a strong current cycle.

3. DEVELOPMENT OF PROTOTYPE OSCILLATORS

In the next two sections, prototype models are developed, starting from the two unstable features.

Prototypes should be as simple as possible, but enough species and reactions ought to be included so that physically not realistic stuations (such as infinite concentrations or nongeneric bifurcations) are avoided.

For convenience we use a diagramatic notation of currents introduced by Clarke /5/. The species symbols are connected by arrows which denote the reactions. The stochiometric coefficients and the kinetic exponents are encoded by the number of barbs and feathers on these arrows: The total number of feathers (barbs) at a reactant (product) denotes the stoichiometric coefficient of its consumption (formation), while the number of left feathers is equal to the kinetic exponent of the reactant.

3.1 Category 1: Critical Current Cycles

The simplest critical current cycle is shown in fig. 1a. It gives rise to an instability, if the autocatalytic species X is consumed by an exit reaction with another species Y (fig. 1b). This unstable feature by itself gives rise to a single saddle point, leading to unlimited growth of one species. To avoid this, two stable currents are added which impose bounds to the concentrations of X and Y. There are several possibilities for such stable currents, simple examples are included in fig. 1c. Such a network exhibits bistability (stemming from a cusp). In order to obtain oscillations, a negative feedback loop has to be introduced into the unstable current via another species Z, which can be achieved in two qualitatively different ways: Either Y is formed by the feedback species Z (fig. 1d) or Z is consumed by the critical current cycle controlling the autocatalytic production of X (fig. 1e). These two subcategories are denoted 1B and 1C, respectively, because the former are usually batch, the latter CSTR oscillators. For the 1C-type two further subcategories 1CX and 1CW are distinguished, the unstable currents of which are shown in figs. 1f and g: in the first case there is an inflow of X in a stable current; when this is not the case another essential species W ('recovery species') is required for the occurrence of oscillations /8,9/.

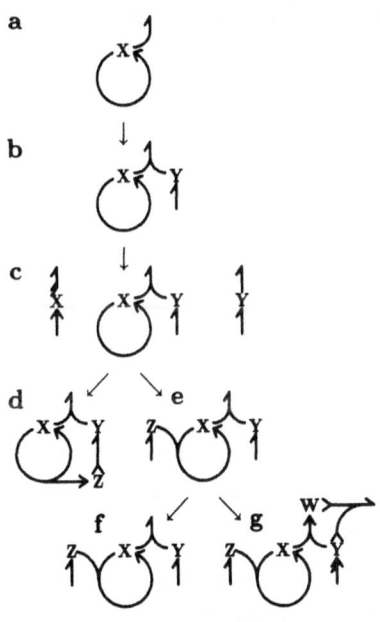

Fig. 1:

Development of the CCC prototypes (category 1): *a)* current diagram of a critical current cycle (first order autocatalysis of X), *b)* addition of an exit species Y, *c)* additions of stable currents which limit X and Y, *d* and *e)* distinction of two subcategories: *d)* Y is formed by Z≡1B type, *e)* Z is consumed by the autocatalysis≡1C type, *f* and *g)* further subdivision of the 1C type: *f)* the recovery of X is assured by an inflow of X in a stable current (not shown)≡1CX type, *g)* the recovery of X is achieved indirectly through the recovery species W≡1CW type.

3.2 Category 2: Strong Current Cycles

The basic unstable feature of a category 2 oscillator (shown in fig. 2a) allows for un-limited growth of X, which can be avoided by inclusion of the reverse reaction. The resulting network in fig. 2b exhibits bistability (stemming from a cusp bifurcation). Oscillations can be obtained, if a negative feedback similar to the 1C-type is intro-duced (fig. 2c). No subtypes need to be distinguished in this category.

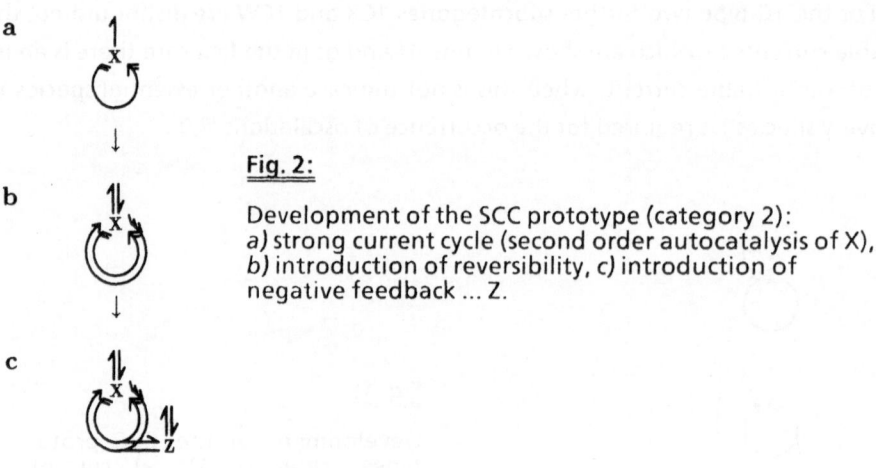

a

b

Fig. 2:

Development of the SCC prototype (category 2):
a) strong current cycle (second order autocatalysis of X),
b) introduction of reversibility, c) introduction of
negative feedback ... Z.

c

4. DISCUSSION

The categorization of oscillators presented in the previous section is probably not exhaustive. On the other hand it has been applied to 25 oscillating mechanisms /8/ taken from the literature and all fall into the 4 categories described.

Operational methods (also applicable experimentally) have been suggested /8,9/ to assign an oscillator to its category and to identify the role each species plays in the mechanism, namely nonessential, autocatalytic, exit, feedback or recovery species. These methods may be useful in mechanistic investigations.

The present study was restricted to simple limit cycle oscillations. The extension to more complex dynamical behavior (such as mixed-mode oscillations, chaos) will be the subject of future work.

REFERENCES

/1/   J. J. Tyson, J. Chem. Phys. **62**, 1010 (1975).

/2/   U. F. Franck, Ber. Bunsenges. Phys. Chem. **84**, 334 (1980).

/3/   U. F. Franck, in "Temporal Order", p. 2-12, L. Rensing and N. I. Jaeger, Eds., Springer, Berlin 1985.

/4/   I. R. Epstein, K. Kustin, D. D. DeKepper and M. Orban, Scient. Am. **248**, 96 (1983).

/5/   B. L. Clarke, Adv. Chem. Phys. **43**, 1 (1980).

/6/   B. L. Clarke, J. Chem. Phys. **75**, 4970 (1981).

/7/   B. L. Clarke, Cell Biophys. **12**, 237 (1988).

/8/   M. Eiswirth, A. Freund and J. Ross, Adv. Chem. Phys. (in press).

/9/   M. Eiswirth, A. Freund and J. Ross, J. Phys. Chem. (submitted).

International Series of Numerical Mathematics, Vol. 97, © 1991 Birkhäuser Verlag Basel

ENVELOPE SOLITON CHAOS MODEL FOR MECHANICAL SYSTEM

M.S. El Naschie[*] and T. Kapitaniak[**^]

[*]Sibley School of Mechanical and Aerospace Engineering, Cornell University, Ithaca, USA

[**]Department of Applied Mathematical Studies and Center for Nonlinear Studies, University of Leeds, U.K.

The paper considers the possibility of purely spatial chaos of envelope soliton localization in long elastic strings. Typical similarities and differencies between strange chaotic attractors in deterministic systems and random behaviour are also discussed.

1. INTRODUCTION

There has been an increased interest in complexity and spatial chaos in recent years [1-5]. In particular, since the discovery of loop solitons and their interaction [6,7] as well as the possibility of a purely spatial chaos [8], there is a renewed interest in the Euler elastica and its applications [9].

El Naschie established the connection between the loop soliton and the Milke-Holmes chaotic elastica using a dynamical version of the Euler elastica [5,9]. He also drew attention to the possibility of interpreting the instability waves in curved compressed thin material surfaces (i.e. shells) as envelope soliton turbulence [9,10]. Thompson and Virgin were the first to publish a numerical confirmation of Milke and Holmes theoretical results using an elementary but neat model [11].

In the present work we give numerical confirmation for the conjecture made in [5,10] that elastic material surfaces, such as shells exhibit under certain conditions purely spatial and statical soliton chaos.

2. INSTABILITY WAVES IN AN ELASTIC STRUCTURE - ENVELOPE SOLITON

Consider the following nonlinear partial differential equation which may be used to describe the propagation of buckling waves in an elastic medium such as the axsymmetrical deformation of an axially compressed cylindrical shell

$$\alpha W'''' + \sigma W'' + c_1 W - c_2 W^2 + \rho \ddot{W} = 0 \tag{1}$$

For a radial strain obeying a logarithmic law, this equation was used in [9, 12] to study the instability waves due to buckling.

Now depending on the number of slow spaces and slow time, different reduced differential equations for the complex amplitude of deflection A may be obtained. For instance, using

$$x = x_0 \ , \ x_1 = \varepsilon x_0 \ , \ x_2 = \varepsilon^2 x_0 \ , \ t = t_0 \ , \ t_1 = \varepsilon t_0 \ , \ t_2 = \varepsilon^2 t_0$$

one finds the following Ginzburg-Landau type equation [13].

$$\alpha_1 A'' - \alpha_2 A + i\alpha_3 A' + i\alpha_4 \dot{A} + \alpha_5 A |A|^2 = 0 \tag{2}$$

where $i=\sqrt{-1}$, $(')=d/dx$, $(\ \dot{}\ )=d/dt$.

On the other hand the P. D. E may be drastically reduced to an O. D. E by reducing stretching to only $x_1 = \varepsilon x$. This leads to the following stationary nonlinear Shrödinger equation [13]

$$\alpha_1 A'' - \alpha_2 A + \alpha_5 A |A|^2 = 0 \tag{3}$$

This equation is easily integrated by elementary methods and gives the soliton solution

$$A = (6/\sqrt{19}) \ \text{sech} \ (19/12)x_1 \tag{4}$$

for $\alpha = \sigma = c_2 = c_3 = \rho = 1$. The homoclinicity of this solution may be established easily as shown in [9].

An optimum choice of the number of slow spaces and slow times which restores the dynamical character of the problem is however when we take

$$x_1 = \varepsilon x_0 \ , \ t_1 = \varepsilon t_0 \ , \ t_2 = \varepsilon^2 t_0 \ .$$

This leads to the nonlinear Shrödinger equation

$$\alpha_1 A'' - \alpha_2 A + i\alpha_4 \dot{A} + \alpha_5 A |A|^2 = 0 \tag{5}$$

with the well known solution [13]

$$A(x,t)\big|_{t=t_0} = a\ \text{sech}(bx)\ \cos(cx) \tag{6}$$

where a, b and c are constant.

Either way we expect spatial forcing to yield spatial envelope soliton chaos. Thus we consider first the periodically forced equation

$$A'' + k_1 A' - k_2 A + k_3 A^3 = k_4 \cos k_5 s . \tag{7}$$

The results of the numerical integrations for different parameter values which are : k_1=0.01, k_2=0.25 , k_3=19/(4)(18) , k_5=1 and different values of k_4 are shown in Figure 1(a). They fully confirm the expectations expressed earlier in [5,9,12].

Subsequently the forcing by band-limited white noise

$$A'' + k_1 A' - k_2 A + k_3 A^3 = \mathcal{A} \sum_{i=1}^{N} \cos(\nu_i s + \gamma_i) \tag{8}$$

where $\mathcal{A}$ is constant ν_i and γ_i are random variables [14]; ν_i and γ_i are distributed between the band frequencies ν_{min} and ν_{max}, γ_i are independent random variables with uniform distribution on the interval [0, 2Π] is considered .

In Figure 1(b) we show a spatial plot of the same system as in Figure 1(a) once more, however this time with band-limited white noise forcing.

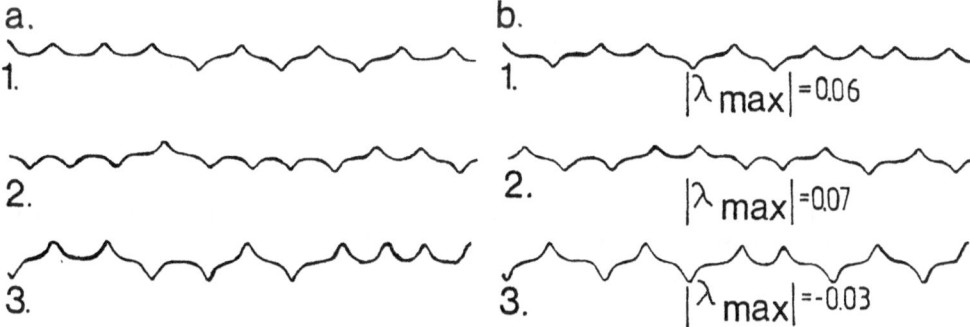

a.

1.

2.

3.

b.

1. $|\lambda \text{max}| = 0.06$

2. $|\lambda \text{max}| = 0.07$

3. $|\lambda \text{max}| = -0.03$

Figure 1: Examples of spatial ptot: (a) eq. (7), k_1=0.01, k_2=0.25, k_3 =19/(4x18), k_5=1, A(0)=1.37649, Ȧ(0)=0, (1) k_4=0.001, (2) k_4 =0.002, (3) k_4 =0.003, (b) eq.(8), k_1 –k_3 as in (a), (1) $\mathcal{A}$=0.001, (2) $\mathcal{A}$=0.002, (3) $\mathcal{A}$=0.003, $|\lambda_{max}|$– the most probable value of the distribution of maximum Lyapunov exponents.

In all above examples the soliton strange behaviour would thus have a similar spatial representation. However in many cases we have no sensitive dependence on initial conditions - Figure 1(b3) (see also [14]). The examples of Poincare maps for periodically forced equation (7) and system forced by band-limited white noise (8) are shown in Figure 2.

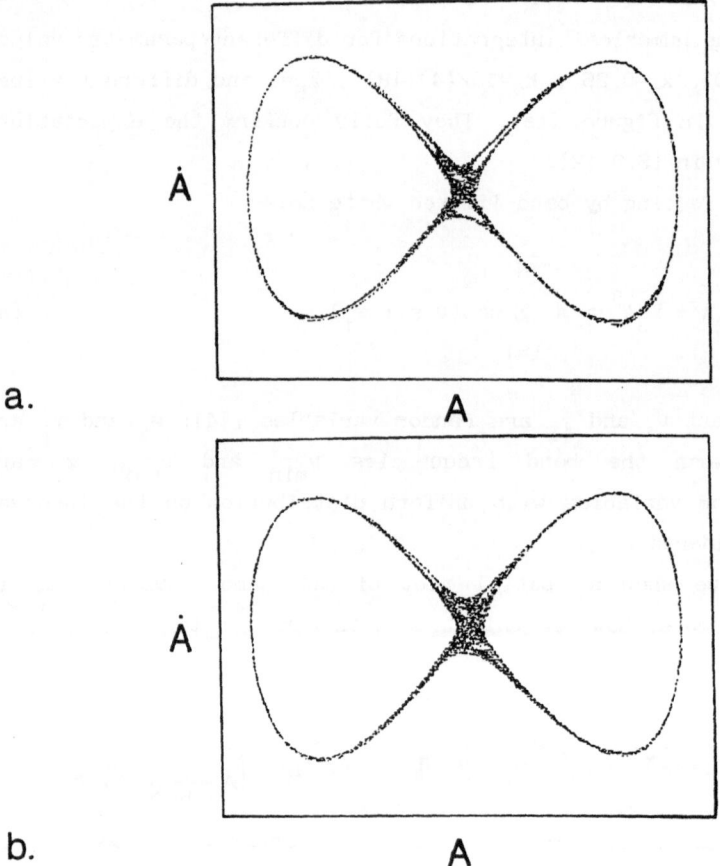

Figure 2. The Poincare maps of the eq. (7) - (a) and eq.(8) - (b), parameters like in Figure 1(a1) and 1(b1)

3. Conclusions

In this work we stress the view expressed probably for the first time by Y. Ueda [15] that soliton and chaos should not be regarded as contradictory. We showed that buckling deformation of shells may take the form of envelope soliton and may be perturbed by shape imperfection into spatial chaos. The conclusion drawn here admittedly from an extremely simplified model, is nevertheless of fundamental importance since we must recognize the existence of internal stochasticity due to the nonlinear deterministic dynamics of a structure besides the external stochasticity due to random imperfection. More examples of soliton chaos models can be found in [17].

Comparison of the deterministic system (7) and stochastic one (8) shows that it is not easy if at all possible to distinguish between chaotic and stochastic behaviour, especially that in our case the noise is small enough and response is still smooth enough to develop typical differences which allow to distinguish between these types of behaviour [16].

^ Permanent address: Institute of Applied Mechanics, Technical University of Lodz , Stefanowskiego 1/15,   90-924 Lodz, Poland.

REFERENCES

1. S. Wolfram , Nature 311, 419-424 (1984)
2. B.M. Herbs and W.-H. Steeb, Z. Naturforsch, 43a, 727-733 (1988)
3. P. Reichert and R Schilling, Phys. Rev. 25B, 917-922 (1984)
4. H.G Purwins, G Klempt and J Berkemeier, Festkorperprobleme 27, 27 -
    -61 (1987)
5. M.S.El Naschie and S.Al Athel, Z.Naturforshung, 449, 645-650, (1989)
6. Y.H. Ichikawa and N. Yajima , In Statistical Physics and Chaos
    in Fusion Plasmas, C.W. Haston and L.E. Reichl (eds), 79-90, Wiley,
    New York (1984)
7. K. Konno and A. Jeffrey, In Advanced in Nonlinear Waves,
    vol.1, L. Debnath (ed.), 162-183, Pitman, London (1984)
8. A. Mielke and P. Holmes, Arch. Rat. Mech. Anal., 101, 4,
    319-348 (1988)
9. M.S. El Naschie, J. Phys. Soc. Japn., 58, 12, 4310-4321 (1989)
10. M.S. El Naschie, K.S.U. J.of Engineering Science, 14, 437-444 (1988)
11. J.M.T. Thompson and L.N. Virgin, Phys. Lett. 134A, 491-496 (1988)
12. M. S. El Naschie, S. Al Athel and A.C. Walker , In
    Proceedings IUTAM Symposium Nonlinear Dynamics in Engineering, W.
    Schiehlen (ed.) 67-74, Springer, Berlin (1990)
13. P. G. Drazin and R.S. Johnson, Solitons, Cambridge University
    Press, Cambridge, (1989)
14. T. Kapitaniak, Chaos in Systems with Noise, World Scientific
    Singapore (1988), second edition (1990)
15. Y. Ueda and A. Noguchi, J. Phys. Soc. Japan, 52, 713-715, (1983).
16. D. Sigeti and W. Horsthemke, Phys. Rev. A35, 2276 (1987)
17. M.S. El Naschie and T. Kapitaniak, Phys. Lett. 147A, 275-281 (1990).

International Series of Numerical Mathematics, Vol. 97, © 1991 Birkhäuser Verlag Basel

Rolling Motion of Ships Treated as Bifurcation Problem

Falzarano J.*   Steindl A.[t]   Troesch A.[‡]   Troger H.[t]

Abstract

The behavior of a vessel slowly turning in waves is studied as a nonlinear bifurcation problem making use of the bifurcation package BIFPACK ([9]). For two different shipmodels the oscillations of the ship are studied. Both external and parametric excitation effects resulting from the wave motion are considered.

1 Introduction

Vessel operators are frequently forced to change vessel heading in order to reduce undesirable rolling motions excited from waves. This rolling motion may result from two different types of excitation. One — and this is easy to understand — follows from an external excitation, from waves travelling in the direction vertical to the ship motion ([1,2]), that is, from beam seas. The frequency of the ship roll motion is equal to the exciting frequency or is a subharmonic oscillation ([1]). The other type of excitation is parametric and follows from waves travelling in the same direction as the ship ([3,4]), that is, from head or following seas. The physical explanation for the parametric excitation is that the hydrostatic spring (see equation (1) below) is time dependent if waves of about the length of the ship travel along the ship. In this paper we intend to give an analysis of the roll behavior of a vessel slowly turning in waves. By slowly turning we understand that the heading angle ϑ of the direction of the ship motion, which will be the distinguished or bifurcation parameter, is varied quasistatically.

Further we compare the results of two different mechanical models one with one degree of freedom and the other with three degrees of freedom.

2 Mechanical models and equations of motion

If we restrict the analysis to a *single* degree of freedom model the investigation reduces to a Duffing equation for the problem with external excitation ([1,2,3]) following from beam seas or an equation with parametric excitation following from head or following seas. In the general case of quatering seas these excitations are acting simultaneously. The equation of motion can be written in the form

$$(I + A)\ddot{\varphi} + B\dot{\varphi} + D\dot{\varphi}|\dot{\varphi}| + (1 + \varepsilon GM \cos \vartheta \cos \omega t)(C_1\varphi + C_3\varphi^3 + C_5\varphi^5) = F \sin \vartheta \cos(\omega t + \gamma) \, . \quad (1)$$

Here is A the added mass, εGM the time varying hydrostatic force, C_1, C_3, C_5 are the coefficients of the hydrostatic restoring force (spring) and F is the linear hydrostatic exciting force. In this model all coefficients are assumed to be independent of the exciting frequency of the waves.

---

*The University of New Orleans, New Orleans, USA
[t]Technische Universität Wien, Wien, Austria
[‡]The University of Michigan, Ann Arbor, USA

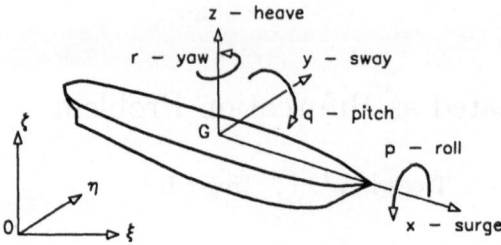

Figure 1: Vessel with six degrees of freedom

The second model has three degrees of freedom. They are the roll angle φ, the yaw angle α and the sway velocity v (Fig.1). The choice of these degrees of freedom follows from a linearization of the full nonlinear equations of motion because then roll, yaw and sway decouple from heave, pitch and surge. Further the frequency dependence of the coefficients is included. This dependence on the excitation frequency is calculated from linear hydrodynamics. Hence the added mass matrix $\mathbf{A}$, the linear matrix $\mathbf{B}$ damping and the linear displacement terms in the equations of motion are frequency dependent. To these linear equations of motion a nonlinear roll damping term $\mathbf{B}_q$ and a nonlinear roll restoring force GZ are added. Now the equations take the following form

$$(\mathbf{M} + \mathbf{A}(\omega))\ddot{\boldsymbol{x}} + \mathbf{B}(\omega)\dot{\boldsymbol{x}} + \mathbf{B}_q(\omega, \dot{\boldsymbol{x}}) + \mathbf{GZ}(\boldsymbol{x}) + \varepsilon\mathbf{GM}(\omega)\cos\omega t\boldsymbol{x} = \mathbf{F}(\omega)\cos(\omega t + f_p) , \quad (2)$$

with $\ddot{\boldsymbol{x}} = (\dot{v}, \ddot{\varphi}, \ddot{\alpha})^T$. The 3×3 matrices and 3-vectors not yet explained are: $\mathbf{M}$ is the mass and inertia matrix, $\varepsilon\mathbf{GM}\cos\omega t$ is the time varying hydrostatic force and $\mathbf{F}$ is the linear hydrodynamic exciting force. The matrices or vectors $\mathbf{M}, \mathbf{A}, \mathbf{B}, \mathbf{B}_q$ and $\mathbf{F}$ are calculated by the program SHIPMO ([6]), $\mathbf{GZ}$ is calculated with the help of STAAF ([7]) and $\varepsilon\mathbf{GM}$ with a program developed in ([8]). The application of these various programs to calculate the coefficients is very important because these coefficients depend significantly on wave length, ship velocity and heading angle. As these quantities are not constant in a turning manoeuvre they have to be adapted to the actual parameter values at each step of the calculations.

The second order system (2) can be transformed into a four dimensional system of first order by introducing a new variable vector y defined by $y_1 = v$, $y_2 = \dot{\varphi}$, $y_3 = \dot{\alpha}$, $y_4 = \varphi$ and adding the trivial relation $\dot{y}_4 = y_2$. This set of equations is

$$\dot{y} = \mathbf{L}y + by_2|y_2| + c(y_4, t) + f(t) \tag{3}$$

where all coefficients are still depending on the wave frequency ω_e. As the vessel turns with constant forward speed under variation of ϑ, the encounter frequency ω_e changes due to the following relation

$$\omega_e = \omega_0 - Vk\cos\vartheta \tag{4}$$

where ω_0 is the wave frequency, V is the vessel speed, $k = 2\pi/\lambda$ is the wave number and λ the wave length. The heading angle $\vartheta = 0$ if the ship is travelling in following seas and $\vartheta = \pi$ if the ship is travelling into head seas.

3   Method of Analysis and Numerical Results

The numerical calculations are performed with the program BIFPACK ([9]) for the fishing trawler Patti B. It is interesting to note that this ship capsized twice in operation resulting in the death of six seamen.

3.1 One degree of freedom model

Depending on the value of the heading angle ϑ we obtain pure parametric excitations for $\vartheta = 0$ and π and pure external excitation for $\vartheta = \pi/2$.

For the results obtained with this model the speed V of the ship is assumed to be zero, that is, we neglect the effect given by (4).

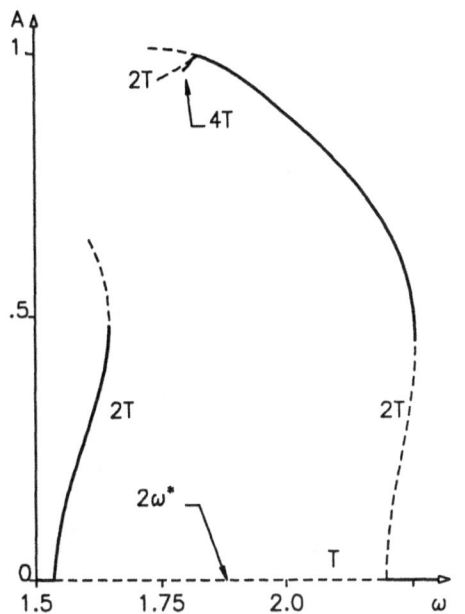

Figure 2: Amplitude $\mathbf{A}$ of the roll motion versus excitation frequency ω for wave amplitude $\eta = 2.5\,\text{ft}$ and heading angle $\vartheta = 0$

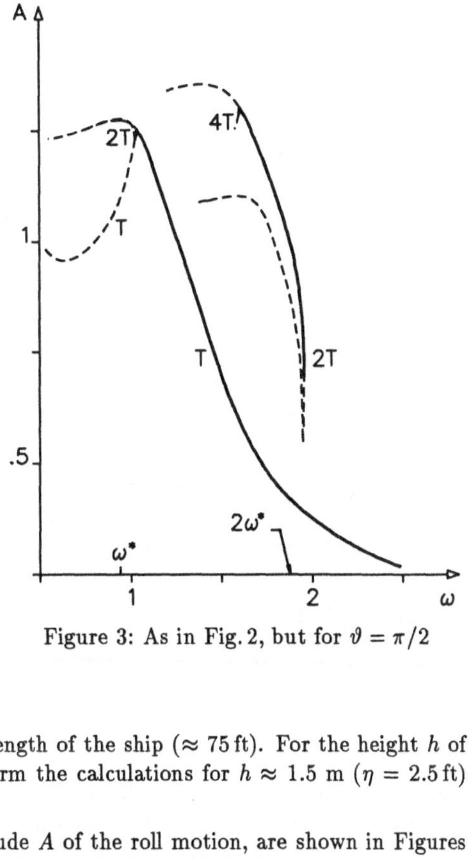

Figure 3: As in Fig. 2, but for $\vartheta = \pi/2$

The wave length is considered to be equal to the length of the ship ($\approx 75\,\text{ft}$). For the height h of the waves measured from crest to trough we perform the calculations for $h \approx 1.5\,\text{m}$ ($\eta = 2.5\,\text{ft}$) where η is the amplitude of the exciting waves.

The results of the calculations, namely the amplitude A of the roll motion, are shown in Figures 2–4. The eigenfrequency of the ship's roll oscillation is denoted by ω^* in the figures.

Fig. 2 shows the case of pure parametric excitation ($\vartheta = 0$). There exists a region on the ω-axis about $2\omega^*$ where a roll oscillation is excited. Outside this region no roll motion exists. The frequency of the oscillation is one half of the frequency of the exciting waves ($2T$-periodic oscillation). We note that stable parametrically excited oscillations of considerable amplitude are possible. In addition for reducing the excitation frequency a period doubling bifurcation sequence starts which can lead to a chaotic ship roll motion. This can be understood from looking at the phaseplane of the statically inclined ship with $\varepsilon = 0$, $F = 0$, $B = D = 0$ in (1). The parameters C_1, C_3, C_5 in (1) are such that in the phase plane a saddle point connection exists similar to that for a pendulum. Hence for the excited ship motion transversal heteroclinic orbits can exist and chaotic motions are possible. This szenario is similar to that studied in [5] for a parametrically excited pendulum. Fig. 3 shows the behavior of the ship under pure external excitation ($\vartheta = \pi/2$). The main stable branch is a T-periodic oscillation which for decreasing excitation frequency again undergoes a period doubling bifurcation. It is further interesting to note that in addition in a small frequency domain a stable subharmonic oscillation exists. In Fig. 5 the superposition of the external and parametric oscillation for $\vartheta = \pi/4$ is presented.

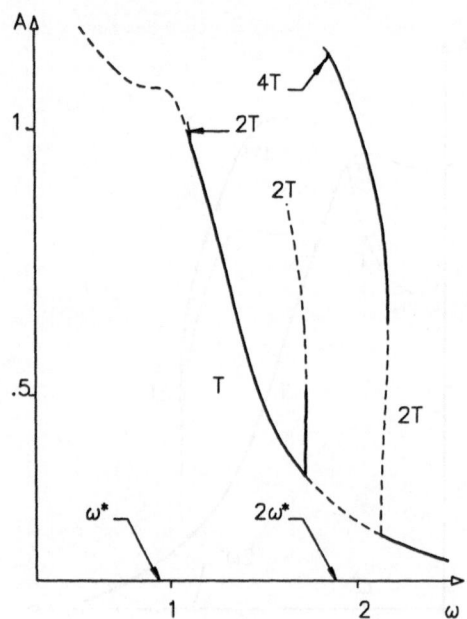

Figure 4: As in Fig. 2, but for $\vartheta = \pi/4$

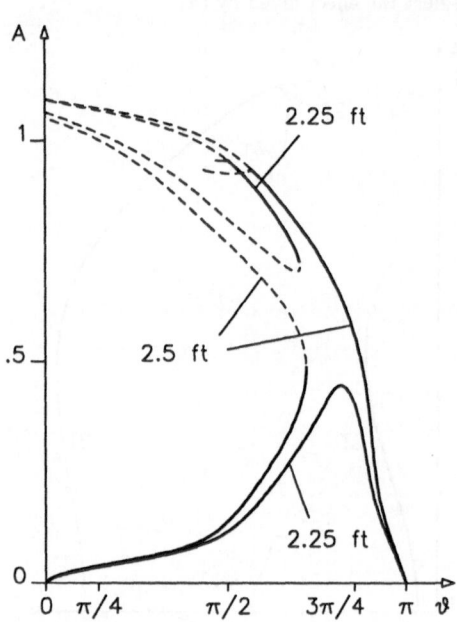

Figure 5: Amplitude A of the roll motion versus the heading angle ϑ for two wave amplitudes $\eta_1 = 2.25$ ft and $\eta_2 = 2.5$ ft

3.2   Three degrees of freedom model

For this model we focus our numerical investigation on the frequency range where only the external excitation is relevant. That is where the encounter frequency is close to the eigenfrequency of the roll motion of the ship. In this case no parametric excitation effects occur. In Fig. 5 the amplitude of the roll motion in its dependence on the heading angle ϑ is presented for the two wave amplitudes $\eta_1 = 2.25$ ft and $\eta_2 = 2.5$ ft. For $\eta = 2.25$ the physically relevant solution curve is connected and smooth and returns to zero after a turn of the ship of 180 degrees. The amplitude of the oscillations of the ship reaches a maximum value which generally is not at $\vartheta = \pi/2$, but at that value of ϑ, where the encounter frequency ω_e (4) is close to the eigenfrequency of the roll motion obtained for the linear ship model. Thus a different roll behavior is obtained in the course of the variation of ϑ depending on whether the ship turns off from following or head seas.

Even more pronounced is this difference in the behavior if the wave amplitude $\eta = 2.5$ ft is considered. In this case we see that starting in following seas at $\vartheta = 0$ the roll motion has a jump at about $\vartheta = 5\pi/8$ and then the roll motion reduces to zero at $\vartheta = \pi$. However if the ship starts to turn from head seas ($\vartheta = \pi$) we see that the amplitude increases steadily and at quite large values secondary bifurcations start which again will lead to a chaotic motion possibly leading to capsizing.

Jumps and secondary bifurcations of the yaw and sway velocity occur at the same frequency values as in Fig. 5.

4 Summary

The three degrees of freedom model proves the practically well known result that turning into head seas or following seas is qualitatively quite different. Further the interesting result can be deduced from the three degrees of freedom model that jump effects or secondary bifurcations occur in all degrees of freedom simultaneously.

Acknowledgement

This work was partly supported by the Austrian Science Fonds under project P 7003.

References

1. Carlo A., Francescutto A., Nabergoi R., *Ultraharmonics and Subharmonics in the Rolling Motion of a Ship: Steady-State-Solution*, International Shipbuilding Progress, vol. 28 (1981) 234–251.

2. Nayfeh A. H., Khdeir A. A., *Nonlinear Rolling of Ships in Regular Beam Seas*, International Shipbuilding Progress, vol. 33 (1986) 40–49.

3. Oakley O. H. Jr., Paulling J. R., Wood P. D., *Ship Motions and Capsizing in Astern Seas*, 10th ONR Symposium on Naval Hydrodynamics, 1974.

4. Bovet D., Jonas E., Johnson R., *Recent Coast Guard Research into Vessel Stability*, Marine Technology SNAME, New York 1974.

5. Troger H., *Über chaotisches Verhalten einfacher mechanischer Systeme*, ZAMM 62 (1982) T18–T27.

6. Beck R. F., A. W. Troesch, *Department of Naval Architecture and Marine Engineering Student's Documentation and User's Manual for the Computer Program SHIPMO*, Report No. 98-2, Ann Arbor, MI. 1989.

7. *Inter CAD, STAFF (Stability Analysis of Arbitrary Forms) User's Manual*, Annapolis, MD, 1983.

8. Falzarano J. M., *Nonlinear Aspects of Ship Dynamic Stability*, Department of Naval Architecture, The University of Michigan, 1988.

9. Seydel R., *BIFPACK: A program package for calculating bifurcations*, State University of New York at Buffalo, 1985.

International Series of Numerical Mathematics, Vol. 97, © 1991 Birkhäuser Verlag Basel

Normal Forms for Planar Systems With Nilpotent Linear Part

E. GAMERO, E. FREIRE, E. PONCE

Dept. Applied Mathematics, University of Sevilla, E.T.S.I.I.

Avda. Reina Mercedes s/n, 41012–SEVILLA, Spain

ABSTRACT. Normal forms corresponding to a linear degeneracy with a 2nd order nilpotent Jordan block are studied. By means of a recursive algorithm well suited to symbolic computation, we achieve the expressions for its coefficients. Further, from ideas behind the algorithm, we investigate the possibilities of simplifying the classical normal form, obtaining simpler and higher order normal forms than in previous works.

In this work, the linear degeneracy $A = \begin{pmatrix} 0 & 1 \\ 0 & 0 \end{pmatrix}$ is considered. This normal form will be of interest in the study of the corresponding bifurcation behaviour, which is related to a Takens–Bogdanov codimension two bifurcation (or to higher codimension bifurcations when additional degeneracies in nonlinear part appear [6]).

We consider the system:

$$\begin{pmatrix} \dot{x} \\ \dot{y} \end{pmatrix} = A \begin{pmatrix} x \\ y \end{pmatrix} + \begin{pmatrix} f(x,y) \\ g(x,y) \end{pmatrix} = \begin{pmatrix} y \\ 0 \end{pmatrix} + \sum_{k \geq 2} \varphi_k(x,y) \tag{1}$$

where $x, y \in \mathbb{R}$ and $\varphi_k \in \mathcal{H}_k^2$, the space of two–dimensional polynomial vectors, homogeneous of k–degree in two variables.

If we consider the linear operator $L_k^A : \mathcal{H}_k^2 \longrightarrow \mathcal{H}_k^2$ defined by

$$L_k^A(P(x,y)) = DP(x,y)A \begin{pmatrix} x \\ y \end{pmatrix} - AP(x,y),$$

then there exists a sequence of near–identity transformations (see, v.g. [1]) leading to

$$\begin{pmatrix} \dot{x} \\ \dot{y} \end{pmatrix} = \begin{pmatrix} y \\ 0 \end{pmatrix} + \sum_{k \geq 2} \psi_k(x,y)$$

where $\psi_k \in Cor L_k^A$, a complementary subspace to $Im L_k^A$, for all $k \geq 2$.

To obtain this transformed system, we can use the following recursive procedure, based upon the use of Lie transforms (see [5], [2], [4], [3]):

For all $k \geq 1$,

a) Set

$$W_{k,1} = k! \varphi_{k+1} + \sum_{j=0}^{k-2} \binom{k-1}{j} (k-j-1)! [\varphi_{k-j}, U_j]$$

b) For all $l = 2, 3, \dots, k$, build

$$W_{k,l} = W_{k,l-1} + \sum_{j=0}^{k-l} \binom{k-l}{j} [W_{k-j-1,l-1}, U_j]$$

c) Obtain the normal form of $k+1$–degree:

$$\psi_{k+1} \equiv Proy_{CorL_{k+1}^A} \left(\frac{W_{k,k}}{k!}\right)$$

d) Compute $U_{k-1} \in \mathcal{H}_{k+1}^2$ solving the linear equation

$$L_{k+1}^A U_{k-1} = Proy_{ImL_{k+1}^A} (W_{k,k}) \tag{2}$$

e) Reassign

$$W_{k,l} - Proy_{ImL_{k+1}^A} (W_{k,k}) \longrightarrow W_{k,l}$$

for $l = 1, 2, \ldots, k$.

Here, $[\cdot, \cdot]$ is the Lie Bracket defined by $[U, V] = DU \cdot V - DV \cdot U$, and U_j, $j \geq 2$, are the coefficients of the generator of the transformation (see [2]). The sequence $\{W_{k,l}\}$ can be organized in a triangular scheme,

$W_{1,1}$					
$W_{2,1}$	$W_{2,2}$				
$W_{3,1}$	$W_{3,2}$	$W_{3,3}$			
$\vdots$	$\vdots$	$\vdots$	$\ddots$		
$W_{k,1}$	$W_{k,2}$	$W_{k,3}$	$\cdots$	$W_{k,k}$	
$\vdots$	$\vdots$	$\vdots$	$\vdots$	$\vdots$	$\ddots$

For the canonical basis of $\mathcal{H}_k^2$ with reverse lexicografical order, we have

$$L_k^A \equiv \left(\begin{array}{c|c} A_k & -I \\ \hline 0 & A_k \end{array}\right) \quad \text{where} \quad A_k = \begin{pmatrix} 0 & 1 & & & & \\ & 0 & 2 & & & \\ & & 0 & 3 & & \\ & & & \ddots & \ddots & \\ & & & & 0 & k \\ & & & & & 0 \end{pmatrix}$$

and I is the identity matrix of $(k+1)$–order.

From this, we have

$$ImL_k^A = lin \left\{ \begin{pmatrix} y^k \\ 0 \end{pmatrix}, \begin{pmatrix} xy^{k-1} \\ 0 \end{pmatrix}, \ldots, \begin{pmatrix} x^{k-1}y \\ 0 \end{pmatrix}, \right.$$
$$\left. \begin{pmatrix} 0 \\ y^k \end{pmatrix}, \begin{pmatrix} 0 \\ xy^{k-1} \end{pmatrix}, \ldots, \begin{pmatrix} 0 \\ x^{k-2}y^2 \end{pmatrix}, \begin{pmatrix} -x^k \\ kx^{k-1}y \end{pmatrix} \right\}$$

and

$$Ker L_k^A = lin \left\{ \begin{pmatrix} y^k \\ 0 \end{pmatrix}, \begin{pmatrix} xy^{k-1} \\ y^k \end{pmatrix} \right\}$$

Consequently, the solution of the linear equation (2) is $U_{k-1} = U'_{k-1} + U''_{k-1}$, where U'_{k-1} is a particular solution (easy to calculate because of the L^A_{k+1} shape), and

$$U''_{k-1} = \begin{pmatrix} \alpha_{k+1}y^{k+1} + \beta_{k+1}y^{k+1} \\ \beta_{k+1}y^{k+1} \end{pmatrix} \in Ker\ L^A_{k+1}$$

where $\alpha_{k+1}, \beta_{k+1}$ are arbitrary constants.

Moreover,

$$Cor L^A_k = lin\ \left\{ \begin{pmatrix} 0 \\ x^k \end{pmatrix}, \begin{pmatrix} 0 \\ x^{k-1}y \end{pmatrix} \right\}$$

is a complementary subspace to ImL^A_k and a normal form for the system (1) is

$$\begin{pmatrix} \dot{x} \\ \dot{y} \end{pmatrix} = \begin{pmatrix} y \\ 0 \end{pmatrix} + \begin{pmatrix} 0 \\ \sum_{j\geq2} \{a_jx^j + b_jx^{j-1}y\} \end{pmatrix} \tag{3}$$

With these ideas, we have obtained an implementation of the algorithm on REDUCE 3.2. For the system (1), the algorithm provides the following expressions for the coefficients of the normal form:

$$a_2 = g_{xx}/2 \qquad b_2 = g_{xy} + f_{xx}$$
$$a_3 = (3f_{xy}g_{xx} - 3g_{xy}f_{xx} + g_{xxx})/6$$
$$b_3 = (-6\beta_2g_{xx} - f_{yy}g_{xx} + g_{yy}g_{xy} - g_{yy}f_{xx} + 2f_{xy}g_{xy} + 2g_{xxy} + 2f_{xxx})/4$$

Here $a_2 \neq 0 \iff g_{xx} \neq 0$, and in this case it is possible to choose β_2 such that $b_3 = 0$. Further,

$$a_4 = (-40\alpha_2g^2_{xx} + 16\beta_2g_{xy}g_{xx} + 16\beta_2f_{xx}g_{xx} + 4f_{yy}g_{xy}g_{xx} - 16f_{yy}f_{xx}g_{xx} - g^2_{yy}g_{xx} +$$
$$+12g_{yy}f_{xy}g_{xx} - 12g_{yy}g_{xy}f_{xx} + 12g_{yy}f^2_{xx} + 4g_{yy}g_{xxx} + 28f^2_{xy}g_{xx} - 24f_{xy}g_{xy}f_{xx} +$$
$$+24f_{xy}g_{xxx} - -16g_{xy}f_{xxx} - 24f_{xx}g_{xxy} - 4g_{xx}g_{xyy} + 16g_{xx}f_{xxy} + 4g_{xxxx})/96$$
$$b_4 = (-8\alpha_2g_{xy}g_{xx} - 8\alpha_2f_{xx}g_{xx} - 18\beta_2f_{xy}g_{xx} - 4\beta_2g^2_{xy} + 10\beta_2g_{xy}f_{xx} - 4\beta_2f^2_{xx} - 6\beta_2g_{xxx} +$$
$$+3f_{yy}g_{yy}g_{xx} - f_{yy}g^2_{xy} - 2f_{yy}f^2_{xx} - f_{yy}g_{xxx} + g^2_{yy}g_{xy} - 2g^2_{yy}f_{xx} + 3g_{yy}f_{xy}g_{xy} +$$
$$+3g_{yy}g_{xxy} + g_{yy}f_{xxx} + 2f^2_{xy}g_{xy} + 2f^2_{xy}f_{xx} + 6f_{xy}g_{xxy} + 4f_{xy}f_{xxx} + g_{xy}g_{xyy} +$$
$$+2g_{xy}f_{xxy} - 5f_{xx}g_{xyy} - 4f_{xx}f_{xxy} - 3g_{xx}g_{yyy} - 3g_{xx}f_{xyy} + 2g_{xxxy} + 2f_{xxxx})/12$$

In the $a_2 \neq 0$ case, selecting α_2, we can obtain $a_4 = 0$, thereby achieving the result previously obtained by Ushiki [7] in a different way. Moreover,

$$a_5 = (150\alpha_2g_{yy}g^2_{xx} - 2520\alpha_2f_{xy}g^2_{xx} + 2520\alpha_2g_{xy}f_{xx}g_{xx} - 840\alpha_2g_{xx}g_{xxx} -$$
$$-300\beta_3g^2_{xx} - 360\beta^2_2g^2_{xx} - 120\beta_2f_{yy}g^2_{xx} + 120\beta_2g_{yy}g_{xy}g_{xx} -$$
$$-120\beta_2g_{yy}f_{xx}g_{xx} + 240\beta_2f_{xy}g_{xy}g_{xx} + 240\beta_2g_{xx}g_{xxy} + 240\beta_2g_{xx}f_{xxx} +$$
$$+240f^2_{yy}g^2_{xx} - 30f_{yy}g_{yy}g_{xy}g_{xx} - 360f_{yy}f_{xy}f_{xx}g_{xx} + \cdots)/2880$$
$$b_5 = (720\alpha_2\beta_2g^2_{xx} + 120\alpha_2f_{yy}g^2_{xx} - 330\alpha_2g_{yy}g_{xy}g_{xx} - 90\alpha_2g_{yy}f_{xx}g_{xx} -$$
$$-600\alpha_2f_{xy}g_{xy}g_{xx} - 360\alpha_2f_{xy}f_{xx}g_{xx} + 360\alpha_2g^2_{xy}f_{xx} + 360\alpha_2g_{xy}f^2_{xx} -$$
$$-120\alpha_2g_{xy}g_{xxx} - 120\alpha_2f_{xx}g_{xxx} - 240\alpha_2g_{xx}g_{xxy} - 240\alpha_2g_{xx}f_{xxx} +$$
$$+420\beta_3g_{xy}g_{xx} + 420\beta_3f_{xx}g_{xx} + 216\beta^2_2g_{xy}g_{xx} + 216\beta^2_2f_{xx}g_{xx} + \cdots)/576$$

Here, by means of β_3, we can also annihilate a_5 in the case $a_2 \neq 0$.

The algorithm leads us to the following theorem, accounting for the possible simplifications in different cases:

Theorem. *For the system (1), we have:*

a) *if $g_{xx} \neq 0$, a normal form is*

$$
\begin{pmatrix} \dot{x} \\ \dot{y} \end{pmatrix} = \begin{pmatrix} y \\ 0 \end{pmatrix} + \begin{pmatrix} 0 \\ a'_2 x^2 + b'_2 xy + a'_3 x^3 + b'_4 x^3 y + b'_5 x^4 y + a'_6 x^6 + b'_7 x^6 y + \\ + b'_8 x^7 y + a'_9 x^9 + b'_{10} x^9 y + b'_{11} x^{10} y + a'_{12} x^{12} + \cdots \end{pmatrix}
$$

b) *in the case $g_{xx} = 0$, $g_{xy} + f_{xx} \neq 0$ we have a normal form up to 8-degree:*

$$
\begin{pmatrix} \dot{x} \\ \dot{y} \end{pmatrix} = \begin{pmatrix} y \\ 0 \end{pmatrix} + \begin{pmatrix} 0 \\ b'_2 xy + a'_3 x^3 + b'_3 x^2 y + a'_4 x^4 + a'_5 x^5 + \\ + b'_5 x^4 y + a'_6 x^6 + a'_7 x^7 + b'_7 x^6 y + a'_8 x^8 \end{pmatrix}
$$

Moreover, if $a'_3 \neq 0$, we can annihilate also b'_5 and b'_7.

Proof. Consider that the initial system is just in normal form (3) (making a change of variables if necessary). The normal form for this system, obtained by means of the triangular scheme, is

$$
\begin{pmatrix} \dot{x} \\ \dot{y} \end{pmatrix} = \begin{pmatrix} y \\ 0 \end{pmatrix} + \begin{pmatrix} 0 \\ \sum_{j \geq 2} \left\{ a'_j x^j + b'_j x^{j-1} y \right\} \end{pmatrix}
$$

where the coefficients a'_j, b'_j depend upon the parameters α_k, β_k.

The coefficients we have obtained are:

$$
\begin{aligned}
a'_2 &= a_2 & b'_2 &= b_2 & a'_3 &= a_3 & b'_3 &= -3\beta_2 a_2 + b_3 \\
a'_4 &= (-5\alpha_2 a_2^2 + \beta_2 b_2 a_2 + 3a_4)/3 & b'_4 &= (-4\alpha_2 b_2 a_2 - \beta_2 b_2^2 - 9\beta_2 a_3 + 3b_4)/3 \\
a'_5 &= (-42\alpha_2 a_2 a_3 - 5\beta_3 a_2^2 - 6\beta_2^2 a_2^2 + 4\beta_2 a_2 b_3 + 12a_5)/12 \\
b'_5 &= (120\alpha_2\beta_2 a_2^2 - 30\alpha_2 b_2 a_3 - 40\alpha_2 a_2 b_3 + 35\beta_3 b_2 a_2 + 18\beta_2^2 b_2 a_2 \\
&\quad - 20\beta_2 b_2 b_3 - 72\beta_2 a_4 + 24b_5)/24 \\
a'_6 &= (252\alpha_3 b_2 a_2^2 + 1120\alpha_2^2 a_2^3 - 196\alpha_2\beta_2 b_2 a_2^2 - 1344\alpha_2 a_2 a_4 - 630\alpha_2 a_3^2 \\
&\quad - 81\beta_3 b_2^2 a_2 - 399\beta_3 a_2 a_3 - 38\beta_2^2 b_2^2 a_2 - 252\beta_2^2 a_2 a_3 - 120\beta_2 b_2 a_4 \\
&\quad + 108\beta_2 a_2 b_4 + 60\beta_2 b_3 a_3 + 360a_6)/360 \\
b'_6 &= (72\alpha_3 b_2^2 a_2 + 380\alpha_2^2 b_2 a_2^2 + 104\alpha_2\beta_2 b_2^2 a_2 + 1260\alpha_2\beta_2 a_2 a_3 - 144\alpha_2 b_2 a_4 \\
&\quad - 240\alpha_2 a_2 b_4 - 180\alpha_2 b_3 a_3 + 100\beta_4 a_2^2 + 50\beta_3\beta_2 a_2^2 + 9\beta_3 b_2^3 \\
&\quad + 186\beta_3 b_2 a_3 + 50\beta_3 a_2 b_3 - 60\beta_2^3 a_2^2 + 22\beta_2^2 b_2^3 + 198\beta_2^2 b_2 a_3 + 60\beta_2^2 a_2 b_3 \\
&\quad - 132\beta_2 b_2 b_4 - 40\beta_2 b_3^2 - 360\beta_2 a_5 + 120b_6)/120
\end{aligned}
$$

$$a_7' = (40\alpha_4 a_2^3 - 80\alpha_3\beta_2 a_2^3 + 264\alpha_3 b_2 a_2 a_3 + 80\alpha_3 a_2^2 b_3 + 8\alpha_2^2 b_2^2 a_2^2$$
$$+ 1800\alpha_2^2 a_2^2 a_3 + 340\alpha_2\beta_3 a_2^3 + 200\alpha_2\beta_2^2 a_2^3 + 48\alpha_2\beta_2 b_2 a_2 a_3$$
$$+ \cdots + 180 a_7)/180$$

$$b_7' = (70\alpha_4 b_2 a_2^2 - 896\alpha_3\beta_2 b_2 a_2^2 + 210\alpha_3 b_2^2 a_3 + 392\alpha_3 b_2 a_2 b_3 a_3$$
$$- 3360\alpha_2^2\beta_2 a_2^3 + \cdots - 1080\beta_2 a_6 + 360 b_7)/360$$

$$a_8' = (-9720\alpha_4 b_2^2 a_2^2 + 37800\alpha_4 a_2^2 a_3 - 176490\alpha_3\alpha_2 b_2 a_2^3 - 14256\alpha_3\beta_2 b_2^2 a_2^2$$
$$+ \cdots + 1650\beta_5 a_2^3 + 8700\beta_4\beta_2 a_2^3 + \cdots + 60480 a_8)/60480$$

$$b_8' = (-10080\alpha_4\beta_2 a_2^3 - 2160\alpha_4 b_2^3 a_2 + 5040\alpha_4 b_2 a_2 a_3 + 3360\alpha_4 a_2^2 b_3$$
$$- 54207\alpha_3\alpha_2 b_2^2 a_2^2 + \cdots + 15120 b_8)/15120$$

So, in the case $a_2 \neq 0$ ($\Longleftrightarrow g_{xx} \neq 0$), we can adequately select $\beta_2, \alpha_2, \beta_3, \beta_4, \alpha_4$ to annihilate $b_3', a_4', a_5', b_6', a_7'$ respectively.

In the case $a_2 = 0, b_2 \neq 0$ ($\Longleftrightarrow g_{xx} = 0$, $g_{xy} + f_{xx} \neq 0$), it is possible to annihilate b_4', b_6', b_8' by means of $\beta_2, \beta_3, \alpha_2$.

Moreover, a similar argument shows that in the first case, using $\beta_5, \beta_6, \alpha_6, \beta_7, \beta_8$ we can annihilate $a_8', b_9', a_{10}', a_{11}', b_{12}'$. □

It must be noticed the possibilities of applying this algorithm to specific systems, and to study, in each case, how many simplifications are possible.

References

[1] Ashkenazi, M.; Chow, S., *Normal Forms Near Critical Points for Differential Equations and Maps*, IEEE Transactions on Circuits and Systems, pp. 850–862, (1988).

[2] Chow, S.; Hale, J. K., *Methods of Bifurcation Theory*, Springer–Verlag, (1982).

[3] Deprit, A., *Canonical Transformations Depending on a Small Parameter*, Celest. Mechan., 1 pp. 12–32, (1969).

[4] Freire, E.; Gamero, E.; Ponce, E.; G.–Franquelo, L., *An Algorithm for Symbolic Computation of Center Manifolds*, Symbolic And Algebraic Computation, Lecture Notes in Computer Science, 358, Ed. P. Gianni, pp. 218–230, (1988).

[5] Meyer, K. R.; Schmidt, D. S., *Entrainment Domains*, Funkcialaj Ekvacioj, 20, pp. 171–192, (1977).

[6] Rodríguez–Luis; A.J., Freire, E.; Ponce, E., *On a Codimension 3 Arising in an Autonomous Electronic Circuit*, see this volume.

[7] Ushiki, S., *Normal Forms for Singularities of Vector Fields*, Japan Journal of Applieds Mathematics, pp. 1–37, (1984).

International Series of Numerical Mathematics, Vol. 97, © 1991 Birkhäuser Verlag Basel

TWO METHODS FOR THE NUMERICAL DETECTION OF HOPF BIFURCATIONS.

T.J. Garratt [1], G. Moore [2], A. Spence [1],

[1] School of Mathematical Sciences, University of Bath, Bath, AVON, BA2 7AY. U.K.
[2] Department of Mathematics, Imperial College, Queen's Gate, LONDON, SW7 2BZ. U.K.

1 Introduction

This paper is concerned with the detection of Hopf bifurcations in the parameter dependent nonlinear system

$$\frac{d\mathbf{x}}{dt} = \mathbf{f}(\mathbf{x}, \lambda), \quad \mathbf{f} : \mathbf{R}^n \times \mathbf{R} \mapsto \mathbf{R}^n, \quad \mathbf{x} \in \mathbf{R}^n, \quad \lambda \in \mathbf{R}. \tag{1}$$

The set $\Gamma := \{(\mathbf{x}, \lambda) \in \mathbf{R}^{n+1} : \mathbf{f}(\mathbf{x}, \lambda) = 0\}$ represents the steady state solutions of (1) and it is often important to determine the *(linearised) stability* of a branch of Γ. If μ_1 denotes the eigenvalue with largest real part of the Jacobian matrix $A := \mathbf{f_x}(\mathbf{x}, \lambda)$, then a steady state solution is stable (unstable) if $\text{Re}(\mu_1)$ is negative (positive). Also in applications it is desirable to detect a point of Γ where μ_1 is complex and $\text{Re}(\mu_1)$ changes sign as λ varies, called a *Hopf bifurcation point*. If n is small in (1) it is certainly simplest and probably best to find all the eigenvalues of A using the QR algorithm. However for systems arising from spatial discretisations of p.d.e.'s n is typically very large with the Jacobian matrix sparse, and the direct application of the QR algorithm for a nonsymmetric matrix will be very expensive or may not even be feasible. Of course we need not find all the eigenvalues of A since only the eigenvalues which determine the stability are wanted. This paper is about the application of iterative methods for the calculation of the eigenvalues of A with largest real part.

To illustrate our overall numerical approach and to compare two specific methods we consider one of the simplest possible situations. We assume: (H1): For all $(\mathbf{x}, \lambda) \in \Gamma$ the matrix $\mathbf{f_x}(\mathbf{x}, \lambda)$ is nonsingular; (H2): The eigenvalues μ_i $(i = 1, ..., n)$ of $\mathbf{f_x}(\mathbf{x}, \lambda)$ are ordered by decreasing real part i.e. $i > j \Rightarrow \text{Re}(\mu_i) \leq \text{Re}(\mu_j)$, with $\mu_{1,2} = v_1 \pm iw_1$, $\mu_3 = v_3$ and $v_3 < v_1$ where $v_1, v_3, w_1 \in \mathbf{R}$ and $w_1 \neq 0$; (H3): For the initial value of λ, the steady state solution is stable with $\text{Re}(\mu_1) < 0$ and as λ varies only the complex pair $\mu_{1,2}$ crosses the imaginary axis. (We emphasise that our approach is not restricted to this situation but we make these assumptions to present the key ideas as simply as possible). We assume that the steady state solution is found using a 'predict-solve' continuation method using direct factorisation of the Jacobian matrix in the solve stage. At each computed point of Γ we wish to estimate the complex pair $\mu_{1,2}$ and hence assign stability. Obviously if a Hopf bifurcation exists on Γ it will be located roughly and hence may be accurately determined by a standard method.

There are several methods available for finding eigenvalues of nonsymmetric matrices but for reasons of space we only consider one, namely Arnoldi's method [4], which we now outline. Given a normalised $\mathbf{v} \in \mathbf{R}^n$, k steps of Arnoldi's method produces V_k, an orthonormal basis for $\{v, Av, \ldots, A^{k-1}v\}$ and H_k, the $(k \times k)$ upper Hessenberg restriction of A to V_k, with eigensolutions $\{\tilde{\mu}_i, \tilde{y}_i\}_{i=1}^k$ found by the QR algorithm. The $\tilde{\mu}_i$ provide approximations to k eigenvalues of A with corresponding approximate eigenvectors given by $\tilde{u}_i = V_k \tilde{y}_i$. Arnoldi's method only requires matrix-vector multiplications on A and so it is possible to take advantage of sparsity or any special structure of A. Practical experience [5, 6] shows that there is rapid convergence to extremal well separated eigenvalues of A and this leads to the following strategy for the determination of $\mu_{1,2}$: *Perform Arnoldi's method on a transformed matrix $T(A)$ where the transformation T is chosen specifically to ensure that μ_1 is mapped to a well separated extremal eigenvalue of $T(A)$.* For practical reasons $T(A)$ must also be chosen so that there is a known mapping between the eigenvalues of A and $T(A)$, and with matrix-vector multiplications on $T(A)$ easy to compute. One example of this approach is given by Saad [7], where the choice $T(A) = p_s(A)$, a shifted and scaled Chebyshev polynomial of degree s, is used and we refer the reader to [6, 7] for further details. In Section 2 we introduce an alternative choice for $T(A)$, namely a generalised Cayley transform [1].

The aims of this paper are (a) to describe the modified Cayley transform, (b) to outline an algorithm using the above ideas to compute $\mu_{1,2}$ at every computed point of Γ, and (c) to provide a comparison of the numerical performance of two transformations, namely, the Chebyshev polynomial and the generalised Cayley transform.

2   The Generalised Cayley Transform

For $\alpha_1, \alpha_2 \in \mathbf{R}$ assume

$$\text{(a)} \ \alpha_1 \neq \mu_i \ (i = 1, \ldots, n), \qquad \text{(b)} \ \alpha_1 > \alpha_2, \qquad (2)$$

and consider the matrix transform of A given by

$$C(A) := (A - \alpha_1 I)^{-1}(A - \alpha_2 I), \qquad (3)$$

which is a generalisation of the usual Cayley transform $(A - \alpha I)^{-1}(A + \alpha I)$ ([1], p.273) . The following result is straightforward to prove:

Theorem 2.1 *Assume* (2). $(\mu, \mathbf{u})$ *is an eigensolution of* $A \Leftrightarrow (\theta, \mathbf{u})$ *is an eigensolution of* $C(A)$ *where*

$$\theta = t(\mu) := (\mu - \alpha_2)/(\mu - \alpha_1), \qquad \mu = t^{-1}(\theta) := (\alpha_1 \theta - \alpha_2)/(\theta - 1). \qquad (4)$$

(Note that both denominators are nonzero by (2).) In the sequel we shall write $\theta_i = (\mu_i - \alpha_2)/(\mu_i - \alpha_1)$ with μ_i satisfying (H2), i.e. the ordering of the θ_i is determined by the ordering of the μ_i. The usefulness of the transform (3) is that eigenvalues of A lying to the right (left) of the line $\text{Re}(\mu) = \frac{1}{2}(\alpha_1 + \alpha_2)$ are mapped to eigenvalues of $C(A)$ lying outside (inside) the unit circle. This is stated formally as:

Theorem 2.2 $\text{Re}(\mu) \leq (\geq) \frac{1}{2}(\alpha_1 + \alpha_2) \iff |\theta| \leq (\geq) 1.$

In applications arising from discretisations of p.d.e.'s we find that many eigenvalues of A have large negative real part and that only a few eigenvalues near the imaginary axis are of interest in a study of stability. The former eigenvalues transform under (3) to eigenvalues of

$C(A)$ very close to unity and (small) changes to α_1, α_2 have little effect on this property. Thus we choose α_1 and α_2 according to the transformation properties of the latter eigenvalues. Under (H2) one option is to choose α_1, α_2 to place θ_3 on the unit circle and to maximise the distance of $\theta_{1,2}$ from the unit disk since then Arnoldi's method applied to $C(A)$ can be expected to converge rapidly to $\theta_{1,2}$. The following result holds.

Theorem 2.3 *Let $\mu_{1,2}$, μ_3 satisfy* (H2) *and take α_1 and α_2 in* (3) *as*

$$\alpha_1 = v_3 + \gamma, \quad \alpha_2 = v_3 - \gamma, \quad \gamma = \sqrt{(v_1 - v_3)^2 + w_1^2}. \tag{5}$$

Then $\theta_3 = -1$, $|\theta_1| = 1 + d$, $d = (w_1 - v_3 + v_1 - \gamma)/(v_3 - v_1 + \gamma) > 0$, with d a maximum with respect to α_1 and α_2 (subject to $\alpha_1 + \alpha_2 = 2v_3$).

Remark: To compute $y := C(A)x$, one has to solve $(A - \alpha_1 I)y = (A - \alpha_2 I)x$, and so direct factorisation of $(A - \alpha_1 I)$ is required. This would probably be acceptable in a continuation context using direct factorisation in the 'solve' stage.

3 Practical implementation in a continuation framework

In this section we give an algorithm for determining the stability of points on Γ and for the detection of Hopf bifurcations. Assuming that there is no *a priori* information about the eigenvalues of the Jacobian $A = \mathbf{f}_\mathbf{x}(\mathbf{x}_0, \lambda_0)$ at the first point $(\mathbf{x}_0, \lambda_0)$ of Γ, we apply Arnoldi's method to A alone to compute estimates of some of its eigenvalues from which to construct an initial transformation (these initial transformations are constructed as in Theorem 2.3 for $C(A)$, or as described by Saad [6] for $p_s(A)$). If necessary, Arnoldi's method is applied to $T(A)$ repeatedly using a suitable starting vector until an accurate approximation of $\mu_{1,2}$ is obtained and thus the stability of $(\mathbf{x}_0, \lambda_0)$ ascertained. As better approximations to the eigenvalues become available in this process, improved transformations, say $T^{(2)}(A), T^{(3)}(A), \dots$ can be constructed (with $T^{(1)}(A) = T(A)$), all with the aim of accelerating convergence. New starting vectors are found using the technique in Saad [6]. This scheme can be repeated for all computed points of Γ, with, as would be expected, full use being made of previously computed eigenvalues, eigenvectors and transformations. An algorithm incorporating some of these ideas is:

Algorithm 3.1 Arnoldi's method accelerated by a transformation for continuation.

1. *Compute $(\mathbf{x}_0, \lambda_0) \in \Gamma$ and the corresponding Jacobian $A := \mathbf{f}_\mathbf{x}(\mathbf{x}_0, \lambda_0)$.*

2. *Compute k eigenvalue approximations $\tilde{\mu}_i$ of A using Arnoldi's method starting with a random initial vector $\mathbf{v} \in \mathbf{R}^n$.*

3. *For $j=1, \dots$ do (i) – (iii) until $\rho_1 \leq \epsilon$:*

   (i). *Obtain a transformation $T^{(j)}(A)$ such that $t(\tilde{\mu}_1)$ is extremal and well separated from $t(\tilde{\mu}_i), i = 3, \dots, k$.*

   (ii). *Compute k approximate eigensolutions $(\tilde{\theta}_i, \tilde{\mathbf{u}}_i)$ of $T^{(j)}(A)$ using Arnoldi's method and evaluate the corresponding eigenvalues $\tilde{\mu}_i = t^{-1}(\tilde{\theta}_i)$ of A.*

   (iii). *Obtain a new starting vector $\mathbf{v}$ and compute $\rho_1 := \|(A - \tilde{\mu}_1)\tilde{\mathbf{u}}_1\|_2/|\tilde{\mu}_1|$.*

4. *Compute the next $(\mathbf{x}_p, \lambda_p) \in \Gamma$ and corresponding Jacobian $A := \mathbf{f}_\mathbf{x}(\mathbf{x}_p, \lambda_p)$.*

5. *Let $j = 1$ and goto 3.*

4 Numerical Experiments

Algorithm (3.1) was tested on a system arising from a pair of equations modelling a tubular reactor (equations (1)-(5) [2]), which are discretised using the simplest central differences to produce $\mathbf{f}(\mathbf{x}, \lambda)$ for a range of values of n, denoted n_i in the table. The numerical continuation package PITCON [3] was used to compute the steady state solutions over a range of $\{0.23, 0.31\}$ for the parameter λ, the Damkohler number. The values of the model parameters were taken as in Fig. 1 in [2], which shows a steady state bifurcation diagram with two Hopf bifurcations produced by a complex conjugate pair of eigenvalues crossing and then recrossing the imaginary axis. Algorithm (3.1) was implemented for $T(A) = p_s(A)$ and $T(A) = C(A)$ and an estimate of $\mu_{1,2}$ was evaluated at *every* computed steady state solution point, with $\epsilon = 10^{-2}$. All computations were performed using a SUN-4/470 workstation in double precision. The first point to make is that (except for one case discussed below) both methods worked and produced the same results as [2], obtained using the QR algorithm applied to A. To compare the relative efficiency of the two methods we present in a table numerical results obtained during the computation of the branch Γ. For each n_i we give the number of points of Γ found by PITCON (p_i); the total CPU time (in seconds) to compute Γ (PTI); the number of Arnoldi steps used ($k = 5, 10, 20$); the total number of matrix-vector multiplications (MV); the total CPU time to compute $\mu_{1,2}$ (ETI); and finally the CPU time used if the QR algorithm were applied directly to A (QTI). Note that the degree, s, of the Chebyshev polynomial was chosen from the set of possible values $\{5, 10, 20, \ldots, 90, 100\}$. In the table for each value of n_i and k, we give only the run with the degree of the polynomial that was most efficient in total CPU time. For $n_i = 800$, $k = 5$, the run with $T(A) = p_s(A)$ was abandonded at the very first steady state solution when over 10000 matrix-vector operations had been performed without an accurate estimate of $\mu_{1,2}$ being obtained. The most striking observations from the table are first, that the Cayley transform out-performs the Chebyshev polynomial in *all* cases, regardless of the number of Arnoldi steps or degree of the Chebyshev polynomial, and second, that both transformation methods completely out-perform the QR algorithm. Finally we note that for this example the least squares polynomial approach of Saad [7] produced very similar results to the Chebyshev polynomial method, but we omit the details.

5 Conclusions

The Cayley transform out-performs the Chebyshev polynomial in the numerical experiments on a model of the tubular reactor, where no more than 20 Arnoldi steps were used. From this limited information we suggest that for cases where direct factorisation of the Jacobian is possible the Cayley transform approach will be more efficient than the Chebyshev polynomial. The two methods can be extended to detect Hopf bifurcations when more than one pair of complex conjugate eigenvalues cross the imaginary axis along a branch of Γ, that is to compute estimates of the first r, say, eigenvalues with largest real part.

Acknowledgements: Thanks are due to S. Ashby, T. Manteuffel and Y. Saad for making various items of software available to us and to Y. Saad for helpful advice. T.J. Garratt is suported by a SERC CASE award and AEA Technology, Harwell, UK.

n_i	p_i	PTI	k	Cayley $T(A) = C(A)$		Chebyshev $T(A) = p_s(A)$			QR
				MV	ETI	MV	ETI	s	QTI
50	16	0.33	5	235	1.19	2725	3.34	30	13.66
			10	330	1.80	1930	3.30	10	
			20	660	4.88	2260	6.91	5	
100	22	0.90	5	275	2.06	3855	9.51	20	100.65
			10	470	3.85	3370	9.89	10	
			20	900	10.01	3240	15.58	5	
200	29	1.80	5	365	4.80	12485	27.30	50	727.44
			10	610	8.64	9470	23.33	20	
			20	1180	20.92	7940	27.33	10	
400	68	7.67	5	775	19.32	38525	167.82	70	10896.72
			10	1430	36.78	30670	133.65	40	
			20	2780	94.93	31260	177.28	20	
800	88	17.93	5	875	41.51	—	—	—	91692.08
			10	1670	54.60	100810	880.08	70	
			20	3300	144.30	104180	939.36	60	

References

[1] J.N. Franklin. *Matrix theory.* Prentice-Hall, New Jersey, 1968.

[2] R. F. Heinemann and A. B. Poore. *Multiplicity, stability, and oscillatory dynamics of the tubular reactor. Chem. Eng. Sci.,* 36:1411–1419, 1981.

[3] W. C. Rheinboldt and J. Burkardt. *A locally parameterised continuation process. ACM Transactions of Math. Software,* 9:215–235, 1983.

[4] Y. Saad. *Variations on Arnoldi's method for computing eigenelements of large unsymmetric matrices. Lin. Alg., Appl.,* 34:269–295, 1980.

[5] Y. Saad. *Projection methods for solving large sparse eigenvalue problems.* In A. Ruhe and B. Kagstrom, editors, *Lecture Notes In Mathematics, Matrix Pencils Proceedings,* pages 121–144. Springer-Verlag, Berlin, 1982.

[6] Y. Saad. *Chebyshev acceleration techniques for solving nonsymmetric eigenvalue problems. Math. Comp.,* 42:567–588, 1984.

[7] Y. Saad. *Numerical solution of large nonsymmetric eigenvalue problems. Computer Physics Communications,* 53:71–90, 1989.

International Series of Numerical Mathematics, Vol. 97, © 1991 Birkhäuser Verlag Basel

Automatic Evaluation of First
and Higher-Derivative Vectors[1]

Andreas Griewank

Mathematics and Computer Science Division

Argonne National Laboratory
Argonne, IL 60439-4801, U.S.A.

Abstract

The numerical analysis of parameter-dependent nonlinear systems almost always involves first-derivative matrices and often also requires the evaluation of higher derivatives along certain singular directions. It is shown here how these selected derivative values can be obtained without differencing and at a reasonable cost under the realistic assumption that the nonlinear system is given in the form of a computer program for its evaluation.

1. Introduction and Basic Assumptions

Over the past two decades great advances have been made regarding the theoretical and numerical analysis of nonlinear systems

$$f(x) = 0 \quad with \quad f : \mathbf{R}^n \to \mathbf{R}^m.$$

Often such finite-dimensional systems belong to a family of discretizations for an underlying differential or integral equation. Apart from the implicit understanding that the number n of independent variables and the number m of dependent variables may be very large, we will not have to consider this discretization aspect explicitly. In particular, sparsity is taken into account implicitly. Typically, the argument $x \in \mathbf{R}^n$ combines a *state* vector with various design and control parameters, plus possibly the time. We will not distinguish the role of these independent variables, because derivatives of the same kind and order are usually required with respect to all of them.

Usually the difference $m - n$ is a small positive integer equal to the largest codimension of any singularity in the region of interest. Sometimes the number n may also be increased by introducing *imperfection* parameters in order to make certain singularities generic. In such cases, it is best to label all potentially varying parameters as independent variables in the first place, even though some of them may be constant for most of the calculation. As we will see, the complexity estimate for the evaluation of the Jacobian matrix

$$J(x) = f'(x) \in \mathbf{R}^{m \times n}$$

[1]This work was supported by the Applied Mathematical Sciences subprogram of the Office of Energy Research, U.S. Department of Energy, under Contract W-31-109-Eng-38.

in the so-called reverse mode of automatic differentiation is in fact completely independent of the number n of independent variables. In other words, there is essentially no computational penalty for considering many input parameters as variable. This fact may be quite helpful for a sensitivity analysis after solutions have been obtained with sufficient accuracy.

Especially if the finite-dimensional system is obtained by an adaptive discretization or involves piecewise smooth approximations, e.g., splines, the vector function f may have cracks, kinks (discontinuities in $J(x)$), or jumps in higher derivatives. Indeed, in many applications, $f(x)$ will not even be globally defined at all $x \in \mathbf{R}^n$. To avoid any unnecessary complications, we will assume throughout that $f(x)$ is sufficiently often differentiable near all arguments of interest. The resulting computational techniques are likely to work as long as the underlying discretization converges in the appropriate Sobolev norm so that the discontinuities in all relevant derivatives become smaller and smaller as the grid is refined.

The chain rule based technique of *automatic differentiation* has benn used for more than 30 years, mostly in the so-called forward mode. For a recent survey see [4]. We restrict our attention to derivative vectors, because a direct accumulation of derivative matrices or tensors involves more complicated data structures and raises combinatorial optimization problems [5].

2. Required Derivatives

Wherever some submatrix of the Jacobian $J(x)$ has full row rank m, the solution set $f^{-1}(0)$ forms locally a smooth $(m - n)$-dimensional manifold. The same is true for the *slices* obtained by fixing the values of variables whose partials are not contained in a particular nonsingular submatrix. Any of these manifolds may be locally parametrized in terms of a subset of the independent variables or some other local coordinates. In either case the corresponding local chart is an implicit function that is often approximately evaluated by using a few steps of Newton's method. Typically Newton steps are calculated by first forming and then factoring the relevant square Jacobian, which we may assume to be a submatrix of $J(x)$. While this may be quite costly in general, the Jacobian often has a regular sparsity pattern or a priori known spectral properties that allow the computation of an approximate Newton step at a reasonable cost. Even when such structural information is not available, one may be able to obtain Newton steps comparatively cheaply by solving an extended system without forming the Jacobian at all. This approach was outlined in [5] and will not be pursued further here.

When a particular submatrix of the Jacobian is singular, the implicit function theorem no longer applies, and the structure of the corresponding slice depends on the size of the rank drop as well as the properties of the second derivative tensor $\nabla^2 f(x) \in \mathbf{R}^{m \times n \times n}$. Fortunately one does not need to evaluate the full tensor but merely its restriction to the null spaces of the relevant submatrix and its transpose, respectively. Let us assume for notational simplicity that at some point x_* the latter is spanned by a single vector $0 \neq u \in \mathcal{R}^n$. Correspondingly, the nullspace of the submatrix itself is spanned by the columns $\{v_i\}_{i=0...m-n}$ of an $n \times (m-n+1)$ matrix V. Then the structure of the solution set for $x \approx x_*$ depends on the symmetric matrix

$$H(x) = [u^T \nabla^2 F(x) v_i \, v_j]_{j=0,1,...m-n}^{i=0,1...m-n}.$$

Provided $H(x_*)$ is nonsingular, it is well known that x_* is an isola formation or a bifurcation point, depending on whether $H(x_*)$ is definite or not. Otherwise x_* is a cusp singularity in that $H(x_*)$ has a nontrivial nullvector $c \in \mathcal{R}^{m-n+1}$, and one has to look into the next higher derivative, namely, the cubic term

$$h(x) \equiv u^T \nabla^3 f(x)(x) v^3 \equiv \frac{\partial^3}{\partial t^3} u^T f(x + tv) \bigg|_{t=0}$$

where $v = Vc \in \mathbf{R}^n$. Similarly, higher singularities (See e.g. [10]) are characterized by the vanishing of one or more components or dot products of partial derivative vectors of the form

$$b(x) \equiv \nabla^d f(x) v_1 v_2 ... v_d = \frac{\partial^d f(x + t_1 v_1 + t_2 v_2 + \cdots + t_d v_d)}{\partial t_1 \partial t_2 \cdots \partial t_d} \bigg|_{t_i=0}. \tag{1}$$

Moreover, in order to rapidly approximate bifurcation points, cusp singularities or other higher order degeneracies by Newton's method, one also needs the gradients of the scalar quantities that vanish at the singularity in questions. As shown, for example, in [7] for the cusp case, these gradients are linear combinations of n-vectors of the form

$$a(x) \equiv u^T \nabla^{d+1} f(x) v_1 v_2 ... v_d = \nabla_x \left[\frac{\partial^d u^T f(x + t_1 v_1 + t_2 v_2 + \cdots + t_d v_d)}{\partial t_1 \partial t_2 \cdots \partial t_d} \right]_{t_i=0}. \tag{2}$$

Expressions of this form that are relevant at Hopf points, Takens-Bogdanov singularities and other bifurcations of dynamical systems are given in [1]. In both (1) and (2) the direction vectors v_i may well be restricted to a (state) subspace but as we discussed above the extra level of differentiation in (2) compared to (1) can be carried out with respect to any potentially varying parameter as well.

We will refer to the n-vector a given in (2) as an **adjoint** of the m-vector b defined in (1). Generally, an adjoint of an arbitrary vector function b is obtained by forming the dot product of b with some weight vector u and then differentiating the resulting scalar with respect to the full vector of independent variables x. For $d = 0$ and with u ranging over all Cartesian basis vectors in $\mathbf{R}^n$ one may obtain from (2) the Jacobian $J(x)$ row by row. When m is significantly larger than n, this scheme is likely to be more efficient than the alternative of picking up successive columns b as defined in (1) with $d = 1$ and v_1 ranging over the Cartesian basis vectors in $\mathbf{R}^n$. However, this rule of thumb need not always apply, and there are mixed modes that yield the Jacobian even more cheaply, at least in certain cases.

Obviously the accuracy with which the *direct* and *adjoint* derivative vectors b and a as defined in (1) and (2) can be approximated by divided differences diminishes rapidly as the degree d is increased. This situation holds in particular when f is evaluated by a substantial applications code. Moreover, the approximation of the adjoint vector a by differencing would be quite costly, as it involves at least n evaluations of f. For this reason many authors restrict the evaluation of such gradients to subspaces through the use of Brown/Brent-like modifications [3] of Newton's method. By automatic differentiation any vector of the form a can be obtained at a cost that is independent of n relative to the cost of evaluating f itself.

3. Directional Derivatives and Mixed Partials

Whenever the d vectors v_i in (1) are identical to a fixed vector $v \in \mathbf{R}^n$, we will call b the directional derivative of f along v. Dividing this b by $d!$, we obtain a Taylor coefficient denoted by $y^{(d)}(v)$ so that for fixed x

$$y(t) = f(x + tv) = \underbrace{y^{(0)} + t \cdot y^{(1)}(v) + t^2 \cdot y^{(2)}(v) + \ldots + t^d \cdot y^{(d)}(v)}_{f_d(t \cdot v) \equiv} + \mathcal{O}(t^{d+1}). \qquad (3)$$

Here we have assumed that f is at least d times Lipschitz-continuously differentiable. We introduced this truncated Taylor series because its coefficients $y^{(j)}(v)$ can be evaluated directly much more easily than the more general mixed derivatives b given in (1). Moreover the latter can be obtained from the univariate Taylor coefficients by the following Lagrange interpolation technique.

Abbreviating $w \equiv t \cdot v$, we find that the function $f_d(w)$ defined in (3) is a multivariate polynomial of degree d in $w \in \mathbf{R}^n$. The jet f_d represents exactly the up to d-th derivatives of f at the fixed base point x. Provided d is sufficiently large, we can expect that the solution set of the algebraic bifurcation equations $f_d(w) = 0$ is locally homeomorphic to that of the original system $f(x) = 0$. Fortunately it is usually sufficient to analyze $f_d(w)$ on a subspace spanned by a the $\hat{n} << n$ columns of a matrix V. While we consider the individual columns of V as constant, their number—like the degree d—may be successively enlarged.

Now suppose we want to obtain the coefficients or some other computationally convenient representation of the polynomial

$$\phi_d(z) \equiv f_d(Vz) \quad . \qquad (4)$$

Being a d-th degree polynomial in $\hat{n}$ variables, ϕ_d has $\hat{n} + d$ choose $d + 1$ unknown coefficients. To tie down these degrees of freedom, one needs at least the same number of values $\phi_d(g) = f_d(Vg)$ at suitable sample points $g \in \mathbf{R}^{\hat{n}}$. To make do with that minimal number and to obtain a convenient Lagrangian interpolation formula, we may let g range over all integer vectors whose $\hat{n}$ components g_j are nonnegative and sum to d or less. There is an obvious 1-1 correspondence between these lattice vectors and the homogeneous terms with multiindex g in the expansion of $\phi_d(z)$. Moreover, with e the $\hat{n}$-vector of ones and g any of the lattice points one can easily check that the polynomial

$$L_g(z) = \prod_{\substack{i \leq d \\ i > e^T z}} (i - e^T z) \cdot \prod_{i=0}^{i < g_1} (z_1 - i) \cdots \prod_{i=0}^{i < g_j} (z_j - i) \cdots \prod_{i=0}^{i < g_{\hat{n}}} (z_{\hat{n}} - i) \qquad (5)$$

has degree d and vanishes at all grid points except g itself. Thus we obtain for any argument z the vector polynomial identity

$$\phi_d(z) = \frac{1}{d!} \sum_{\substack{0 \leq g}}^{\substack{e^T g \leq d}} \begin{pmatrix} e^T g \\ g_1 \; \cdots \; g_{\hat{n}} \end{pmatrix} \cdot \phi_d(g) \cdot L_g(z) r, \tag{6}$$

where the first term under the sum is the multinomial coefficient representing $d!/L_g(g)$.

When d and $\hat{n}$ are small integers, this Lagrange formula provides easy access to the values and derivatives of f_d at arbitrary arguments. It should also be noted that an additional column in V and thus an additional component in z can be accommodated without resampling on the previous subspace. Similarly, one can reuse the old sample values after an increase in the degree d. By differentiating (6) with respect to the vector z and then evaluating at the origin $z = 0$, all partials of f with respect to the columns of V can be expressed as linear combinations of the sample values $\phi_d(g)$. Consequently, adjoints of the mixed partials can also be obtained as adjoints of the vectors $\phi_d(g)$ considered as a function of x. Hence we conclude that the problem of calculating mixed partials and their adjoints can be reduced to that of calculating directional derivatives (or more precisely univariate Taylor coefficients) and their adjoints.

It is also possible to directly calculate multivariate Taylor series of f on the range of some V by automatic differentiation. As of now we prefer instead the repeated evaluation of univariate Taylor coefficients because it allows for much simpler and smaller data and program structures. Also, if the subspace needs to be enlarged it would be impossible to update the multivariate Taylor series unless an enormous amount of intermediate data has been stored. This assertion will become clearer in the following section.

4. Propagation of Taylor Series and their Adjoints

Virtually all vector functions $y = f(x) : \mathbf{R}^n \to \mathbf{R}^m$ of practical interest are evaluated by computer programs whose execution consists of a sequence of scalar assignments, that is, a loop of the form

Original Program

$$For \; i = n + 1, n + 2, \ldots, n + p + m$$

$$s_i = f_i \langle s_j \rangle_{j \in \mathcal{J}_i}$$

Here each *elementary function* f_i depends on already computed scalar quantities s_j with j belonging to an index set

$$\mathcal{J}_i \subset \{1, 2, \ldots, i - 1\} \quad for \quad i = n + 1, n + 2, \ldots, n + p + m.$$

The nonnegative integer p represents the number of intermediate variables, which we expect to be much larger than both n and m for seriously nonlinear problems. Thus we can combine the variables s_i into the three vectors:

$$
\begin{aligned}
x &\equiv (s_1, s_2, \ldots \quad \ldots, s_{n-1}, s_n) & (\textit{independent}) \\
z &\equiv (s_{n+1}, \ldots \quad \ldots, s_{n+p}) & (\textit{intermediate}) \\
y &\equiv (s_{n+p+1}, \ldots \quad \ldots, s_{n+p+m}) & (\textit{dependent})
\end{aligned}
$$

On a parallel machine, components of z and y that do not directly or indirectly depend on each other can be evaluated concurrently. On a sequential machine the cost of executing all assignments is largely independent of the variable ordering.

In mathematical terms f is the composition of $p + m$ elementary or *library* functions f_i, which are assumed to be analytic in the interior of their domains. For example, this is clearly the case when all f_i represent either elementary arithmetic operations, namely, $+$, $-$, $*$, and $/$ or nonlinear system functions of a single argument, for example, logarithms, exponentials, and trigonometric functions.

Rather than restricting ourselves to unary and binary elementary functions, we may allow any number of arguments. However, we do need the property that given a truncated Taylor series

$$
s_j(t) = s_j^{(0)} + s_j^{(1)}t + s_j^{(2)}t^2 + \ldots + s_d^{(d)}t^d + \mathcal{O}(t^{d+1}) \tag{7}
$$

for all $j \in \mathcal{J}_i$, the corresponding coefficients of $s_i(t) = f_i(s_j(t))_{j \in \mathcal{J}_i}$ can be computed at an $\mathcal{O}(d^2)$ cost. Here the computational cost may be measured in any reasonable way accounting in particular also for reads and writes from and to memory. Using the sum, product, and quotient rule of differentiation, one can easily check that this condition is met for all arithmetic operations. Similarly, one may use the fact that all standard functions are quadratures or solutions to linear ODE's in order to show that Taylor series can be propagated from their arguments to their values with quadratic complexity in d. In the Appendix we list the required calculations for the natural logarithm and its adjoint in detail. Obviously, the same complexity requirement holds also for multivariate linear and diagonally quadratic functions.

Combining the first $(d + 1)$ Taylor coefficients of s_i into a vector

$$
S_i \equiv \left(s_i^{(0)}, s_i^{(1)}, s_i^{(2)}, \ldots, s_i^{(d)} \right)^T
$$

we may thus expand the original program to a recursion on Taylor series:

Forward Sweep

$For\ i = n + 1, n + 2, \ldots, n + p + m$

$$
S_i = F_i \langle S_j \rangle_{j \in \mathcal{J}_i}
$$

where the vector functions $F_i : \mathbf{R}^{d+1} \to \mathbf{R}^{d+1}$ are uniquely determined by the underlying scalar functions f_i. The transition from f_i to F_i can be interpreted as an extension of real or complex arithmetic to an algebra of truncated Taylor series ([13]). In any case we can ensure that for some constant $c > 0$ the evaluation costs of f_i and F_i satisfy the simple relation

$$Cost\{F_i\langle S_j\rangle_{j\in\mathcal{J}_i}\} \,/\, Cost\{f_i\langle s_j\rangle_{j\in\mathcal{J}_i}\} \quad \leq \quad c\cdot(1+d)^2.$$

Now let us consider the composite vector function F that maps a truncated Taylor series for a path $x(t)$ into the corresponding approximation to $y(t) = F(x(t))$. Since this function can be evaluated by one forward sweep, it follows from the last inequality by summation over i that at (least on a serial machine)

$$Cost\{F(X)\} \,/\, Cost\{f(x)\} \leq c\cdot(1+d)^2$$

Here $X = \left[x_i^{(j)}\right]$ may be thought of as a matrix with n rows X_i and $(d+1)$ columns $x^{(j)}$. Similarly $Y = \left[y_i^{(j)}\right]$ forms a matrix with m rows Y_i and $(d+1)$ columns $y^{(j)}$.

To recover from these Taylor series the derivative vectors discussed in Section 2, we may restrict $x(t)$ to be linear, so that only the first two columns $x \equiv x^{(0)}$ and $v \equiv x^{(1)}$ of X are nonzero. Then the resulting $(d+1)$ columns of $Y = F(X)$ are exactly the coefficient vectors occurring in equation (3). Therefore the software package ADOL-C briefly described in the following section contains a problem-independent routine **forward** for calculating Y given x and v, and after the structure of the composite function f has been recorded during an execution of the original program.

Finally, let us consider the problem of computing for fixed x, v, and some $u \in \mathbf{R}^m$ the adjoint values

$$\bar{s}_i^{(j)} \quad \equiv \quad \frac{\partial < u^T y^{(d)} >}{\partial s_i^{(j)}}$$

The adjoints $\bar{s}_i^{(j)}$ describe the sensitivity of a weighted sum of the highest Taylor coefficients $y_i^{(d)}$ with respect to the intermediate coefficients $s_i^{(j)}$. In particular, we find for $j = 0$ and $i \leq n$ that

$$\bar{x}_i \equiv \bar{s}_i^{(0)} = \frac{\partial}{\partial x_i} u^T \nabla^d f(x) v^d$$

is the i-th component of the vector a defined in (2) with all v_i identical. It can also be seen that the other columns $\bar{x}^{(j)}$ of the matrix $\bar{X}$ represent the vectors a defined in (2) with d replaced by $d - j$. The salient fact about the scalar coefficients $\bar{s}_i^{(j)}$ is that like the underlying $s_i^{(i)}$ they can be computed recursively with the original order of dependency exactly reversed. More specifically, $\bar{s}_i^{(j)}$ may depend on all $\bar{s}_{\tilde{i}}^{(\tilde{j})}$ with $\tilde{i} \geq i$ and $\tilde{j} \geq j$. It follows from the chain rule that each vector $\bar{S}_j$ can be computed from the vectors $\bar{S}_i$ with $j \in \mathcal{J}_i$ by using the Jacobians ∇F_i of the corresponding mappings F_i at the given point. After initializing all $\bar{S}_i$ except the components of $\bar{Y}$ to zero, one must execute the following loop:

Reverse Sweep

$$For\ i= n+p+m\ ,\ n+p+m-1\ ,\ \ldots\ ,\ n+1$$

$$\bar{S}_j^T\ +=\ \bar{S}_i^T\left[\tfrac{\partial}{\partial S_j}F_i\right]\quad for\ \ all\ \ j\in\mathcal{J}_i$$

where $a+ = b$ means *increment a by b* as in the programming language C. Since the mappings F_i arise from single elementary operations and functions, they are highly structured, and one can perform the adjoint evaluations without actually forming the Jacobian matrix of the F_i. In fact, the adjoint evaluations associated with any F_i can usually be carried out at no more than twice the computational effort of evaluating F_i itself (see the Appendix). However, one does have to store the current arguments S_j of each F_i during the forward sweep in order to access this information on the reverse sweep. The size of this data set is proportional to the original run time and may therefore grow extremely large. Fortunately, the data are accessed strictly sequentially so that they can be stored and retrieved from disk without severe run-time penalties. The amount of data that need to be accessed at random can be limited to $2(d+1)$ times the core memory requirement of the original program. The routine **reverse** described below calculates $\bar{X}$ given u and implicitly x, v.

If multivariate rather than univariate Taylor series are propagated during the forward sweep, the amount of information that has to be stored in preparation for the reverse sweep grows significantly. The same amount of storage is needed if one does not plan a reverse sweep but merely wishes to increase the degree d or the dimension $\hat{n}$ of the subspace recursively, without regenerating terms that were already calculated during earlier sweeps. If the final $\hat{n}$ and d are a priori known and of significant size, it might be efficient to implement the functions F_i with *fast* methods [2], for example, Fourier transforms. This sophisticated approach would reduce the complexity per elementary operation to an order of $d \log d$ rather than the order d^2 of the naive method based on explicit polynomial multiplication. Fortunately, the quadratic term is initially small compared to the affine part and the general overhead, so that the run time growth appears to be essentially linear for $d < 9$.

5. Implementation by Overloading

The mathematical techniques sketched above have been implemented in various ways. Apart from computational efficiency and accuracy, user convenience must be a leading design objective. The first question is how the user specifies the nonlinear system $f(x) = 0$. Sometimes $f(x)$ may have a concise algebraic representation; sometimes it may be generated automatically, for example, by a finite element package. In large-scale problems it seems likely that the vector function $f(x)$ will at some stage be represented by an evaluation code in a high-level programming language like

Fortran, Algol, or C. Assuming that this is the case, such given code is the natural object for automatic differentiation. Early on, some researchers advocated and practiced the modification of the original code by hand (for a survey, see [11]). While this is obviously rather tedious and error prone for problems of any size, an experienced programmer can probably write an *adjoint code* that is smaller and more efficient than automated alternatives. In fact, certain features (e.g., implicit functions and other subprograms) are handled inefficiently or not at all by the automatic differentiation packages that we are aware of.

Several researchers developed precompilers that analyze a given code (usually Fortran 77) and add the extra instructions that are needed to compute the required derivatives (see, e.g., [8], and [9]). The resulting expanded code is then compiled with a standard compiler into an object code that can simultaneously evaluate derivatives if called with appropriate options. A more direct route to the same result can be taken in languages that allow the overloading of operators and functions for arguments of user-defined types. With the help of problem-independent header files, one may then coax the compiler itself to issue the required extra instructions. Essentially, all the user has to do is to redeclare his floating-point variables to be of a different type (see, e.g., [14]). Unfortunately, the overloading capability is currently not available in Fortran, but it will be included in Fortran 90. Using the public domain converter **f2c** of AT&T, codes written in Fortran 77 can be translated into C, which is essentially a subset of C++ where overloading is possible. The routines for performing subsequent reverse and forward sweeps may then again be called from a Fortran program (e.g., an optimization package). We have traveled this circuitous route on a weather model based on the shallow water equation, as reported in the following section. For the automatic differentiation of the C code, we used our C++ package ADOL-C, which is documented in [6] and can be obtained free of charge from the author.

To convey an idea of how ADOL-C is used, let us consider the following univariate example

```
#include "adouble.h"
#include "adutils.h"
#include <stdio.h>
adouble power(adouble x, int n) {
adouble z=1;                          'This function computes integral'
if (n > 0){                           'powers by recursive bisection. '
    int nh =n/2; z = power(x,nh);
    z*= z; if (2*nh != n) z *= x;
    return z;}
else {
    if (n==0) return z;
    else return 1/power(x,-n);} }
```

```
main() {
int n; scanf("%d",&n);
double xp,yp;
adouble x,y;
trace_on();                              'Initiate recording of calculation'
x <<= xp=0.5;                            'Select independent variable'
y = power(x,n);                          'Main crunchy section unaltered'
y >>= yp;                                'Select dependent variable'
trace_off();                             'Terminate recording of calculation'
forward(&res,1,3,&xp);                   'Forward sweep of degree 3'
reverse(&yp,3);                          'Reverse sweep of degree 3'
forward(&res,0,4,&xp);                   'Forward sweep of degree 4'
printf("%f = ? %f ",yp/4,res); }         'Check derivative consistency'
```

The letters and words printed in boldface had to be added in order to make the original program fit for differentiation. The quoted header files contain the definition of the new type **adouble** to which all floating-point variables that are considered as differentiable functions must be redeclared. Independent and dependent variables are selected with the special assignment operators <<= and >>=. All calculations between the function calls **trace_on** and **trace_off** are internally recorded on a *tape*. Subsequently this tape can be used to calculate direct and adjoint derivatives of arbitrary order. In this univariate case the reverse sweep has no advantage, but its results for some d can be compared to those from the forward sweep with d incremented by one. The example shows that even recursive subprogram calls are no problem. Similarly, one may have branches as in the function **power**, provided they are not conditioned on the values of **adoubles**.

On problems with many independent and dependent variables, the modifications of the original C code are exactly the same, except that there must of course be several special assignments of the form <<= and >>= . Also, the derivative evaluation routines **forward** and **reverse** need more arguments, i.e. their calls must be of the form

```
void forward(Y,rev,deg,x,v)
double Y[m][];          // columns as in (1) for d ≤ deg
int rev ;               // flag for subsequent reverse
int deg ;               // maximal degree of derivatives
double x[] ;            // independent variable vector
double v[] ;            // tangent vector in domain
```

```
void reverse(Xbar,deg,u)
double Xbar[n][] ;              //columns as in (2) for d ≤ deg
int deg ;                      // maximal degree of derivative -1
double u[] ;                   // adjoint weight vector
```

where m and n are the (internally stored) numbers of dependent and independent variables respectively. A call to **reverse** must always be preceded by a corresponding call to *forward* with **rev=1**. As we have stressed before, the run time for either routine is bounded by a constant multiple of $(1 + d)^2$ relative to the run time of the underlying undifferentiated evaluation program.

The package includes convenient drivers for the evaluation of gradients and Hessians. A driver that calculates full derivative tensors on subspaces using the Lagrangian formula (6) is currently under development.

6. Numerical Results

To verify the practical possibility of calculating higher-order derivatives for sizable problems, we consider the classical Bratu problem and a shallow water model for weather modeling [12].

For the Bratu problem we chose a cylindrical geometry with homogeneous Dirichlet boundary conditions and an Arrhenius term with two parameters. These were considered as independent variables together with the temperature distribution. Rather than actually calculating the cubic turning point of this system, we simply calculated direct and adjoint derivative vectors of degree 0 to 8 at an arbitrary point. Under the assumption of rotational symmetry, the problem is reduced to a rectangle and then discretized by using bilinear elements on an orthogonal but unevenly spaced grid. The original Fortran code was hand translated into C and then modified for automatic differentiation as indicated in the previous section.

Grid	10 × 20		20 × 40		30 × 60		40 × 80	
Tape	59392		272384		641024		1163264	
Deaths	7081		32441		76201		138361	
Original	0.085		0.41		1.0		1.7	
d = 0	5.6	8.6	4.8	8.3	4.9	8.4	4.9	8.4
d = 2	12.6	24.8	12.4	23	11.6	22.8	12.4	23.8
d = 4	24.6	49.4	23.1	45.2	22.2	45.2	23.2	45.2
d = 6	40.8	82.4	37.8	75	36.6	73.6	37.6	77.2
d = 8	61	121.4	55.4	111.9	55.5	113.4	55.9	111.6
Mode	for	rev	for	rev	for	rev	for	rev

The first two entries called *tape* and *deaths* in each column determine the sequential storage requirement for forward and reverse sweeps as $tape + (d + 1)deaths$ bytes and $tape + 2(d + 1)deaths$

bytes, respectively. The factor 2 in the storage requirement for **reverse** can be avoided if adjoint derivative vectors are computed only in single-precision accuracy. As always, the amount of randomly accessed core memory is roughly $(d+2)$ times that of the original code, which is in this case negligible. The entries in the third row represent the run time of the undifferentiated code on a Sun 3 in seconds. All entries underneath are ratios between the run time of forward or reverse sweeps and the corresponding times in the third row. As one can see, the reverse sweeps take some $50-100\%$ longer than forward sweeps, and the times for both grow roughly linear in $(d+1)$.

A similar picture emerges for the second example, the discretization of three coupled hyperbolic PDEs with two spatial dimensions. Here we considered only one $11 \times 11 \times 30$ grid. The 3483 independent variables represent the initial and boundary conditions on all but the terminal face of the box. Starting from the initial face the shallow water equation is integrated over 30 time steps. The ostensible purpose of the calculation is to minimize a weighted sum of errors between calculated and observed values at the 3630 grid points by varying the independent variables representing the initial conditions. For this purpose the original Fortran code contained a separate subroutine for the generation of artificial data in common blocks. Only the routine for the subsequent evaluation of the sum of square residual was translated into C by using the converter **f2c** of AT&T. After the minor modifications discussed above, the C++ code was compiled and then linked with the object codes generated from the Fortran parts of the program. Our utilities **forward** and **reverse** were also called from Fortran, which works because their parameters were deliberately restricted to compatible data types. The results of the calculation are again listed as run-time ratios relative to the five seconds taken by the execution of the original code.

d	0	1	2	4	6	8
Forward	3.357	5.046	6.914	11.36	18.77	25.82
Reverse	5.691	8.874	12.411	20.93	34.88	48.55

The total storage requirement for these runs varied between 2.5 and 25 megabytes. Again we see that the run time grows essentially linearly in the degree $d \leq 8$ with reverse sweeps taking almost twice as long as the corresponding forward sweeps. Overall we conclude that direct and adjoint derivative vectors can be computed at a reasonable cost in terms of run time. The extra storage requirement for the reverse sweeps is very substantial but manageable. The reader may wander why we did not discuss the accuracy of the numerical derivative obtained. The reason is that on all test problems for which handcoded derivatives are availables their numerical values are virtually identical with those obtained by ADOL-C. It should be noticed that automatic differentiation does not incur any truncation errors and would yield exact results if the calculations could be carried out with infinite precision. Further investigations and software development should target the automatic differentiation of implicit functions and numerical approximations to Poincaré maps.

7. Appendix

As a typical example of a univariate system function we examine the natural logarithm as an elementary function, namely, $s_i = f_i(s_j) \equiv log(s_i)$. Assuming that the scalars s_j and consequently s_i are in fact smooth functions of a parameter $t \in \mathbf{R}$, we find by differentiation that for all $k \geq 1$

$$[s_j(t)\, s_i(t)']^{(k-1)} \equiv s_j(t)^{(k)},$$

where $s(t)^{(k)}$ denotes the k-th Taylor coefficient of $s(t)$ with respect to t. By identifying terms on both sides, one finds the recurrence

$$\tilde{s}_i^{(k)} = \frac{1}{s_j}\left[s_j^{(k)} - \sum_{l=1}^{k-1}\tilde{s}_i^{(l)}s_j^{(k-l)}\right] \quad and \quad s_i^{(l)} = \frac{1}{l}\tilde{s}_i^{(l)} \quad for \ 1 \leq l \leq d.$$

This simple algorithm evaluates the vector function $S_i = F_i(S_j)$ for $f_i = log$ using d divisions and $d(1+d)/2$ multiplications followed by additions or subtractions. The corresponding adjoint operation can be effected by the recurrences $\bar{\tilde{s}}_i^{(k)} = \frac{1}{k}\bar{s}_i^{(k)}$ for $1 \leq k \leq d$ and

$$\bar{\tilde{s}}_i^{(l)} \quad -= \quad \frac{1}{s_j}\left[\sum_{k=l+1}^{d} \bar{\tilde{s}}_i^{(k)}s_j^{(k-l)}\right], \quad for \quad l = d-1,\ldots,1$$

$$\bar{s}_j^{(l)} \quad += \quad \frac{1}{s_j}\left[\bar{s}_i^{(l)} - \sum_{k=l+1}^{d} \bar{\tilde{s}}_i^{(k)}\tilde{s}_i^{(k-l)}\right], \quad for \quad 0 \leq l \leq d$$

When d is of some size, these adjoint recurrences involve almost exactly twice as much work as the original ones for the evaluation of F_i. The logarithm is in this respect typical of the other transcendental functions and the multiplication operation. As a consequence, reverse sweeps tend to take almost twice as long as the corresponding forward sweeps, which was certainly observed in our numerical results.

8. Acknowledgments

The author is indebted to B. Fiedler and P. Kunkel for advice, to I. M. Navon for providing the shallow water code, and to S. Reese for conducting the numerical experiments. The work on this project was began during a three week visit at the Konrad Zuse Zentrum in Berlin.

References

[1] W.J. Beyn (1991). *Numerical Methods for Dynamical Systems*, To appear in the Proceedings of the SERC Summer School in Lancaster, England.

[2] R. P. Brent and H. T. Kung (1978). *Fast Algorithms for Manipulating Power Series*, J. of the ACM, Vol. 25, No. 4, pp. 581–595.

[3] K. Brown (1969). *A Quadratically Convergent Method for Solving Nonlinear Equations*, SIAM J. Numer. Analysis, Vol. 6, pp. 560–569.

[4] A. Griewank (1989). "On Automatic Differentiation," in *Mathematical Programming: Recent Developments and Applications,* ed. M. Iri and K. Tanabe, Kluwer Academic Publishers, pp. 83–108.

[5] A. Griewank (1990) *Direct Calculation of Newton Steps without Accumulating Jacobians,* Preprint MCS-P127-0290, Argonne National Laboratory, Argonne, IL 60439. To appear in the Proceedings of the SIAM Workshop on Large-Scale Optimization, at Cornell University.

[6] A. Griewank, D. Juedes, and J. Srinivasan (1990). *ADOL-C, A Package for the Automatic Differentiation of Algorithms Written in C/C++,* Preprint MCS-180-1190, Argonne National Laboratory, Argonne, IL 60439.

[7] A. Griewank and G. W. Reddien (1989). *Computation of Cusp Singularities for Operator Equations and Their Discretizations,* J. Comp. and App. Math., Vol. 48, pp. 591–601.

[8] J. E. Horwedel, B. A. Worley, E. M. Oblow, and F. G. Pin (1988). *GRESS Version 0.0 Users Manual,* ORNL/TM 10835 , Oak Ridge National Laboratory, Oak Ridge, TN 37830.

[9] K. Kubota and M. Iri (1990). *Padre 2, version 1—User's Manual,* Research Memorandum RMI 90-01, Faculty of Engineering, University of Tokyo, Hongo 7-3-1, Bunkyo-ku, Tokyo.

[10] P. Kunkel (1988). *Quadratically convergent methods for the computation of unfolded singularities* SIAM J.Numer. ANAL. Vol. 25, No 6. pp. 1392-1408.

[11] H. Kagiwada, R. Kalaba, N. Rosakhoo, and K. Spingarn (1986). Numerical Derivatives and Nonlinear Analysis, Vol. 31, *Mathematical Concepts and Methods in Science and Engineering,* ed. A. Miele, Plenum Press, New York.

[12] I. M. Navon and U. Muller (1979). *FESW—A Finite-Element Fortran IV Program for Solving the Shallow Water Equations,* Advances in Engineering Software Vol. 1, pp. 77–84.

[13] L.B. Rall (1990). *Differentiation Arithmetics.* in Computer Arithmetic and Self-validating Numerical Methoids, Notes and Reports in Mathematics in Science and Engineering, Vol.7, pp. 73-90, Academic Press, Boston 1990.

[14] L. B. Rall (1984). *Differentiation in PASCAL-SC: Type GRADIENT,* ACM TOMS, Vol. 10, pp. 161–184.

International Series of Numerical Mathematics, Vol. 97, © 1991 Birkhäuser Verlag Basel

On the Stability of a Spinning Satellite in a Central Force Field

Ardéshir Guran,

University of Toronto, Canada

Introduction

It is known that the stability or instability of motion of a rigid axi-symmetric satellite spinning around its axis of symmetry can be determined by knowing its inertia-moment-ratio and its winding number [1,2]. This classical result is based on a linear stability analysis of the equation of motion which allows very small angular deviations of the satellite's axis of symmetry from the space-fixed Z-direction (see Figure 1.).

The present paper sets out to study the full non-linear equations of motion of a spinning satellite without confining its motion to small attitude angles and small perturbations which are necessary in linear stability analysis. To that end, using a normalized Hamiltonian formalism, the nonlinear differential equations of motion of the satellite are obtained. It is shown that the motion of a spinning satellite in a central force field defines a motion on a three dimensional manifold. Various dynamic behaviours e.g. periodic, quasiperiodic, and chaotic are diagnosed via the Poincaré map technique. The effect of oblateness on the motion of the satellite is also studied.

Equations of Motion in a Normalized Hamiltonian Formalism

Consider an axisymmetric spinning satellite in a circular orbit (Figure 1), the kinetic and the potential energy of the satellite in terms of Cardan angles ϕ_1, ϕ_2, ϕ_3 of the first kind [3] are

$$
\begin{aligned}
T &= \frac{A}{2}[\dot{\phi}_1^2 cos^2\phi_2 + \dot{\phi}_2^2 + \Omega^2(sin^2\phi_1 + cos^2\phi_1 sin^2\phi_2) - \\
&\quad 2\dot{\phi}_1\Omega cos\phi_1 sin\phi_2 cos\phi_2 + 2\dot{\phi}_2\Omega sin\phi_1] + \frac{C}{2}\omega_0^2 \\
V &= \frac{3}{2}(C - A)\Omega^2 sin^2\phi_2
\end{aligned}
\tag{1}
$$

where A, A, C are the principal moments of inertia, and ω_0, Ω are the total spin and the orbital angular velocity of the satellite respectively.

Defining the generalized momenta $p_i = \frac{\partial L}{\partial \phi_i}$, and passing to dimensionless variables

$$\tau = t\sqrt{\frac{H}{A}}$$

$$\lambda_i = \frac{p_i}{\sqrt{HA}} \tag{2}$$

the normalized Hamiltonian reads [4]

$$h = \frac{1}{2}\left(\frac{\lambda_1^2}{\cos^2 \phi_2} + \lambda_2^2\right) + \lambda_1\sqrt{\gamma}\cos\phi_1\tan\phi_2 - \lambda_1\sqrt{\gamma}\mu_1\mu_2\frac{\sin\phi_2}{\cos^2\phi_2} -$$

$$\lambda_2\sqrt{\gamma}\sin\phi_1 - \gamma\mu_1\mu_2\frac{\cos\phi_1}{\cos\phi_2} + \gamma\frac{\mu_1^2\mu_2^2}{2}\tan^2\phi_2 +$$

$$\frac{\gamma}{2}\mu_1\mu_2^2 + \frac{3}{2}\gamma(\mu_1 - 1)\sin^2\phi_2 \tag{3}$$

where $\mu_1 = \frac{C}{A}$ gives the degree of oblateness of the gyro and $\mu_2 = \frac{\omega_0}{\Omega}$ is a frequency ratio or winding number which tells us how many turns the satellite around itself completes while it goes around its master once.

Note that $\gamma = \frac{4\Omega^2}{H}$ measures the strength of gravitational energy relative to the total energy of the satellite. The equations of motion of the satellite have the canonical form

$$\dot{\lambda}_1 = \sqrt{\gamma}\lambda_1\sin\phi_1\tan\phi_2 + \sqrt{\gamma}\lambda_2\cos\phi_1 - \gamma\mu_1\mu_2\frac{\sin\phi_1}{\cos\phi_2}$$

$$\dot{\lambda}_2 = -\lambda_1^2\frac{\sin\phi_2}{\cos^3\phi_2} - \lambda_1\lambda_1\sqrt{\gamma}\frac{\cos\phi_1}{\cos^2\phi_2} + \lambda_1\sqrt{\gamma}\mu_1\mu_2\frac{1+\sin^2\phi_2}{\cos^3\phi_2}$$

$$\qquad +\gamma\mu_1^2\mu_2^2\frac{\sin\phi_2}{\cos^3\phi_2} + \gamma\mu_1\mu_2\cos\phi_1\frac{\sin\phi_2}{\cos^3\phi_2} - 3\gamma(\mu_1 - 1)\sin\phi_2\cos\phi_2$$

$$\dot{\phi}_1 = \frac{\lambda_1}{\cos^2\phi_2} + \sqrt{\gamma}\cos\phi_1\tan\phi_2 - \sqrt{\gamma}\mu_1\mu_2\frac{\sin\phi_2}{\cos^2\phi_2}$$

$$\dot{\phi}_2 = \lambda_2 - \sqrt{\gamma}\sin\phi_1 \tag{4}$$

subject to the constraint $h = 1$

Dynamic Behaviour via Poincaré Map

It is clear that the system of equations (4) is integrable for $\gamma = 0$. In this case the angular momentum λ_1 is a second conserved quantity, besides the Hamiltonian. The system is also integrable for the case of $\gamma = \infty$ where the motion is confined to small attitude angles ϕ_1 and ϕ_2 and the Hamiltonian takes a quadratic form.

To study the motion for various bifurcation parameters a total of 200 points have been calculated for each of 23 various initial conditions in the $\lambda_1 \phi_1$ plane. A fourth-order Runge-Kutta integrator has been used. Figure 2 shows the Poincaré maps for a prolate satellite. For small γ's Poincaré maps are fairly well covered by invariant tori. As we go on increasing the bifurcation parameter γ some tori break up into chaotic trajectories. When γ is further increased both attitude angles ϕ_1, ϕ_2 are confined to small angles, and the Hamiltonian becomes effectively a quadratic function of the state variables ϕ_1, ϕ_2, and more regularity emerges. For an oblate satellite the chaotic trajectories emerge in higher values of bifurcation parameter. This is illustrated in Figure 3.

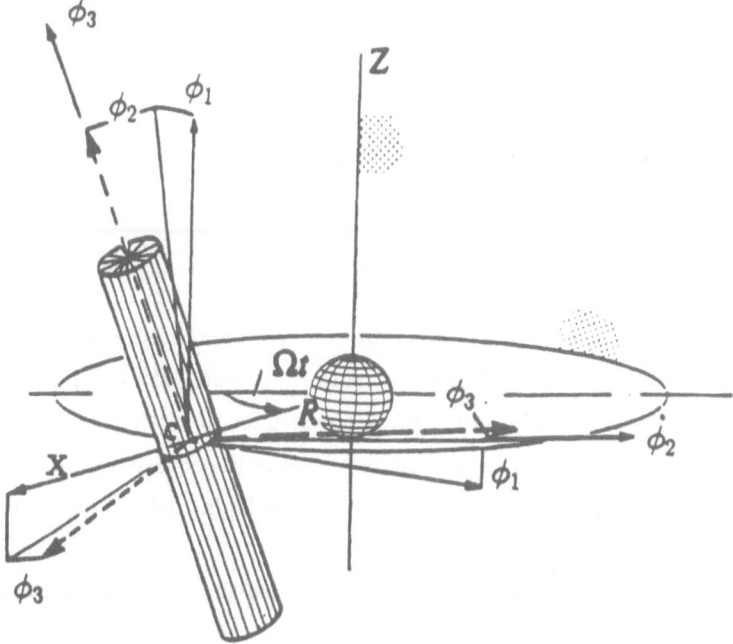

Figure 1 Configuration of the spinning satellite in a circular orbit (ϕ_1, ϕ_2, ϕ_3 are Cardan angles of the first kind).

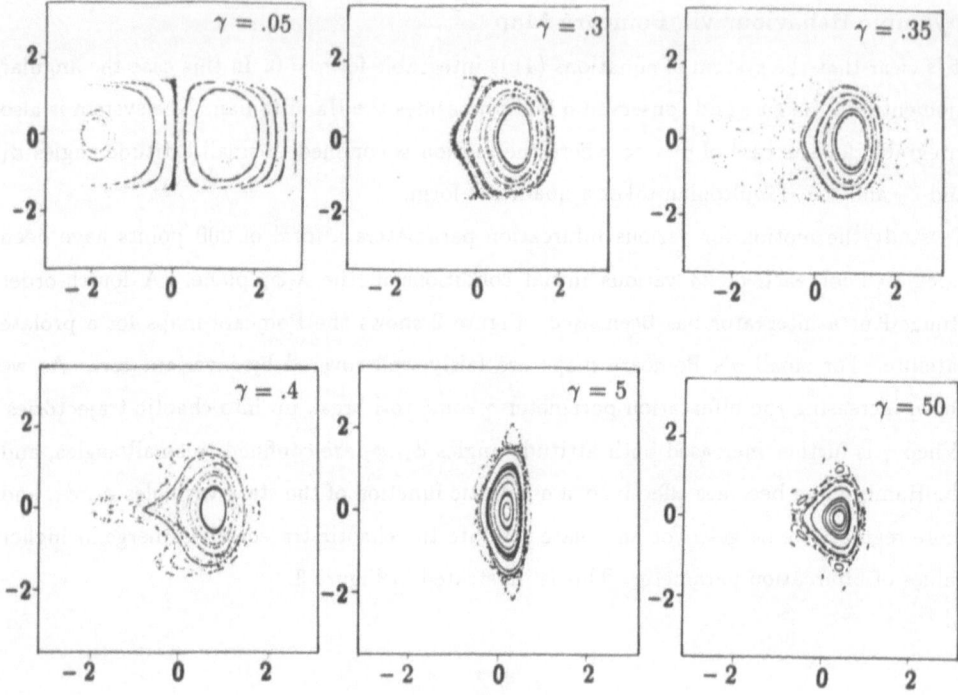

Figure 2 Poincaré sections λ_1 versus ϕ_1 at $\phi_2 = 0$, $\dot{\phi}_2 > 0$ for a prolate satellite in locked rotation ($\mu_1 = 2$, $\mu_2 = 1$, γ as indicated)

Concluding Remarks

The two integrable cases $\gamma = 0$ and $\gamma = \infty$ are so different in nature that the transition between these two kinds has to first destroy one of them before the other can emerge. It is shown that the mixed behaviour of the satellite which exhibits both regular and chaotic motion is located between two limiting integrable cases of the equations of motion. This is analogous to the results reported by Richter and Scholz [5] in their studies of the motion of a planar double pendulum.

The present communication is meant to illustrate the rich and beautiful complexity of the motion in one of the simplest problems of satellite dynamics.

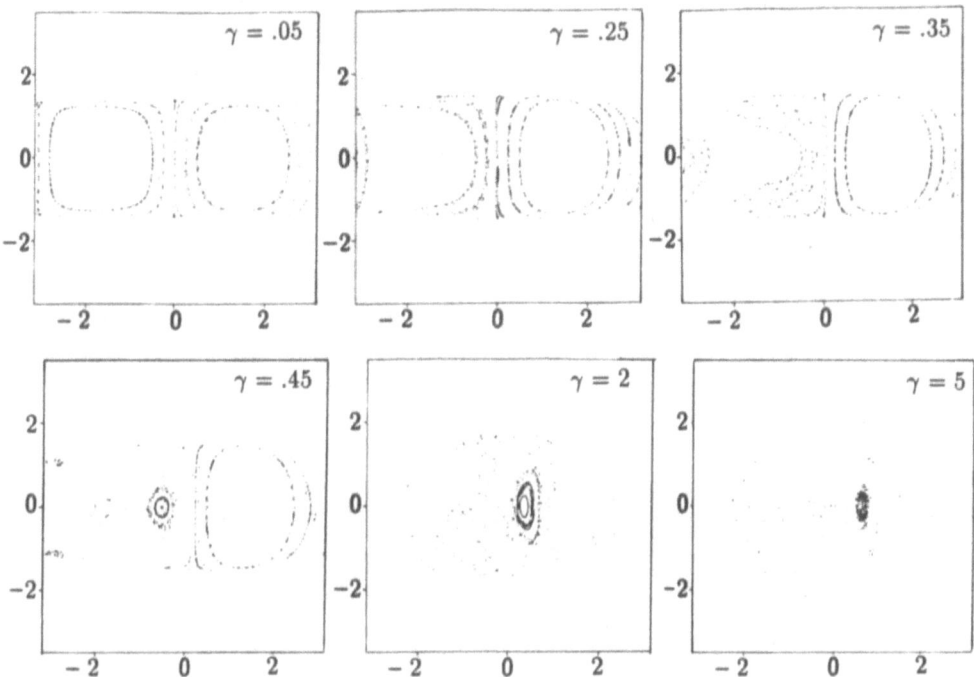

Figure 3 Poincaré sections λ_1 versus ϕ_1 at $\phi_2 = 0$, $\dot{\phi}_2 > 0$ for an oblate satellite in locked rotation ($\mu_1 = .5$, $\mu_2 = 1$, γ as indicated)

References

1. Thomson, W.T.,"Spin Stabilization of Attitude against Gravity Torque", *Journal of the Astronautical Sciences*, Vol. 9, No. 1, pp. 31-33, 1962.

2. Kane, T.R., Marsh, E.L., Wilson, W.G., "Letter to the Editor", *Journal of the Astronautical Sciences*, Vol. 9, pp. 108-109, 1962.

3. Magnus, K., *Kreisel*, Springer-Verlag, Berlin, 1971.

4. Guran, A., *Contributions to the Study of Instabilities in a Class of Conservative Systems*, Toronto, 1990.

5. Richter, P.H., and Scholz, H.J. *The Planar Double Pendulum*, Film C1574, Göttingen; Publ. Wiss. Film, Sekt. Techn. Wiss/Naturw. Ser.9, Nr.7/C1574, 1985.

Prof. Dr. Ardéshir Guran, 5 King's College Road, Mechanical Engineering
University of Toronto, Toronto, Ontario, CANADA, M5S-1A4

International Series of Numerical Mathematics, Vol. 97, © 1991 Birkhäuser Verlag Basel

CODIMENSION TWO BIFURCATION IN AN APPROXIMATE MODEL FOR DELAYED ROBOT CONTROL

G. Haller and **G. Stépán**

Department of Applied Mechanics
Technical University of Budapest
H-1521 Budapest, Hungary

INTRODUCTION

Delay in robot control may result in unexpectedly sophisticated dynamics, which is not possible to explain without the time lag built in the system. But once we have a delayed system, even with a small degree of freedom, it can produce phenomena which are well known in the theory of nonlinear oscillations, but they are not likely to occure in simple, smooth and undelayed mechanical systems.

In this paper we investigate the delayed positioning of a single degree of freedom robotic arm. We reveal the strongly nonlinear behaviour of the system which goes through a generalized Hopf bifurcation of codimension two.

1. THE MECHANICAL MODEL

The mechanical system under investigation can be seen in Fig. 1 (cf. [1]).

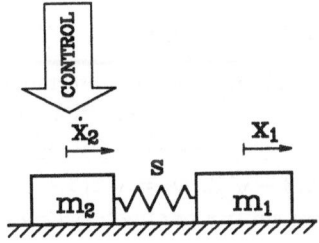

Fig. 1. The mechanical model

It is clear from the picture that we want to position the mass m_1 by the proper movement of m_2 through a flexible element with stiffness s. We use the feedback

$$\dot{x}_2(t) = f\big(x_1(t - \tau)\big), \tag{1.1}$$

where the parameter $\tau > 0$ characterizes the delay in the response of the control system. We assume that f is odd and analytical in a neighbourhood of the origin. Therefore, it can be written in the form

$$f\big(x_1(t - \tau)\big) = -kx_1(t - \tau) + \varepsilon x_1^3(t - \tau) + O\big(|x_1(t - \tau)|^5\big),$$

where $k = -f'(0) > 0$ is the amplification factor or gain, and $\varepsilon > 0$ is a parameter assumed small. With the notation $\alpha = \sqrt{s/m_1}$, the equations of motion are

$$\ddot{x}_1(t) + \alpha^2(x_1(t) - x_2(t)) = 0, \qquad (1.2)$$

$$\dot{x}_2(t) = -kx_1(t - \tau) + \varepsilon x_1^3(t - \tau) + O(|x_1(t - \tau)|^5).$$

In the sequel we will focus on the stability and bifurcations of the $x_1 = x_2 \equiv 0$ trivial solution of (1.2). Our investigations will be based on the assumption that the delay τ is small.

After approximating the right hand side of (1.2) with its 4-jet in τ and introducing the dimensionless time $\tilde{t} = t/\tau$, we get the system

$$y_1' = y_2,$$
$$y_2' = y_3,$$
$$y_3' = y_4, \qquad (1.3)$$
$$y_4' = -24y_1 + 24(1 - \frac{1}{KT})y_2 - 12y_3 + 24(\frac{1}{6} - \frac{1}{KT^3})y_4$$
$$+ 12\frac{E}{K}(2y_1^3 - 6y_1^2 y_2 + 6y_1 y_2^2 + 3y_1^2 y_3 - 2y_2^3 - 6y_1 y_2 y_3 - y_1^2 y_4),$$

where

$$y_1(\tilde{t}) \equiv x_1(\tau \tilde{t}) = x_1(t), \quad y_2 = y_1', \quad y_3 = y_2', \quad y_4 = y_3',$$

$$T = \alpha\tau, \quad K = \frac{k}{\alpha}, \quad E = \frac{\varepsilon}{\alpha},$$

and the prime denotes differentiation with respect to $\tilde{t}$. Clearly, the $(0,0,0,0)$ equilibrium of (1.3) coincides with $x_1(t) \equiv 0$. The characteristic equation of the linear part of (1.3) reads

$$\frac{1}{24}KT^3\lambda^4 + (1 - \frac{1}{6}KT^3)\lambda^3 + \frac{1}{2}KT^3\lambda^2 + T^2(1 - KT)\lambda + KT^3 = 0, \qquad (1.4)$$

from which the conditions of stability can be determined. Simple calculations show that at the point $T_c = \sqrt{6}, K_c = 1/\sqrt{6}$ an interaction of two Hopf bifurcations occurs, resulting in a codimension two bifurcation with two pairs of pure imaginary eigenvalues $\lambda_{1,2,3,4} = \pm i\omega_{1,2}$ with $\omega_1 = \sqrt{2(3 - \sqrt{3})}, \omega_2 = \sqrt{2(3 + \sqrt{3})}$. To determine the unfolding of the bifurcation we choose two control parameters μ_T and μ_K, where

$$\mu_T = T - T_c, \quad \mu_K = K - K_c.$$

2. NORMAL FORM CALCULATION

At the critical point $\mu_T = 0, \mu_K = 0$ system (1.3) reduces to

$$y_1' = y_2,$$
$$y_2' = y_3,$$
$$y_3' = y_4, \qquad (2.1)$$
$$y_4' = -24y_1 - 12y_3 + 12\sqrt{6}E(2y_1^3 - 6y_1^2 y_2 + 6y_1 y_2^2 + 3y_1^2 y_3 - 2y_2^3$$
$$- 6y_1 y_2 y_3 - y_1^2 y_4),$$

In its eigenbasis, system (1.3) takes the form

$$\dot{z}_1 = i\omega_1 z_1 + h^1(z_1, \bar{z}_1, z_2, \bar{z}_2),$$
$$\dot{z}_2 = i\omega_2 z_2 + h^2(z_1, \bar{z}_1, z_2, \bar{z}_2), \tag{2.2}$$

where

$$h^j = \sum_{k+l+m+n=3} h^j_{klmn} z_1^k \bar{z}_1^l z_2^m \bar{z}_2^n, \quad j = 1, 2 .$$

For $\mu_1 \neq 0, \mu_2 \neq 0$ the normal form of (2.2) near the critical point reads

$$\dot{w}_1 = \lambda_1(\mu_T, \mu_K)w_1 + a(\mu_T, \mu_K)w_1^2\bar{w}_1 + b(\mu_T, \mu_K)w_1 w_2 \bar{w}_2 + O(|w|^5),$$
$$\dot{w}_2 = \lambda_2(\mu_T, \mu_K)w_2 + c(\mu_T, \mu_K)w_1\bar{w}_1 w_2 + d(\mu_T, \mu_K)w_2^2\bar{w}_2 + O(|w|^5),$$

with

$$\lambda_j(\mu_T, \mu_K) = \alpha_j(\mu_T, \mu_K) + i\omega_j(\mu_T, \mu_K), \quad j = 1, 2 ,$$

or in polar coordinates (see eg. [2]):

$$\dot{r}_1 = r_1(\alpha_1(\mu_T, \mu_K) + \text{Re } a(\mu_T, \mu_K)r_1^2 + \text{Re } b(\mu_T, \mu_K)r_2^2) + O(r^5),$$
$$\dot{r}_2 = r_2(\alpha_2(\mu_T, \mu_K) + \text{Re } c(\mu_T, \mu_K)r_1^2 + \text{Re } d(\mu_T, \mu_K)r_2^2) + O(r^5),$$
$$\dot{\phi}_1 = \omega_1(\mu_T, \mu_K) + O(r^2),$$
$$\dot{\phi}_2 = \omega_2(\mu_T, \mu_K) + O(r^2). \tag{2.3}$$

For small values of μ_T and μ_K, as in [2], we study the truncated planar system

$$\dot{r}_1 = r_1(\alpha_{1T}\mu_T + \alpha_{1K}\mu_K + ar_1^2 + br_2^2),$$
$$\dot{r}_2 = r_2(\alpha_{2T}\mu_T + \alpha_{2K}\mu_K + cr_1^2 + dr_2^2), \tag{2.4}$$

where

$$\alpha_{jT} = \frac{\partial \alpha_j(\mu_T, \mu_K)}{\partial \mu_T}\Big|_{(0,0)}, \quad \alpha_{jK} = \frac{\partial \alpha_j(\mu_T, \mu_K)}{\partial \mu_K}\Big|_{(0,0)}, \quad j = 1, 2 ,$$

$$a = \text{Re } a(0, 0), \quad b = \text{Re } b(0, 0), \quad c = \text{Re } c(0, 0), \quad d = \text{Re } d(0, 0).$$

After computing the coefficients of (2.4) we get the equations

$$\dot{r}_1 = r_1\left(\left(\frac{\sqrt{6}}{2} - \sqrt{2}\right)\mu_T + \sqrt{6}\mu_K - 3\sqrt{6}Er_1^2 - 6\sqrt{6}Er_2^2\right),$$
$$\dot{r}_2 = r_2\left(\left(\frac{\sqrt{6}}{2} + \sqrt{2}\right)\mu_T + \sqrt{6}\mu_K - 6\sqrt{6}Er_1^2 - 3\sqrt{6}Er_2^2\right). \tag{2.5}$$

Using (2.5) we obtain the following:

1. If $\mu_K > \left(\frac{1}{\sqrt{3}} - \frac{1}{2}\right)\mu_T$ then there exists a periodic solution

$$\tilde{r}_1 = \sqrt{\frac{1}{3E}\left(\frac{1}{2} - \frac{1}{\sqrt{3}}\right)\mu_T + \mu_K}, \tag{2.6}$$

which is stable if $\mu_K > (\sqrt{3} - \frac{1}{2})\mu_T$

2. If $\mu_K > -(\frac{1}{\sqrt{3}} + \frac{1}{2})\mu_T$ then there exists a periodic solution

$$\tilde{r}_2 = \sqrt{\frac{1}{3E}(\frac{1}{2} + \frac{1}{\sqrt{3}})\mu_T + \mu_K}, \tag{2.7}$$

which is stable if $\mu_K > -(\sqrt{3} + \frac{1}{2})\mu_T$.

3. If both periodic solutions exist and are stable, there exists an invariant two-torus

$$\bar{r}_1 = \frac{1}{3}\sqrt{\frac{1}{E}(\frac{1}{2} + \sqrt{3})\mu_T + \mu_K}, \quad \bar{r}_2 = \frac{1}{3}\sqrt{\frac{1}{E}(\frac{1}{2} - \sqrt{3})\mu_T + \mu_K}, \tag{2.8}$$

which is unstable.

In the view of the computations above, the unfolding of system (2.1) can be seen in Fig. 2. The coexistence of the two periodic solutions and the invariant torus has been numerically confirmed in the appropriate sector of Fig. 2, as it is indicated in Fig. 3. The limit cycles and the torus were computed from (2.6), (2.7) and (2.8), while the two solutions approaching different limit cycles were produced by numerical simulation.

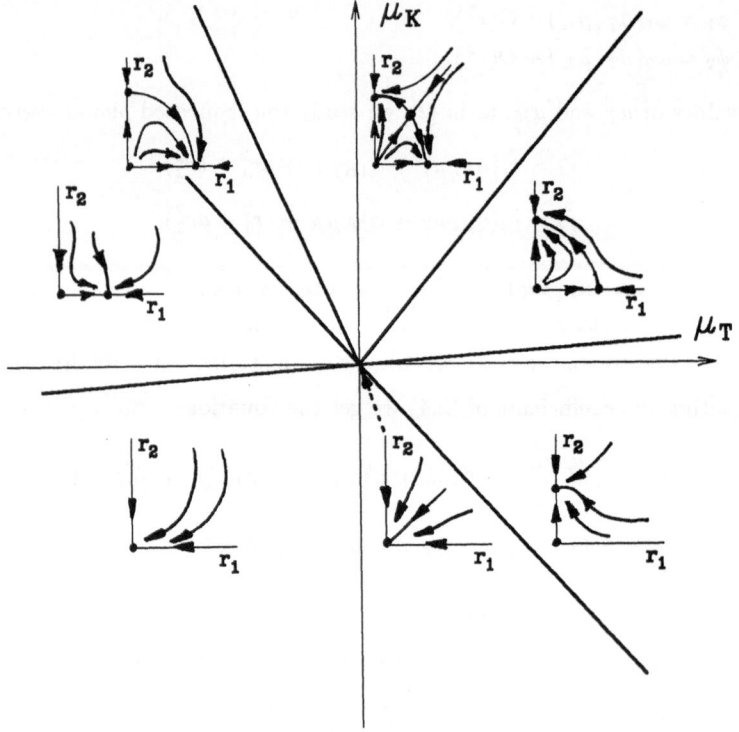

Fig. 2. The unfolding of system (3.5)

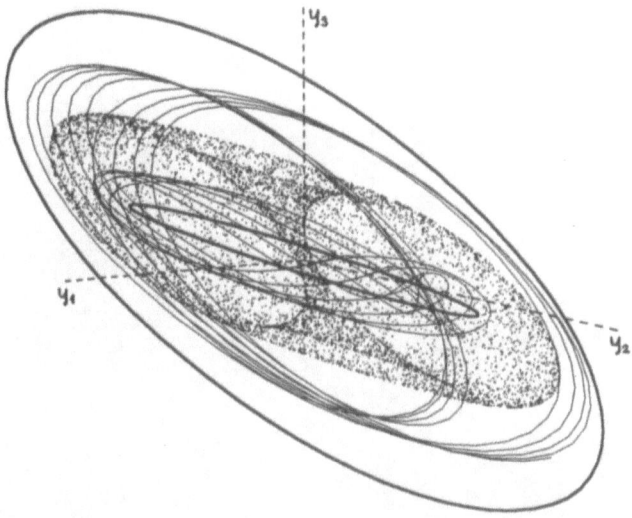

Fig. 3. Coexistence of a torus and two limit cycles
(Projection from the four dimensional phase space
onto the coordinate space $[y_1, y_2, y_3]$.)

ACKNOWLEDGEMENT

This research was supported by the *Hungarian Scientific Research Fundation OTKA*
5-207.

REFERENCES

[1] G. Stépán, Retarded Dynamical Systems: Stability and Characteristic Functions. Longman, Harlow
Essex (1989)

[2] J. Guckenheimer and P. Holmes, Nonlinear Oscillations, Dynamical Systems and Bifurcations of
Vector Fields. Springer-Verlag, New York (1983)

International Series of Numerical Mathematics, Vol. 97, © 1991 Birkhäuser Verlag Basel

LACUNARY BIFURCATION OF MULTIPLE SOLUTIONS
OF NONLINEAR EIGENVALUE PROBLEMS

by

Hans–Peter Heinz

Fachbereich Mathematik der Joh.–Gutenberg–Universität Mainz

Postfach 3980, D–6500 Mainz

1. Introduction. In order to describe the type of nonlinear eigenvalue problems we are going

to discuss, consider a densely defined closed linear operator T in a real Hilbert space H and let H_1 be the Hilbert space which consists of the domain of T together with the graph norm. Also, let H_1^* be the dual space of H_1 and denote the dual operator corresponding to $T: H_1 \to H$ by $T': H \to H_1^*$. Since H_1 is dense in H, we may view H as a subspace of H_1^*, and then the scalar product $(\cdot,\cdot)$ on H and the dual pairing $<\cdot,\cdot>$ on $H_1 \times H_1^*$ coincide on $H_1 \times H$. Moreover, the bounded linear operator

$$T'T: H_1 \to H_1^*$$

extends the positive self–adjoint operator $S := T^*T$, whose spectral properties will play a crucial role. (Detailed accounts of all this can be found e.g. in [4] and [7].)

Now consider the problem

$$T'Tu + \xi(\lambda)N(u) = \lambda u \qquad (1.1),$$

where λ is a real parameter, $\xi: \mathbb{R} \to \mathbb{R}$ a function, $N: H_1 \to H_1^*$ a nonlinear continuous gradient operator such that $N(0) = 0$, $<N(u),u> \geq 0$ $\forall u \in H_1$ and such that N has vanishing Fréchet derivative at $u=0$. We are interested in the case where the spectrum of S presents gaps, as happens, for example, when S is derived from Hill's equation or from a

Schrödinger equation with periodic potential (cf. [2]). Recently, T. Küpper, C.A. Stuart and the author have discussed conditions under which solutions of (1.1) bifurcate from boundary points of these spectral gaps (cf. [3–9]. In the present note we outline a further development of the approach from [3–6], which is based on critical point theory for strongly indefinite functionals ([1,10]). The essential new ingredient is a critical point theorem due to Benci, Capozzi and Fortunato [1]. This result enables us to treat multiple solutions and to streamline the whole theory. Results concerning the operator equation (1.1) will be presented in Section 2, while in Section 3 we indicate an application to nonlinear perturbations of periodic Schrödinger equations. A detailed account of this material will be given elsewhere.

2. The Main Results in the Abstract Setting. Our point of departure is the following

theorem:

Theorem 1. Let E be a real Hilbert space and $J \in C^1(E;\mathbb{R})$ an even functional satisfying

the Palais–Smale condition as well as the following hypotheses:

(i) $J(u) = \frac{1}{2}(Lu,u)_E - \varphi(u) \quad \forall u \in E$, where $(\cdot,\cdot)_E$ denotes the scalar product on E,

where $L:E \to E$ is a bounded self–adjoint linear operator possessing a bounded inverse, and

where $\varphi \in C^1(E;\mathbb{R})$, $\varphi(0) = 0$, and the gradient of φ is a compact operator in E.

(ii)   There exist two L–invariant closed linear subspaces V,W of E as well as

constants $c_0 > 0, \rho > 0, c_\infty \in \mathbb{R}$ such that

$\quad\quad$ (α) $V + W = E$, $m := \dim (V \cap W) < \infty$,

$\quad\quad$ (β) $J(u) \geq c_0 \quad \forall u \in V$ such that $\|u\| = \rho$,

$\quad\quad$ (γ) $J(u) \leq c_\infty \quad \forall u \in W$.

Then J has at least m distinct antipodal pairs $(u_j,- u_j)$ of critical points such that

$J(u_j) \in [c_0,c_\infty]$ (in particular, $u_j \neq 0$) for $j = 1,...,m$.

This is a corollary of the $\mathbb{Z}_2$–version of Thm. 1.4 in [1], as indicated by the remarks

surrounding that theorem.

Now let us return to eq. (1.1). Let $\sigma(S)$ denote the spectrum of S, and let

$G :=]a,b[$ be an interval such that $a,b \in \sigma(S)$, but $G \cap \sigma(S) = \phi$ (i.e. a "spectral gap").

The bifurcation we are interested in occurs at $\lambda = b$ (resp. at $\lambda = a$) when ξ is continuous

at b and $\xi(b) < 0$ (resp. when ξ is continuous at a and $\xi(a) > 0$). For simplicity of

notation we only treat the case $\xi \equiv -1$ here (the general case differs from this only by

minor technicalities). Thus the equation to be discussed reads

$$T'Tu - N(u) = \lambda u \quad\quad\quad\quad\quad (2.1).$$

We use exactly the same general assumptions as in [5]. They are as follows:

(H1) $N = \varphi'$, where $\varphi \in C^1(H_1, \mathbb{R})$, $\varphi(0) = 0$.

$\quad\quad$ Moreover, at $u = 0$, N is Fréchet differentiable with $N'(0) = 0$, $N(0) = 0$.

(H2) φ is even and convex.

(H3) $\exists \gamma \geq \eta > 1$ such that

$$0 \leq \frac{<N(u),u>}{\gamma + 1} \leq \varphi(u) \leq \frac{<N(u),u>}{\eta + 1} \quad \forall u \in H_1.$$

(H4) $N: H_1 \to H_1^*$ is a compact operator.

(H5) $\exists K > 0, c \geq 0$ such that $\|N(u)\|_* \leq K\varphi(u) + c \quad \forall u \in H_1$.

(Here $\|\ \|_*$ denotes the norm on H_1^* dual to the Hilbert space norm on H_1).

For the treatment of eq. (2.1) under these hypotheses it is convenient to renorm H_1

and H_1^* in the following way: By adding a suitable constant to both sides of (2.1) we arrange that $S \geq I$, take $T := S^{1/2}$ and use the norm

$$\|u\|_1 := \|Tu\| \qquad (u \in H_1)$$

on H_1 (here and in the sequel, the symbol $\| \ \|$ always denotes the norm of H). This changes neither the space H_1 nor the operator $T'T$ nor does it affect the validity of (H1)–(H5). We have only passed to an equivalent norm on H_1.

For fixed $\lambda \in \mathbb{R}$ the solutions of (2.1) are evidently the critical points of the functional $J_\lambda \in C^1(H_1, \mathbb{R})$ given by

$$J_\lambda(u) := \frac{1}{2}\|Tu\|^2 - \varphi(u) \qquad (u \in H_1).$$

If $\lambda \in G$, the Palais–Smale condition is satisfied by J_λ, as has been shown in [5], Lemma 2.1. To find suitable spaces V and W, consider the spectral resolution $(E(\lambda))_{\lambda \in \mathbb{R}}$ associated with the self–adjoint operator S, and put

$$H_G := E(a)\,H \subseteq D(S) \subseteq H_1, \qquad V := H_1 \cap H_G^\perp .$$

Given a closed linear subspace X of V, set $X_1 := \left\{ x \in X \mid \|x\|_1 = 1 \right\}$ and $W := H_G \oplus X$, moreover

$$h(\lambda,x) := \left[\frac{a(1-\lambda\,\|x\|^2)}{\lambda - a} \right]^{1/2}$$

and

$$\mu(\lambda;x) := \inf\{\, \varphi(x+w) \mid w \in H_G,\ \|w\|_1 \leq h(\lambda,x)\}$$

for $x \in X_1$, $\lambda \in G$. Suppose

$$\inf_{x \in X_1}\ \mu(\lambda,x) > 0 \qquad (2.2).$$

Then it can be shown (essentially as in [4], Lemma 4.6) that the quantity

$$c_\infty(\lambda;X) := \sup_{x \in X_1}\ \max\left\{ \frac{\eta - 1}{2(\eta+1)^{(\eta+1)/(\eta-1)}} (1-\lambda\,\|x\|^2)^{\frac{\eta+1}{\eta-1}}\, \mu(\lambda x)^{-\frac{2}{\eta-1}}, \right.$$

$$\left. \frac{\gamma - 1}{2(\gamma+1)^{(\gamma+1)/(\gamma-1)}} (1-\lambda\,\|x\|^2)^{\frac{\gamma+1}{\gamma-1}}\, \mu(\lambda,x)^{-\frac{2}{\gamma-1}} \right\}$$

is an upper bound for J_λ on W. The remaining assumptions of Thm. 1 are easily verified (with $c_0 > 0$ arbitrarily small), and we obtain

Theorem 2. Suppose (H1) – (H5) are satisfied. With the above notations and conventions, consider $\lambda \in G$, $m \in \mathbb{N}$, and let $X \subseteq V$ be an m–dimensional linear subspace for which (2.2) holds. Then there exist m distinct antipodal pairs $(u_j, -u_j)$ of non–trivial solutions of (2.1) such that $0 < J_\lambda(u_j) \le c_\infty(\lambda, X)$ for $j = 1,...,m$.

Bifurcation results can be derived from this when (H5) is replaced by the stronger requirement

(H6) $\exists K_0 > 0$ such that $\qquad \|N(u)\|_* \le K_0 \left[\langle N(u), u \rangle^{\frac{\eta}{\eta+1}} + \langle N(u), u \rangle^{\frac{\gamma}{\gamma+1}} \right] \; \forall \; u \in H_1$.

Using this, the proof of Proposition 4.1 in [4] can be mimicked to yield

<u>Theorem 3.</u>   Let (H1) − (H3) and (H6) be satisfied, and let $(\lambda_n)_n \subseteq G$, $(u_n)_n \subseteq H_1$ be sequences such that

      (i)   for every n, $\lambda = \lambda_n$ and $u = u_n$ satisfy (2.1),

      (ii) $\lim\limits_{n \to \infty} \lambda_n = b$, $\lim\limits_{n \to \infty} J_{\lambda_n}(u_n) / (b - \lambda_n) = 0$.

Then $\lim\limits_{n \to \infty} \|u_n\|_1 = 0$ and, in particular, b is a bifurcation point of (2.1) with respect to the H_1−topology.

The last two theorems show us what we have to do in order to establish existence resp. bifurcation results in applications:

A)   If, for some $\lambda \in G$, an m−dimensional subspace $X \subseteq V$ can be found for which (2.2) holds, then (2.1) has m pairs of non−trivial solutions for that value of λ.

B)   If sequences $(\lambda_n)_n$, $(X_n)_n$ can be found such that

      (i) $\lambda_n \in G \; \forall n$, $\lim\limits_{n \to \infty} \lambda_n = b$,

      (ii)   each X_n is an m−dimensional subspace of V for which (2.2) holds with $\lambda = \lambda_n$, and

      (iii) $\lim\limits_{n \to \infty} c_\infty(\lambda_n, X_n) / (b - \lambda_n) = 0$,

then b is a bifurcation point from which systems of m solution pairs $(\lambda_n, \pm u_{nj})$ ($j = 1,...,m$) emanate.

Examples given in [5,8] show that, even for $m = 1$, both A) and B) require additional assumptions. The simplest one of these is

$$\varphi(u) > 0 \qquad \forall \, u \in H_1 - (0) \tag{2.3}.$$

Using an argument based on weak compactness, one can prove that (2.3) implies (2.2) for <u>every</u> $\lambda \in G$ and <u>every</u> finite−dimensional subspace X of V. Thus we obtain

<u>Theorem 4.</u> Suppose (H1) − (H5) and (2.3) are satisfied. Then, for every $\lambda \in G$ there exist infinitely many distinct pairs $(\lambda, \pm u_j)$ ($j \in \mathbb{N}$) of non−trivial solutions of (2.1).

However, (2.3) is unsuitable for B), because B) (iii) cannot be guaranteed. We now discuss a more subtle condition, which is a generalization of condition $(T(\delta))$ from [5,6]. To state it, fix $\lambda_o \in \mathbb{R}$, $\delta > 0$, $m \in \mathbb{N}$.

Definition. We say that condition $(T(\delta, m))$ is satisfied at λ_o provided there exists an m–dimensional real normed space Z and a sequence of linear maps $L_n : Z \to H_1$ ($n \in \mathbb{N}$) such that

(i) $\lim\limits_{n \to \infty} \|L_n \zeta\| = \|\zeta\|$ $\forall \zeta \in Z$,

(ii) $\exists K_1 > 0$ such that

$$\|(T'T - \lambda_o I) L_n \zeta\|^2 \leq K_1 \, | \, \langle (T'T - \lambda_o I) L_n \zeta, L_n \zeta \rangle \, | \qquad \forall \zeta \in Z, \, n \in \mathbb{N}$$

(iii) $\exists C > 0$ such that

$$0 < \varphi(L_n \zeta) \leq C \qquad \forall \zeta \in Z \text{ with } \|\zeta\| = 1$$

(iv) $\left[\inf\limits_{\|\zeta\|=1} \varphi(L_n \zeta) \right]^{-\delta} \sup\limits_{\|\zeta\|=1} | \, \langle (T'T - \lambda_o I) L_n \zeta, L_n \zeta \rangle \, | \to 0$ as $n \to \infty$.

Cumbersome as this condition may seem, it is very suitable for applications, as we shall see in Section 3. Here we indicate how it can be used to construct subspaces of V as required by A) resp. B). Thus, suppose $(T(\delta, m))$ is satisfied at $\lambda_o = b$. Choose $l \in G$, $r > b$, and let $P := E(r) - E(l)$ be the spectral projection corresponding to the interval $]l, r]$. It can be shown that the convergence $\|L_n \zeta\| \to \|\zeta\|$ is uniform on bounded subsets of Z, and this implies that there is n_o such that for $n \geq n_o$ the spaces

$$X_n := P L_n(Z) \subseteq V$$

have dimension m. Put $X_n^1 := \{ x \in X_n \mid \|x\|_1 = 1 \}$. It can be shown (essentially as in Lemma 3.1 of [5]) that in case $\delta \geq 2 / (\eta + 1)$ there is $C_1 > 0$ and $n_1 \geq n_o$ such that

$$0 < \varphi(x) \leq C_1 \qquad \forall x \in \bigcup_{n=n_1}^{\infty} X_n^1 \tag{2.4}$$

and

$$\varepsilon_n := \left[\inf\limits_{x \in X_n^1} \varphi(x) \right]^{-\delta} \sup\limits_{x \in X_n^1} | \, \langle (T'T - bI) x, x \rangle \, | \to 0 \quad \text{as} \quad n \to \infty \tag{2.5}$$

For $\lambda \in G$, $\|x\|_1 = 1$, and $h(\lambda, x) \leq 1$ it easily follows from (H1) – (H3) that

$$\mu(\lambda, x) \geq 2^{-\gamma} \varphi(x) - M \, h(\lambda, x)^{\eta+1} \tag{2.6}$$

with a constant M independent of x and λ. Now consider $\lambda_n \epsilon G$ such that

$$b - \lambda_n = b \cdot \sup_{x \in X_n^1} [1 - b\|x\|^2].$$

Since $\langle T'Tx, x \rangle = \|Tx\|^2 = \|x\|_1^2$, (2.5) yields

$$1 - \lambda \|x\|^2 \le 2 \, \varepsilon_n \, \varphi(x)^\delta \tag{2.7}$$

for $\lambda_n \le \lambda < b$, $x \epsilon X_n^1$, and hence $\lim_{n \to \infty} \lambda_n = b$ by (2.4), (2.5). Moreover, (2.7) enables

us to find an upper bound for $h(\lambda, x)$ on $[\lambda_n, b[\times X_n^1$. Using this bound in (2.6) we find

$$\mu(\lambda, x) \ge 2^{-\gamma} \varphi(x) - M_1 \, \varepsilon_n^{\frac{\eta+1}{2}} \, \varphi(x)^{\frac{\delta(\eta+1)}{2}} \tag{2.8}$$

for $\lambda_n \le \lambda < b$, $x \epsilon X_n^1$, $n \ge n_2 \ge n_1$, where $M_1 > 0$ is independent of n, λ, x. Since

$\delta \ge 2/(\eta+1)$ by assumption, we only have to choose n large enough to satisfy (2.2).

Moreover, (2.7), (2.8) and the choice of λ_n yield the estimate

$$\frac{c_\infty(\lambda_n, X_n)}{b - \lambda_n} \le \text{coust} \cdot \sup_{x \in X_n^1} \max \left\{ \left[\frac{2 \, \varepsilon_n}{2^{-\gamma}\varphi(x)^{1-\delta} - M_1 \, \varepsilon_n^{(\eta+1)/2} \, \varphi(x)^{(\eta-1)\delta/2}} \right]^{\frac{2}{\eta-1}}, \right.$$

$$\left. \left[\frac{2 \, \varepsilon_n}{2^{-\gamma}\varphi(x)^{1-\delta} - M_1 \, \varepsilon_n^{(\eta+1)/2} \, \varphi(x)^{(\eta-1)\delta/2}} \right]^{\frac{2}{\gamma-1}} \right\}.$$

If $\delta \ge 1 > \frac{2}{\eta+1}$, this bound tends to zero as $n \to \infty$, as is easily seen from (2.4), (2.5). Thus,

in this case, the sequences $(\lambda_n)_{n \ge n_2}$, $(X_n)_{n \ge n_2}$ are as required in B).

The upshot of this is that one can prove the following theorem, which is probably the most important of our results:

Theorem 5. Suppose (H1) – (H4) are satisfied. Then:

a) If also (H5) holds and $(T(\delta, m))$ is satisfied at b for some $m \epsilon \mathbb{N}$ and $\delta \ge \frac{2}{\eta+1}$, then there exists $\hat{\lambda} \epsilon G$ such that, for every $\lambda \epsilon [\hat{\lambda}, b[$ there are at least m distinct pairs $(\lambda, \pm u_j)$ $(j = 1, ..., m)$ of non–trivial solutions of (2.1).

b) If (H6) holds and $(T(\delta, m))$ is satisfied at b for some $m \epsilon \mathbb{N}$ and $\delta \ge 1$, then b is a bifurcation point. More precisely, there exist $2m$ sequences $((\lambda_n, \pm u_{nj}))_n$ $(j=1, ..., m)$ of non–trivial solutions of (2.1) such that $\lim_{n \to \infty} \lambda_n = b$, $\lambda_n \epsilon G$ $\forall n$, and $\lim_{n \to \infty} \|u_{nj}\|_1 = 0$ $\forall j$.

As was mentioned at the beginning of this section, similar results with b replaced by a hold for other cases of eq. (1.1), e.g. for

$$T'Tu + N(u) = \lambda u \ .$$

Hypotheses (H1) – (H6) remain unchanged, but of course $(T(\delta,m))$ is now required to hold at $\lambda_0 = a$.

3. An Application. Consider the nonlinear Schrödinger equation

$$- \Delta u + q(x)u - f(x,u) \ = \ \lambda u \tag{3.1}$$

on $\mathbb{R}^N$, where $q \in L^\infty (\mathbb{R}^N)$ is real–valued, and where the function $f{:}\mathbb{R}^N \times \mathbb{R} \to \mathbb{R}$ satisfies the following conditions:

(C1)   Carathéodory conditions

(C2) $f(x, -s) = - f(x,s) \ \forall \ s{\in}\mathbb{R}$ a.e. in $\mathbb{R}^N$, and $f(x,\cdot)$ is nondecreasing on $\mathbb{R}$ a.e. on $\mathbb{R}^N$

(C3) $\exists \ \gamma, \eta$ with $1 < \eta \le \gamma < \frac{N+2}{N-2}$ and such that a.e. on $\mathbb{R}^N$ the function $s \to f(x,s)s^{-\eta}$

is nondecreasing on $] \, 0,\infty[$ and the function $s \to f(x,s)s^{-\gamma}$ is nonincreasing on $]0,\infty[$.

(C4) $\lim\limits_{R\to\infty} ess \sup\limits_{|x| \ge R} |f(x,1)| = 0$, and $f(\cdot,1) \in L^\infty (\mathbb{R}^N)$.

Let $H := L^2(\mathbb{R}^N)$ be the basic Hilbert space, and denote by H^k $(k{\in}\mathbb{Z})$ the real Sobolev space $W^{k,2}(\mathbb{R}^N)$. There is a unique self–adjoint operator S in H such that $D(S) = H^2$ and $Su = -\Delta u + qu \quad \forall \ u{\in}D(S)$. By adding a suitable constant to q and λ we arrange $q(x) \ge 1$ a.e., so that $S \ge I$, and then we can write (3.1) in the form (2.1) with $T := S^{1/2}$ and $N{:}\, H^1 \to H^{-1} = (H^1)^*$ given by

$$\langle \ N(u),v \ \rangle = \int f(x,u(x)) \ v(x) \ dx \qquad\qquad (u,v{\in}H^1) \ .$$

The spaces H_1 (resp. H_1^*) from Section 2 agree with H^1 (resp. H^{-1}) up to equivalence of

norms. Conditions (C1) – (C4) ensure that N is well–defined and satisfies (H1) – (H6) , as is verified by standard arguments (of [4–6]). In particular, the potential φ is given by

$$\varphi(u) := \int F(x, u(x)) \ dx \qquad\qquad (u{\in}H^1),$$

where $F(x,s) := \int_0^s f(x,t) \ dt$. Thus (2.3) is satisfied if $f(x,1) \ne 0$ a.e. ((C3) is used in the

proof!), and then Thm. 4 guarantees the existence of infinitely many solutions $u_j{\in}H^1$ for

each λ belonging to a spectral gap $G =]a,b[$ of S.

Thm. 5 can be applied to the case where q is periodic , i.e. there is a basis $\{a_1,...,a_N\}$ of

$\mathbb{R}^N$ such that

$$q(x{+}a_k) = q(x) \ \text{a.e.} \qquad\qquad (k = 1,...,N) \ .$$

Then it is known (cf. [2]) that for $\lambda_0{\in}\sigma(S)$ there exist bounded solutions $u \ne 0$ of the

equation

$$- \Delta u + q(x) u = \lambda_0 u \qquad\qquad (3.2)$$

having a special form ("Bloch waves") from which it is clear that they are almost periodic.

Since the data are real, we can take real (or imaginary) parts, thus obtaining real–valued

almost periodic solutions $\Psi \neq 0$ of (3.2).

Now let $\lambda_0 = b$, the right end point of a spectral gap G of S, and fix a real–valued

almost periodic solution $\Psi \neq 0$ of (3.2). It is well known that every almost periodic

function f on $\mathbb{R}^N$ has a unique mean value, which we denote by $M(f)$. For arbitrary $m \in \mathbb{N}$,

choose an m–dimensional space Z consisting of rapidly decreasing real C^∞–functions

(e.g. the span of m distinct Hermite functions), endow Z with the L^2–norm, and define

$L_k : Z \to H^2$ for $k \in \mathbb{N}$ by

$$(L_k \zeta)(x) := M(\Psi^2)^{-1/2} k^{-N/2} \Psi(x)\zeta(x/k)$$

for $x \in \mathbb{R}^N$, $\zeta \in Z$. These operators are suitable for the verification of $(T(\delta,m))$, but of course

we need additional assumptions to find lower bounds for $\varphi(L_k \zeta)$ (note that $(T(\delta,m))$ is

never satisfied for $f \equiv 0$). To state such an assumption, fix two numbers $\sigma > 0, \tau > 0$. We

say f satisfies condition $(D(\sigma,\tau))$ if there exist positive constants A, s_0, d and, a point

$x_0 \in \mathbb{R}^N$ with $|x_0| > d$ such that

$$F(x,s) \geq A \, |x|^{-\tau} \, |s|^{\sigma+2} \quad \text{for} \quad 0 \leq s \leq s_0 \quad \text{a.e. on} \quad C := \{ty \, | \, t \geq 1, \, |y-x_0| \leq d\} .$$

(Here $|\cdot|$ denotes the Euclidean norm on $\mathbb{R}^N$.) Using the same techniques as in [6], one

can prove that (C1) – (C4) and $(D(\sigma,\tau))$ imply $(T(\delta,m))$ for arbitrary m and $\delta < \frac{4}{N\sigma+2\tau}$.

Thus, part a) of Thm.5 applies to (3.1) if $(D(\sigma,\tau))$ is satisfied with

$$\tau < \eta+1 \; , \; \sigma < \frac{2}{N}(\eta+1-\tau) \qquad\qquad (3.3),$$

and part b) applies if $(D(\sigma,\tau))$ holds with

$$\tau < 2, \; \sigma < \frac{2}{N}(2-\tau) .$$

The simplest choice for σ is $\sigma = \eta - 1$. In this case (3.3) amounts to

$$\tau < 2, \; \sigma < \frac{2}{N-2}(2-\tau) .$$

The last two conditions are familiar from [6]. Note, however, that m was arbitrary, so that

we obtain the existence (resp. the bifurcation) of an infinity of solutions of (3.1).

Clearly, the same considerations hold for the left end–point a of a spectral gap when the

sign of the nonlinearity is reversed.

REFERENCES

[1]   V. Benci, A. Capozzi, and D. Fortunato, Periodic solutions of Hamiltonian systems with superquadratic potential. Ann.Mat.Pura Appl., Ser.4 143 (1986), p. 1–46

[2]   M.S.P. Eastham, "The Spectral Theory of Periodic Differential Equations". Scottish Academic Press, Edinburgh 1973

[3]   H.–P. Heinz, Bifurcation from the essential spectrum for nonlinear perturbations of Hill's equation, in: "Differential Equations–Stability and Control" (ed. S. Elaydi), Marcel Dekker, New York 1990, pp. 219–226

[4]   H.–P. Heinz, Lacunary bifurcation for operator equations and nonlinear boundary value problems on $\mathbb{R}^N$, preprint

[5]   H.–P. Heinz and C.A. Stuart, Solvability of nonlinear equations in spectral gaps of the linearization, preprint

[6]   H.–P. Heinz, T. Küpper and C.A. Stuart, Existence and bifurcation for nonlinear perturbations of the periodic Schrödinger equation, preprint

[7]   T. Küpper and C.A. Stuart, Bifurcation into gaps in the essential spectrum I, to appear in J. Reine Angew. Math.

[8]   T. Küpper and C.A. Stuart, Bifurcation into gaps in the essential spectrum II, to appear in J. Reine Angew. Math.

[9]   T. Küpper and C.A. Stuart, Gap–bifurcation for nonlinear perturbations of Hill's equation, to appear in J. Reine Angew. Math.

[10] P.H. Rabinowitz, "Minimax Methods in Critical Point Theory with Applications to Differential Equations". CBMS Conference Lectures, AMS, Providence 1986.

International Series of Numerical Mathematics, Vol. 97, © 1991 Birkhäuser Verlag Basel

Branches of Stationary Solutions for Parameter-dependent Reaction-Diffusion Systems from Climate Modeling

Georg Hetzer & Paul G. Schmidt

Division of Mathematics, FAT, Auburn University, AL 36849, USA

We are concerned with a parameter-dependent reaction-diffusion system on a two-dimensional compact connected oriented Riemannian manifold M without boundary (e.g., the two-sphere):

$$(*) \qquad c_j\, \partial_t u_j - \text{div}\, (k_j\, \text{grad}\, u_j) = f_j(\mu; x, u_1, u_2, u_3) \qquad (j = 1,2,3).$$

The unknowns u_j are nonnegative functions of time $t \geq 0$ and position $x \in M$; inertia and diffusion coefficients c_j, k_j are positive and of class $C^2(M)$; the reaction terms f_j (whose precise form will be discussed below) depend smoothly on position x, on the unknowns u_1, u_2, u_3, and on a nonnegative parameter μ.

The system $(*)$ arises from an energy balance climate model due to Jentsch [12, 13, 14]. The quantities u_1, u_2, u_3 can be thought of as annual means of atmospheric temperature, surface temperature, and atmospheric humidity, respectively, each expressed in appropriate energy flux units. Horizontal energy transport is very roughly modeled by linear diffusion operators. Absorbed and emitted radiation fluxes, sensible heat exchange between surface and atmosphere, and latent energy fluxes due to the hydrological cycle (precipitation and evaporation) are incorporated into the reaction terms, as well as some important feedback mechanisms like the greenhouse effect, or the ice-albedo effect. For details about the climatic background, we refer to [9, 12, 13, 14]; for general information about energy balance climate models, see [6] or [17].

A topic which has received much attention since the pioneering work of Budyko [3] and Sellers [20] is the effect of solar radiation variations on the planet's global climates. In models like $(*)$, such variations can be simulated by varying the parameter μ, the so-called solar constant. Thus we are mainly interested in the structure of the set of stationary solutions of $(*)$ in dependence on the parameter μ.

This subject has been studied extensively for one-layer models typically involving a mean atmospheric temperature u as the only dependent variable. Often the problem is further reduced to a spatially one-dimensional reaction-diffusion equation by taking averages along the circles of latitude, or even to an ordinary differential equation by global averaging. In the ODE case, it is not hard to see that the stationary solutions typically form an S-shaped simple curve in the (μ, u) - plane with two (or, more generally, an even number of) turning points (see, e.g., [17], or the discussion in [8]). Solutions on back-bending curve segments being unstable, this suggests the possibility of turning-point catastrophes.

In fact, many quantitative models locate our present climate on the upper stable branch of an S-shaped solution curve, not far from the upper turning point. Thus a relatively small reduction of the solar constant might cause a rapid transition to a deep-freeze climate on the lower stable branch.

Combining computational and analytical tools, similar scenarios have been predicted based on spatially one- or two-dimensional one-layer models (see [5, 16] for early results in the 1D case, and [7, 8] for recent studies of a 2D model). The precise location of the turning points is quite controversial, - it depends heavily on the concrete set-up and on certain crucial parameters; but qualitatively, "S-shapedness" of a principal solution branch appears to be a structurally robust feature of one-layer energy balance models.

The question arises whether this feature will prevail if one proceeds to more sophisticated two- or multi-layer models allowing a much better resolution of coupling and feedback mechanisms. In this paper, we provide a qualitative framework in which "S-shapedness" of a principal solution branch for the system (∗) can be proved and understood analytically. In view of the largely phenomenological nature of the equations, an attempt is made to keep the functional setting as general as possible.

We assume that the reaction terms $f_j = f_j(\mu; x, u_1, u_2, u_3)$ (for $\mu \geq 0$, $x \in M$, $u_1, u_2, u_3 \geq 0$) are given as

$$f_1 = \mu R_1(x, u_1, u_2, u_3) - e_1 u_1{}^4 + g_1(x, u_1, u_3) u_2{}^4 + H(x, u_1, u_2) + P(x, u_1, u_3)$$

$$f_2 = \mu R_2(x, u_1, u_2, u_3) - e_2 u_2{}^4 + g_2(x, u_2, u_3) u_1{}^4 - H(x, u_1, u_2) - E(x, u_2, u_3)$$

$$f_3 = E(x, u_2, u_3) - P(x, u_1, u_3)$$

with sources and sinks (of class C^2) subject to the following conditions:

(H$_1$) R_1, R_2 (short-wave radiation absorbed by atmosphere and surface, resp.) positive and bounded; R_1, R_2 constant w.r.t. (u_1, u_2, u_3) for (u_1, u_2) close to $(0, 0)$ or near (∞, ∞); R_1 nondecreasing w.r.t. u_2, u_3; R_2 nondecreasing w.r.t. u_1, u_3.

(H$_2$) e_1, e_2 (emissivity constants) positive; g_1, g_2 (greenhouse factors) positive and bounded with sup $g_1 < e_2$, sup $g_2 < e_1$; g_1, g_2 constant w.r.t. (u_1, u_2, u_3) for (u_1, u_2) close to $(0, 0)$ or near (∞, ∞), and nondecreasing w.r.t. u_3.

(H$_3$) H (vertical heat flux) bounded; nonincreasing w.r.t. u_1, nondecreasing w.r.t. u_2; $H(x, u_1, u_2)(u_2 - u_1) \geq 0$.

(H$_4$) P (precipitation) decreasing w.r.t. u_1, increasing w.r.t. u_3; $P = 0$ for $u_3 = 0$, $P \to \infty$ as $u_3 \to \infty$ (uniformly w.r.t. x and u_1).

(H$_5$) E (evaporation) bounded; increasing w.r.t. u_2, decreasing w.r.t. u_3; $E = 0$ for $u_2 = 0$.

Note that quite a few of these assumptions concern only the functional behavior for u_1, u_2, or u_3 near 0 or ∞, i.e., outside the range of physical relevance.

For each $\mu \geq 0$, and under even more general hypotheses, the system (∗) generates a solution semiflow Φ_μ with a compact global attractor in the positive cone of a fractional order Sobolev space $W^{2\alpha,p}(M, \mathbf{R}^3) \subset C(M, \mathbf{R}^3)$ with $p > 2$ and $1/p+1/2 < \alpha < 1$ (cf. [9, 10]). Under Hypotheses (H_1) - (H_5), the system becomes cooperative, and Φ_μ can be shown to be strongly monotone with respect to the natural ordering of the phase space. General results of Hirsch [11] and others then imply the convergence of almost all trajectories to the (nonempty) set S_μ of steady-state solutions. S_0 contains only the trivial solution $(0, 0, 0)$; its index is 1, by a homotopy argument (cf. [9]). In any case, S_μ possesses a minimal element $\underline{u}_\mu$, and a maximal element $\overline{u}_\mu$; $\underline{u}_\mu$ is nondecreasing and left-continuous ($W^{2,p}$ or C topology) with respect to the parameter μ (cf. [2]).

For $\mu > 0$, all three components of $\underline{u}_\mu$ are strictly positive (maximum principle). As $\mu \to \infty$, the first and second component of $\underline{u}_\mu$ (atmospheric temperature and surface temperature) diverge to ∞ (uniformly) whereas the third component (atmospheric humidity) approaches a strictly positive limiting solution of the third equation of (∗) (convergence in $W^{2,p}$ and C). Moreover, we have $\underline{u}_\mu, \overline{u}_\mu \to (0, 0, 0)$ (in $W^{2,p}$ and C) as $\mu \to 0+$.

Based on these observations, it can be shown that in fact $\underline{u}_\mu = \overline{u}_\mu$ for μ close to 0 or ∞, i.e., for sufficiently small or sufficiently large values of the solar constant, there is exactly one stationary solution u_μ (say). The proof relies on the maximum principle for weakly coupled elliptic systems, and on a duality argument: A crucial condition on the row sums of the Jacobian of f_1, f_2, f_3 w. r. t. u_1, u_2, u_3 being difficult to verify, we pass to an adjoint problem (thus replacing row sums by column sums), and arrive at the desired uniqueness result by duality.

Now define $S = \{(\mu, u): \mu \geq 0, u \in S_\mu\}$, and consider S a metric subspace of $\mathbf{R} \times C(M, \mathbf{R}^3)$. Let P denote the connected component of $(0, 0)$ in S. A well-known continuation argument (cf. [19]) proves P to be unbounded. Employing a-priori estimates established in [9], we conclude that actually the first projection of P is unbounded, i.e., P covers the parameter space $\mathbf{R}_+$. In particular, $(\mu, u_\mu) \in P$ for μ near 0 or ∞. In a sense, P is the only relevant solution branch: other branches, should they exist, would have to be bounded and of total degree 0.

To derive more information about the structure of P, we have to analyze the spectrum of the linearization of (∗) (with respect to $u = (u_1, u_2, u_3)$) along P. The system being cooperative, it can be shown that at each point $(\mu, u) \in P$, the linearized problem possesses a principal eigenvalue $\lambda_0(\mu, u)$ which is real, algebraically simple, and strictly dominant in the sense that all other eigenvalues have strictly larger real parts (cf. [18] or [15]). For μ close to 0 or near ∞, an argument similar to that employed in the proof of unique solvability yields $\lambda_0(\mu, u) > 0$; in particular, u_μ is asymptotically stable under the semiflow Φ_μ.

Let us now impose the following structural condition, which should be seen in the light of path-following techniques allowing to control the principal eigenvalues numerically in the process of computing a solution branch:

(H_6) Along the principal branch P, all eigenvalues except λ_0 stay in the plane Re $\lambda > 0$,

and λ_0 vanishes at no more than finitely many points $(\mu, u) \in P$.

Basically following the lines of [7], it is then possible to prove that $P \setminus (0, 0)$ is a one-dimensional connected C^2-submanifold of $\mathbf{R} \times C(M, \mathbf{R}^3)$. To this end, the stationary problem corresponding to (∗) is rephrased as a parameter-dependent fixed-point equation $u = G(\mu, u)$, with $G: \mathbf{R} \times C(M, \mathbf{R}^3) \to C(M, \mathbf{R}^3)$ a completely continuous operator of the form $G(\mu, u) = (A + \rho\mathrm{Id})^{-1} (F(\mu, u) + \rho u)$; here A is a linear differential operator associated with the diffusion part of (∗), F is a substitution operator generated by the reaction terms f_j, and ρ is a positive constant.

In the neighborhood of points $(\mu, u) \in P \setminus (0, 0)$ with $\lambda_0(\mu, u) \neq 0$, a C^2-para-metrization of P is easily obtained by means of the implicit function theorem. Near zeros of λ_0, however, modified versions of results due to Amann [1, 2] apply (provided that the constant ρ is suitably chosen) and yield again a local parametrization. In addition, we conclude that turning points occur wherever λ_0 changes sign, and only there. As compared with [7], additional difficulties arise from the fact that the linearization of F with respect to u is not self-adjoint; that makes it harder, e.g., to establish a crucial relationship between $\lambda_0(\mu, u)$ and the spectral radius of $(DG(\mu, \cdot))(u)$.

Employing the classification theorem for one-dimensional connected manifolds, we arrive at the following theorem, valid under Hypotheses (H_1) - (H_5) and the structural condition (H_6):

THEOREM

a) P is the trace of a Jordan curve in $\mathbf{R} \times C(M, \mathbf{R}^3)$: There exists a homeomorphism h from $\mathbf{R}_+$ onto P with $h(0) = (0, 0)$, which is of class C^2 except at 0, and satisfies $h'(s) \neq (0, 0)$ for all $s > 0$.

b) P is "S-shaped": There is an even number of turning points, i.e., points $(\mu, u) = h(s) \in P \setminus (0, 0)$ such that $\mathrm{pr}_1 \cdot h$ attains a strict local extremum at s, and they coincide with those points on P where λ_0 changes sign.

c) Solutions on "forward-bending" curve segments are asymptotically stable, those on "back-bending" segments are unstable: Given $(\mu, u) = h(s) \in P \setminus (0, 0)$, u is asymptotically stable (resp. unstable) under the semiflow Φ_μ provided that $(\mathrm{pr}_1 \cdot h)'(s) > 0$ (resp. < 0).

A further question to ask is what happens if higher eigenvalues are allowed to cross the imaginary axis. If a simple real eigenvalue changes sign at $(\mu, u) \in P$, we can still establish a local parametrization for P near (μ, u) provided that a certain transversality condition holds (cf. [4]); otherwise bifurcation may occur. Since pairs of conjugate-complex eigenvalues may very well exist, Hopf bifurcation cannot be excluded either. General results about strongly monotone semiflows, however, imply that any periodic solution of $(*)$ is necessarily unstable (cf. [11]).

REFERENCES

1. H. Amann, *Multiple positive fixed points of asymptotically linear maps*, J. Funct. Anal. 17 (1974), 174-213.
2. H. Amann, *Fixed point equations and nonlinear eigenvalue problems in ordered Banach spaces*, SIAM Rev. 18 (1976), 620-709.
3. M.I. Budyko, *The effect of solar radiation variations on the climate of the Earth*, Tellus 21 (1969), 611-619.
4. M.G. Crandall & P.H. Rabinowitz, *Bifurcation, perturbation of simple eigenvalues, and linearized stability*, Arch. Rat. Mech. Anal. 52 (1973), 161-180.
5. M. Ghil, *Climate stability for a Sellers-type model*, J. Atmos. Sci. 33 (1976), 3-20.
6. M. Ghil & S. Childress, *Topics in Geophysical Fluid Dynamics: Atmospheric Dynamics, Dynamo Theory, and Climate Dynamics*. Appl. Math. Sciences, Vol. 60, Springer-Verlag, Berlin etc., 1987.
7. G. Hetzer, *The structure of the principal component for semilinear diffusion equations from energy balance climate models*, Houston J. Math. (to appear).
8. G. Hetzer, H. Jarausch & W. Mackens, *A multiparameter sensitivity analysis of a 2D diffusive climate model*, Impact of Computing in Science and Engineering 1 (1989), 327-393.
9. G. Hetzer & P.G. Schmidt, *A global attractor and stationary solutions for a reaction-diffusion system arising from climate modeling*, Nonlinear Analysis, Theor. Meth. Appl. 14 (1990), 915-926.
10. G. Hetzer & P.G. Schmidt, *Global existence and asymptotic behavior for a quasilinear reaction-diffusion system from climate modeling*, J. Math. Anal. Appl. (to appear).
11. M.W. Hirsch, *Stability and convergence in strongly monotone dynamical systems*, J. reine angew. Math. 383 (1988), 1-53.
12. V. Jentsch, *Cloud-ice-vapor feedbacks in a global climate model*, in: *Irreversible Phenomena and Dynamical Systems Analysis in Geosciences* (C. Nicolis & G. Nicolis, eds.), 417-437. Reidel, Dordrecht, 1987.
13. V. Jentsch, *An energy balance climate model with hydrological cycle, Part 1, Model description and sensitivity to internal parameters* (preprint).
14. V. Jentsch, *An energy balance climate model with hydrological cycle, Part 2, Stability and sensitivity to external forcing* (preprint).
15. R. Nagel (ed.), *One-parameter Semigroups of Positive Operators*. Lect. Notes Math., Vol. 1184, Springer-Verlag, Berlin etc., 1986.
16. G.R. North, *Theory of energy-balance climate models*, J. Atmos. Sci. 32 (1975), 2033-2043.
17. G.R. North, R.E. Calahan & J.A. Coakley, *Energy balance climate models*, Rev. Geophys. Space Phys. 19 (1981), 91-121.
18. R.D. Nussbaum, *Positive operators and elliptic eigenvalue problems*, Math. Z. 186 (1984), 247-264.
19. P.H. Rabinowitz, *A global theorem for nonlinear eigenvalue problems and applications*, in: *Contributions to Nonlinear Functional Analysis* (E.H. Zarantonello, ed.), 11-36. Academic Press, New York, 1971.
20. W.B. Sellers, *A global climate model based on the energy balance of the earth-atmosphere system*, J. Appl. Meteorol. 8 (1969), 392-400.

International Series of Numerical Mathematics, Vol. 97, © 1991 Birkhäuser Verlag Basel

A NOTE ON THE DETECTION OF CHAOS
IN MEDIUM SIZED TIME SERIES

by

Johan F. Kaashoek and Herman K. van Dijk

Econometric Institute, Erasmus University Rotterdam

P.O.Box 1738, 3000 DR Rotterdam, The Netherlands

Introduction

The length of macro-economic time series is often restricted, e.g., to 200 observations. Either no more data are available, or economic intervention policy makes a restriction to a limited size necessary. We have examined the presence of "chaotic properties" for medium sized time series by using numerical procedures for calculating the largest Lyapunov exponent and the correlation dimension.

Our economic time series are monthly observations of the natural logarithm of the real exchange rate between yen and dollar (period December 1972 to June 1988). We denote this time series by $JPUS_t$ (see Figure 1). For more details on the data we refer to Schotman and van Dijk (1990) who fitted a linear autoregressive model of order one, $x_t = \alpha x_{t-1} + \epsilon_t$, and obtained as least squares estimate for α a value of 0.982 (s.e. 0.014).

This motivated us to consider two simulated data sets. First, data are generated according to a linear autoregressive process of order one ($\alpha = 0.95$) with disturbances that are white noise; this data set is denoted by $N95_t$. Second, we use again the same linear autoregressive process, however now with disturbances generated according to the logistic map $4\epsilon_t (1 - \epsilon_t)$ minus 0.5; this series is denoted by $CH95_t$.

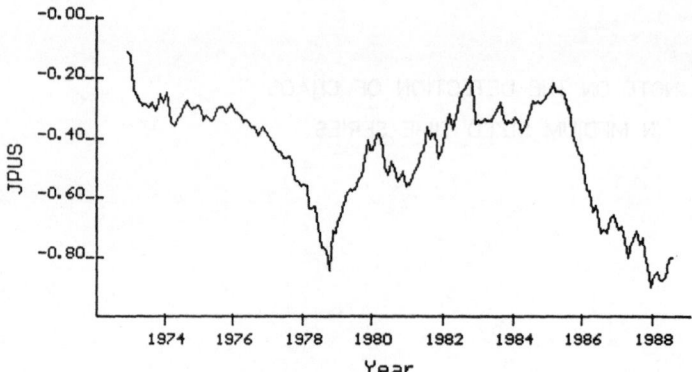

Figure 1. Data series: logarithm of the Yen-Dollar real exchange rate.

Since the pure chaotic series $\epsilon_{t+1} = 4\epsilon_t(1 - \epsilon_t)$ has mean value 0.5 and variance 0.125, we took for the white noise processes the same variance 0.125.

The linear model $x_t = \alpha x_{t-1} + \epsilon_t$ was fitted to the data set JPUS; we did the same for the data sets N95, C95. The least squares residuals are denoted by JPUSR, N95R, CH95R. Moreover, we looked at time series from a pure white noise process (denoted by N_t) and from a pure chaotic one (denoted by CH_t). In total we have eight time series. All simulated data sets consist of 200 observations.

Experiments, results and conclusions

Suppose $a_t \in \mathbb{R}$, $t = 1,..,T_0$ is an observed time series. Suppose the series has a deterministic explanation, i.e. $\exists$ F: $\mathbb{R}^N \to \mathbb{R}^N$ and a smooth observer function h: $\mathbb{R}^N \to \mathbb{R}$ with

$$x_{t+1} = F(x_t), \quad h(x_t) = a_t.$$

Define $a_t^M = (a_t, a_{t+1}, \ldots, a_{t+M-1}) \in \mathbb{R}^M$, then for $M \geq 2N + 1$, the dynamic properties of ψ^M: $\mathbb{R}^M \to \mathbb{R}^M$ defined by $\psi^M(a_t^M) = a_{t+1}^M$ are generically the same as the dynamic properties of F. (See Takens, 1981.) The dimension M is called the embedding dimension.

Our experiments consist of the reconstruction of an attractor (if present) for the given time series and examine the quantities which can be used in detecting chaotic properties, i.e. looking for positiveness of the largest Lyapunov exponent and a correlation dimension less than

the used embedding dimension (a low dimensional attractor). The
Lyapunov exponent is calculated by applying the procedure introduced by
Wolf et. al. (1988); the calculation of the correlation dimension is
based on Grassberger and Procaccia (1983). The results are summarized
in table 1 : Lyapunov exponents (scaled down by ln 2, the theoretical
value for the process $x_{t+1} = 4x_t(1 - x_t)$); table 2 contains results on
correlation dimension which are visualized in figure 2.

Table 1

Largest Lyapunov exponent

dim.	CH	CH95R	CH95	JPUSR	JPUS	N95	N95R	N
1	0.98	1.55	2.46	4.31	1.21	3.19	5.19	5.08
2	0.96	0.97	0.61	0.80	0.31	0.73	1.21	1.40
3	0.97	0.90	0.46	0.56	0.23	0.47	0.77	0.71
4	0.88	0.85	0.38	0.38	0.21	0.34	0.57	0.55
5	0.93	0.82	0.28	0.34	0.17	0.26	0.42	0.46
6	0.91	0.74	0.29	0.25	0.15	0.22	0.31	0.32
7	0.84	0.62	0.19	0.18	0.14	0.17	0.21	0.22
8	0.75	0.49	0.18	0.13	0.13	0.15	0.12	0.15
9	0.67	0.35	0.14	0.10	0.12	0.14	0.09	0.12
10	0.47	0.31	0.11	0.07	0.09	0.12	0.06	0.08

Table 2

Correlation dimension

dim.	CH	CH95R	CH95	JPUSR	JPUS	N95	N95R	N
1	0.83	0.77	0.79	0.92	0.88	0.89	0.90	0.87
2	0.90	0.84	1.18	1.61	1.37	1.64	1.64	1.67
3	0.93	0.89	1.62	2.17	1.72	2.14	2.40	2.40
4	0.88	1.00	1.96	2.57	1.91	2.61	3.04	3.01
5	0.83	1.09	2.33	2.88	2.15	3.05	3.69	3.65
6	0.77	1.17	2.67	3.18	2.26	3.23	4.03	4.28
7	0.65	1.15	2.96	3.39	2.34	3.51	4.49	4.66
8	0.61	1.08	3.18	3.53	2.44	3.81	5.12	4.94
9	0.52	0.93	3.35	3.68	2.48	3.82	5.71	5.11
10	0.41	0.78	3.50	3.74	2.47	3.95	6.04	5.59

The one dimensional pure chaotic process (data set CH) can be traced
back. Till embedding dimension 7, the values for the Lyapunov exponent
and the correlation dimension are comparable with the theoretical
values. The same holds for the series CH95R. See columns 2 and 3 in
tables 1 and 2.

If a time series a_t is a pure white noise process, then for each
embedding dimension, the correlation dimension will be equal to the
used embedding dimension. However, for a pure chaotic process, we
expect that the correlation dimension will stay fixed at some value if
the embedding dimension is high enough. So we expect to find a
different pattern for the correlation dimension with respect to the
series CH (or CH95R, "chaotic" residuals) and N (or N95R, "white noise"
residuals). This is indeed the case. The series CH and CH95R show
clearly that the correlation dimension becomes (is) stabilized at some
level whereas the correlation dimension for the series N and N95R (and
also N95) stays on growing with increasing embedding dimension, see
also figure 2.

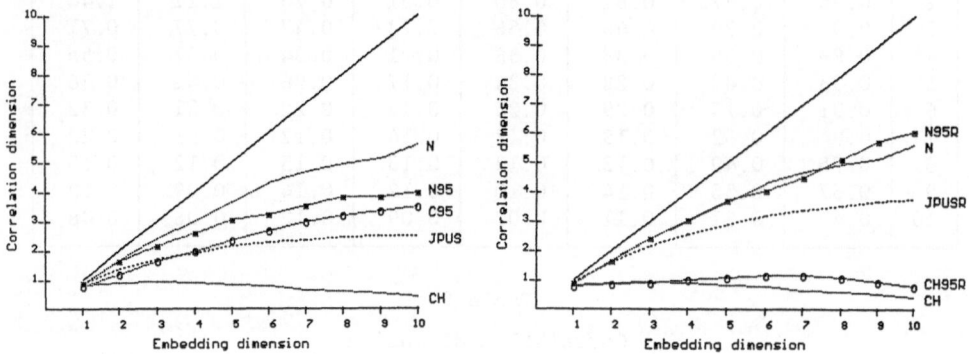

Figure 2. Correlation dimension.

For the strong contaminated series CH95 and N95, the results are far
more different. The Lyapunov exponent decreases monotone whereas the
correlation dimension increases with increasing embedding dimension.
The latter may be correct for the data sets N, N95R and N95 but
according to Brock's asymptotic analysis (residual test) (see Brock,
1986) one would expect that the series CH95 and CH95R have similar
properties. This is hardly the case here. The results on the series
CH95 and CH95R are not comparable.
Finally, the economic series on the real exchange rates. The
correlation dimension for JPUS (and less convincing for JPUSR) shows
some saturation (see column 5 and 6, table 2), however in all series
the growth rate diminishes at higher embedding dimension! But in any

case, the results on the series JPUSR, the residuals of JPUS, differ
much from those on the series N, respectively N95R. The residuals do
not follow the pattern of a pure white noise process. This indicates
that the modeling by a linear autoregressive model of order one is not
totally satisfying.

We want to conclude with some final observations with respect to the
limited sample size. First, the Lyapunov exponent is positive for all
series at all used embedding dimensions but at the same time is found
to be rather strong decreasing at higher dimensions; this is certainly
due to the limited sample size. As we are looking for embedding
dimensions where the Lyapunov exponent is stable, only low dimensional
chaotic processes can be detected in medium sized data sets. See column
2 and 3 in table 1 (series CH, CH95R) where such a stable pattern is
found. Secondly, strong contamination gives almost no possibility for
detection at the given sample size. Brock's residual test fails here.
Only by finding and applying the proper filter while removing the
non-chaotic components, the chaotic properties can be revealed.

References
Brock, W: Distinguishing random and deterministic systems: Abridged
version. Journal of economic theory 40, 168-195, 1986.
Grassberger, P. and I. Procaccia: Measuring the strangeness of strange
attractors, Physica D 9, 189-208, 1983.
Schotman, P.C. and H.K. van Dijk: A Bayesian analysis of the unit root
in real exchange rates, Report 9015/A of the Econometric Institute,
Erasmus University Rotterdam (to appear in Journal of Econometrics)
1990.
Takens, F.: Detecting strange attractors in turbulence, in D.A. Rand
and L.S. Young, eds., Dynamical systems and turbulence.
Springer-Verlag, Berlin 1981.
Wolf, A., J.B. Swift, H.L. Swinney, and J.A. Vastano: Determining
Lyapunov exponents from a time series. Physica 16D, 285-317, 1985

International Series of Numerical Mathematics, Vol. 97, © 1991 Birkhäuser Verlag Basel

An approach for the analysis of spacially localized oscillations.

M.Kirby,[†][§] D. Armbruster,[‡] W. Güttinger [†]

[†]Institut für Informationsverarbeitung
Universität Tübingen, Köstlinstr. 6
7400 Tübingen 1, F.R.G.

[‡]Department of Mathematics, Arizona State University, Tempe, AZ, USA

[§]Department of Mathematics, Colorado State University, Fort Collins, CO, USA

Abstract

The dynamics near a *strange fixed point* arising in the Kuramoto-Sivashinsky equation are investigated as the equation undergoes a Hopf bifurcation and then further bifurcates to a modulated wave. While the representation of the spatial structures generated requires a relatively broad fourier spectrum we show that the essential features are captured by 3 eigenfunctions of the time averaged correlation matrix. At the Hopf bifurcation point, the eigenfunctions correspond to the fixed point and the 2 unstable directions of the center eigenspace.

1   Introduction

The connection of the Kuramoto-Sivashinsky (K-S) partial differential equation to a strictly low-dimensional system of ordinary differential equations is well established. Numerical simulations have demonstrated, e.g. [2], complex behavior and a rich structure of bifurcations, including chaos. In addition, it is one of the few PDE's for which rigorous statements have been made concerning the existence of an inertial manifold. An application of Center-Unstable Manifold theory [1] as well as non-linear approximations of the inertial manifold [3] provide further evidence for the low dimensional behavior of the K-S equation.

The focus of this paper is on characterizing the evolution of the dynamics around a so-called *strange fixed point* [2] through a series of bifurcations. To this end we apply a method known as the Karhunen-Loeve Expansion (K-L) or the Principal Orthogonal Decomposition [5]. It amounts to identifying large-scale structure with maximally correlated flow fields. These fields form a set of orthogonal functions which capture maximal energy and can be viewed as a set of *the most likely instantaneous flows* in a way to be described in Section 2.

Our primary goal in this study is to examine the evolution of the dynamics in terms of the K-L eigenfunctions as the bifurcation parameter is varied. As such, the emphasis differs from previous applications of the procedure which focus on using the K-L basis functions to generate a low-dimensional system of dynamical equations, see [5] and references therein. We propose the methodology as a statistical tool for probing the dynamics of a system; specifically, to obtain information which other approaches may not readily yield.

2   Optimality Criteria

The optimality properties of the eigenfunctions play a central role in this procedure. We briefly summarize them here and refer the reader to [5] for further details. The eigenfunctions have the mathematically equivalent interpretation of maximizing the average energy and correlation of the flow fields in each coordinate direction. Specifically, if $u(x,t)$ represents the spatio-temporal evolution of the PDE in question, then we choose ψ_1 such that

$$\lambda_1 = \lim_{T \to \infty} \frac{1}{T} \int_0^T (\psi_1, u)^2 dt$$

is a maximum with the side constraint that $(\psi_1, \psi_1) = 1$. Proceding inductively, the resulting variational problem leads to a Fredholm type integral equation

$$\int K(x,y)\psi(y) = \lambda\psi(x)$$

where the kernel $K(x,y) = \langle u(x,t)u(y,t) \rangle$ and the brackets denote time average. Consider a truncated expansion of the flow in terms of the $\{\psi_n\}$, namely

$$u(x,t) = \sum_{n=1}^N a_n(t)\psi_n(x).$$

Then we observe that the modes are uncorrelated in time as $\langle a_j(t)a_k(t) \rangle = \lambda_j \delta_{jk}$ and furthermore, the eigenvalue λ_j corresponds to the statistical variance in the j'th coordinate direction, and is maximal. Also, the time averaged truncation error $\epsilon_N = \|\langle \sum_{n=1}^N a_n(t)\psi_n(x) \rangle\| = \sum_{n=N+1}^\infty \lambda_n = 1 - E_N$. The energy captured by N terms $E_N = \sum_{n=1}^N \lambda_n$ is a maximum, thus ϵ_N is a minimum.

In addition, the K-L eigenfunctions minimize a measure of information H, known as Shannon's Entropy, where $H = \sum \gamma_i ln \gamma_i$ and $\gamma_i = \lambda_i / \sum \lambda_i$. Also, for translationally invariant flows the K-L expansion reduces to harmonic decomposition.

3   Procedure and Results

The K-S equation

$$u_t + 4u_{xxxx} + \alpha(u_{xx} + \frac{1}{2}(u_x)^2) = 0$$

was simulated using a 10 complex mode Fourier-Galerkin procedure with initial conditions $u(x,0) = \cos x + \sin 2x + \sin 3x + \cos 4x$ and periodic boundary conditions. Under these circumstances the equation possesses $O(2)$ symmetry.

At $\alpha = 72$ the strange fixed point, shown in Fig. 1., becomes a global attractor possessing a relatively broad fourier spectrum, with significant energy in the first six modes. The fixed point persists until $\alpha = 83.75$ when it becomes unstable and undergoes a Hopf bifurcation resulting in a localized oscillatory pattern (see Fig. 2.).

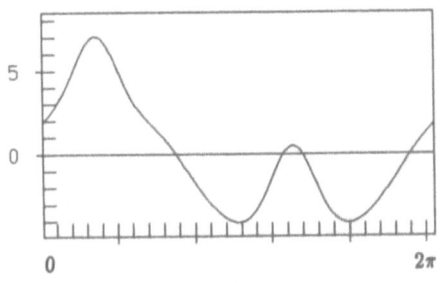

Fig. 1. Strange fixed point, $\alpha = 72.0$.

The smaller hump remains stationary while the large hump oscillates with an amplitude increasing with α. We remark on the existence of a *strange node* on the top of the large hump which can be seen in the contour plot of the fluctuating field $u(x,t) - \langle u(x,t) \rangle$, Fig. 2b. At $\alpha = 86.0$ the solution bifurcates to a modulated wave, with a similar localized oscillation in the frame of reference traveling with the wave. As can be seen in Fig. 3. the amplitude of the oscillations has grown and the small hump has also begun to oscillate slightly. We observe that for the bifurcation parameter values under consideration (save for the case of the traveling wave where a template fitting procedure is applied) there exist a circle of global attractors, .i.e., the flow is not ergodic in the direction of the translational invariance and the eigenfunctions are not sinuoids, as described in Section 2.

To gain insight into the evolution of the dynamics as α varies, we computed the maximally correlated fields at $\alpha = 84.25$, i.e., for the case of local oscillatory motion, and for $\alpha = 87$, after the wave has started to travel. Note that, at $\alpha = 72$ the strange fixed point is itself a maximally correlated field and that all the other eigenfunctions for this case have 0 energy. In the case of the local oscillations, *snapshots* of the flow were taken at fixed intervals and then directly used to compute the eigenfunctions. For the locally oscillatory traveling wave it was necessary to apply a template fitting procedure to track the coherent structure, similar to the one used in [6].

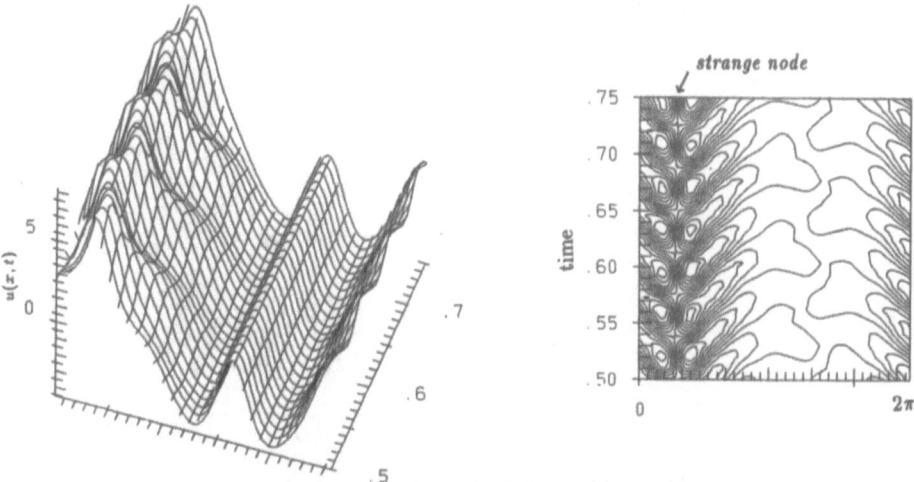

Fig. 2. $\alpha = 84.25$ a) time evolution of u. b) contour plot of fluctuating field $u(x,t) - \langle u(x,t) \rangle$.

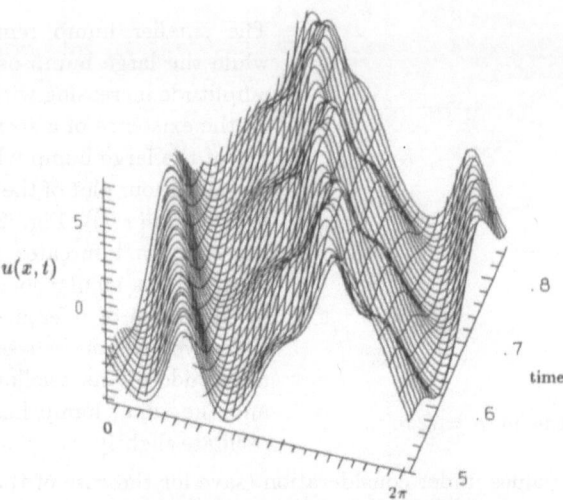

Fig. 3. $\alpha = 87$ time evolution of u

The results of these computations are shown in Fig. 4 and Fig 5.. In each case the first 3 eigenfunctions capture over 99.9% of the flow's energy. The first eigenfunction corresponds to the global attractor from $\alpha = 72$ which is now an unstable steady state. The next 2 eigenfunctions, which describe the fluctuating field (the % energy with respect to the fluctuating field is shown in brackets), are still very close to the unstable directions of the Hopf bifurcation when computed for $\alpha = 84.25$. Note that they capture the existence of the strange node. The 2nd eigenfunction (see middle picture of Fig. 4) shows a sinusoidal oscillation on top of the large hump. At $\alpha = 87$, an additional zero in the region of the small hump in the 3rd eigenfunction (Fig. 5, far right) reflects the fact that the small hump has also begun to oscillate on the modulated wave.

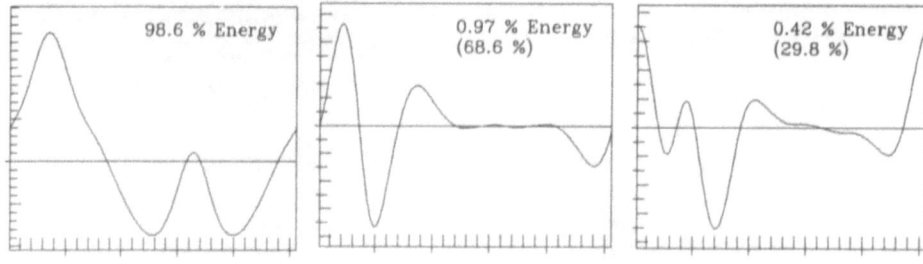

Fig. 4. Eigenfunctions 1-3 for $\alpha = 84.25$

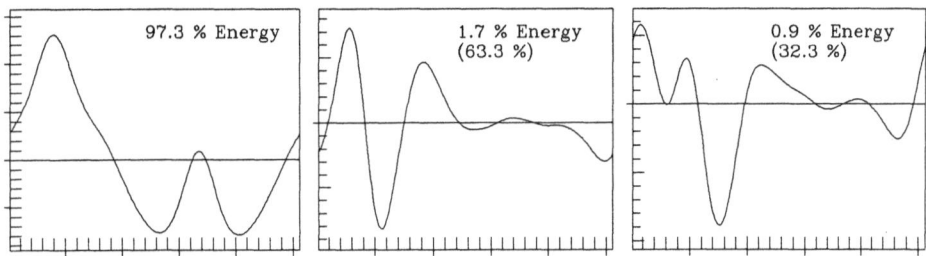

Fig. 5. Eigenfunctions 1-3 for $\alpha = 87$.

4 Discussion

We have demonstrated that the maximally correlated fields reveal the spatio-temporal structure of the flow and how it changes with the bifurcation parameter. Geometrically, one can view the K-L eigenfunctions as spanning the *appropriate plane*, i.e., the one which contains the limit cycle produced by the Hopf bifurcation. At the Hopf bifurcation point this plane coincides with the center eigenspace. As the K-L plane can be computed far away from the bifurcation, we can interpret the K-L procedure as a basis for a global extension of the center manifold reduction. Also, it is clear that this K-L plane is skewed with respect to the Fourier basis as the eigenfunctions themselves have broad fourier spectra. The incorporation of the K-L eigenfunctions into a center unstable manifold reduction is currently being performed [4].

Acknowledgements: The authors would like to thank G. Dangelmayr and B. Nicolaenko for helpful comments and suggestions. M.K. acknowledges the generous financial support of the Alexander von Humboldt Foundation.

5 References

[1] D. Armbruster, J. Guckenheimer and P. Holmes (1989), SIAM J. Appl. Math., **49**, 676
[2] J.M. Hyman, B. Nicolaenko and S. Zaleski (1985), Physica 23 D, pp. 265-292.
[3] M.S. Jolly, I.G. Kevrekidis and E.S. Titi, (1990) *Approximate inertial manifolds for the Kuramoto-Sivashinsky equation: Analysis and computations,* (in press)
[4] M. Kirby and D. Armbruster, *The analysis of spatio-temporal structure,* (in prep.)
[5] L. Sirovich (1987), Parts I-III, (Quarterly of Applied Mathematics, Vol. XLV, **3** 561, October), :plus references therein.
[6] L. Sirovich, M. Kirby and and M. Winter (1990), Phys. Fluids A **2**, 127

International Series of Numerical Mathematics, Vol. 97, © 1991 Birkhäuser Verlag Basel

On the Application of Invariant Manifold Theory, in particular to Numerical Analysis

U. Kirchgraber, F. Lasagni, K. Nipp, D. Stoffer

Department of Mathematics, ETH Zürich

1   Introduction

Consider a suitable space, say R^n for simplicity, and a map $P : R^n \to R^n$; then the pair (R^n, P) defines a discrete (semi-) dynamical system; a (semi-) orbit is a sequence $\{x_i\}$, $i \in N_0$, satisfying $x_{i+1} = P(x_i)$ for all i; since the orbit is defined by x_0 we call $\{x_i\}$ the orbit generated by x_0. A subset $M \subset R^n$ is called (semi-) invariant if $P(M) \subset M$. M is (globally) attractive if $dist(x_i, M) \to 0$ for $i \to \infty$, for each orbit $\{x_i\}$. If M is invariant, attractive and, in addition, a manifold, then M is called an (attractive) invariant manifold.

The basic question in invariant manifold theory is: Which maps P admit invariant manifolds? The available results are essentially of perturbation type, i.e. one starts off with a map P_0 which admits an invariant manifold M_0 and shows that all nearby maps P admit nearby invariant manifolds; actually, to achieve this comparably modest goal, fairly strong attractivity properties for P_0 with respect to M_0 have to be imposed. We give a typical invariant manifold result in Section 2.

The main goal of this paper is to indicate how invariant manifold theory can be used in numerical analysis. Yet we first review briefly a few recent applications from dynamical system theory to provide a more comprehensive picture.

The most classical application is to plane maps with a hyperbolic fixed point leading to the fundamental notion of transversal homoclinic point and the famous Poincaré-Smale theorem. As is well-known, the Smale theorem can be used to prove existence of chaotic behaviour in periodically perturbed plane systems of ODE's; for a description using shadowing, see Kirchgraber and Stoffer (1990). D. Stoffer has recently extended these results to non-periodically perturbed plane systems, cf. Stoffer (1988), Kirchgraber and Stoffer (1989).

Another beautiful application of invariant manifold theory is to universality. Universality means, roughly speaking, that the speed of the period doubling phenomena is essentially independent of the particular system under consideration. From the work of Feigenbaum and Lanford we know that with the period doubling phenomena there is associated a certain map T in the space of systems. This map T admits a hyperbolic fixed point and universality is an easy consequence of the dynamics in the neighborhood of a hyperbolic fixed point, cf. Collet and Eckmann (1980). This is a rather striking example because the link between period doubling and the concept of a hyperbolic fixed point is not obvious at all.

Another subject of current interest are inertial manifolds. Consider the equation (*) $\dot{x} + Ax + R(x) = 0$ where x is in a Hilbert space; A is a linear unbounded selfadjoint positive operator with spectrum satisfying $0 < \lambda_1 \leq \lambda_2 \leq \lambda_3 \leq \ldots, \lambda_n \to \infty$ as $n \to \infty$, and R is a nonlinearity, such that all solutions of (*) eventually enter a fixed ball. Then there is $N_0 \in \mathbb{N}$ and $\Delta > 0$ such that if there is $N > N_0$ satisfying $\lambda_{N+1} - \lambda_N > \Delta$ then the (semi-) flow generated by (*) admits an N-dimensional invariant manifold. The nice feature about this result is the following. If (*) was a finite dimensional system, we could prove a related result provided R was small enough. In the infinite dimensional case we do not need a smallness condition on R. In this case, so to speak, R is small compared to A if the spectrum of A admits the gap described above. For details, cf. Foias, Sell and Temam (1988).

P. Boxler and L. Arnold have recently generalized the concept of center manifold to stochastic differential equations, cf. P. Boxler (1989). It is not surprising that this proves to be useful in treating stochastic bifurcation problems.

2 An Invariant Manifold Theorem

The prototype invariant manifold result is due to the french mathematician J. Hadamard, cf. Hadamard (1901). The following is a (slight) generalization and extension of his result.

Consider four non-negative constants L_{zz}, L_{zy}, L_{yz}, L_{yy} and assume that they satisfy the following condition

$$\sqrt{L_{zy} L_{yz}} < \frac{1}{2} (1 - L_{zz} - L_{yy}) \tag{1}$$

Let $\mathcal{Z}$, $\mathcal{Y}$ be Banach spaces and D a positive constant. Consider the strip

$$\mathcal{U} := \{x = (z, y) | z \in \mathcal{Z}, y \in \mathcal{Y}, |y| \leq D\}$$

and the following class $\mathcal{K}$ of maps

$$P : x = (z, y) \in \mathcal{U} \to P(x) = (z + Z(z, y), e(z, y)) \in \mathcal{U}$$

where Z, e are assumed to satisfy the following Lipschitz conditions: $|Z(z, y) - Z(z', y')| \leq L_{zz} |z - z'| + L_{zy} |y - y'|, |Y(z, y) - Y(z', y')| \leq L_{yz} |z - z'| + L_{yy} |y - y'|$ for all (z, y), $(z', y') \in \mathcal{U}$.

Theorem 1 (Invariant Manifold Theorem) *Let $P \in \mathcal{K}$ be given. Then there is g :* $\mathcal{Z} \to \mathcal{Y}$ *with the following properties.*

1. $\sup_{z \in \mathcal{Z}} |g(z)| \leq \sup_{x \in \mathcal{U}} |e(x)|$

2. g is μ- lipschitzian with $\mu = 2\, L_{yz}/\left(1 - L_{zz} - L_{yy} + \sqrt{(1 - L_{zz} - L_{yy})^2 - 4L_{zy}\,L_{yz}}\right)$

3. $M := \{(z,\, g(z))|z \in \mathcal{Z}\}$ is invariant with respect to P.

4. If $e(x) = L(x)y + Y(x)$ holds with $\ell := \sup_{x \in \mathcal{U}} |L(x)| < 1$ then $\sup_{z \in \mathcal{Z}} |g(z)| \leq (1 - \ell)^{-1} \sup_{x \in \mathcal{U}} |Y(x)|$.

5. If $\mathcal{Z} = \mathbf{R}^r$ and if Z, e are 2π-periodic with respect to the components of z, then this holds true for g as well.

6. Let $x \in \mathcal{U}$ then there is $\hat{x} \in M$ such that the orbits $\{x_i = (z_i,\, y_i)\}$, $\{\hat{x}_i = (\hat{z}_i,\, \hat{y}_i)\}$ generated by x, $\hat{x}$ respectively converge to each other; more precisely, we have

$$|\hat{z}_i - z_i| \leq \chi^i c |y_0 - g(z_0)|, \quad |\hat{y}_i - y_i| \leq \chi^i (1 + \mu c)|y_0 - g(z_0)|$$

where $\chi = L_{yy} + \mu\, L_{zy} \in (0, 1)$ and $c = L_{zy}\nu/(1 - \nu\chi)$ (with $\nu = (1 - L_{zz} - \mu L_{zy})^{-1}$, $\nu\chi \in (0, 1)$).

Let P, $\overline{P} \in \mathcal{K}$. Assume that $\delta_z := \sup_{x \in \mathcal{U}} |Z(x) - \overline{Z}(x)| < \infty$ and put $\delta_e := \sup_{x \in \mathcal{U}} |e(x) - \overline{e}(x)|$ then $\sup_{z \in \mathcal{U}} |g(z) - \overline{g}(z)| \leq C_1\,\delta_z + C_2\,\delta_e$ where $C_1 = (L_{yz} + \mu L_{yy})/Det$, $C_2 = (1 - L_{zz} - \mu L_{zy})/Det$, $Det = (1 - L_{zz} - \mu L_{zy})(1 - L_{yy}) - L_{zy}(L_{yz} + \mu L_{yy})$ (Note that C_1, C_2, Det are positive).

We provide explicit estimates (on the bound of g, its Lipschitz constant, etc.) because this is needed in many applications. Although Theorem 1 is somewhat different from a result given in Section 12 of Kirchgraber and Stiefel (1978), the reader is referred to this reference because the basic ideas of the proof of Theorem 1 can be found there.

3   Applications to Numerical Analysis

An early application of invariant manifold theory to numerical analysis has been given in Kirchgraber (1979) in connection with the justification of ideas of E. Stiefel and J. Baumgarte which aimed at improving the accuracy of the numerical integration of ODE's in the presence of first integrals. Here we briefly describe some more recent results. There are essentially three groups of applications of invariant manifold theory. *Group I* deals with the following question: Given a certain numerical scheme, does it reflect the geometric properties of the underlying system of ODE's? *Group II* deals with what could be called "hidden invariant manifolds", i. e. invariant manifolds present in certain numerical schemes. *Group III*, finally, addresses the question of how to use invariant manifold theory to design efficient algorithms, if possible. We give examples of all three types.

3.1 Numerical Schemes that do (or do not!) Reflect the Underlying Geometry

We briefly describe three results.

a) Consider a plane autonomous system $\dot{x} = f(x)$ which admits $x = 0$ as *critical* equilibrium, i.e. $f(0) = 0$ and $\sigma(df(0)) = \{\pm i\}$. Near $x = 0$ the system is of the type $\dot{x} = Jx + 0(x^2)$ where $J = \begin{pmatrix} 0 & -1 \\ 1 & 0 \end{pmatrix}$. With the help of a near identity map $x = z + 0(z^2)$ we transform the system into Birkhoff normal form (up to terms of order $0(z^4)$); we obtain $\dot{z} = Jz + G_1 |z|^2 z + G_2 |z|^2 Jz + 0(z^4)$. The following result is easy to see. $G_1 < 0 \, (G_1 > 0)$ implies: 0 is asymptotically stable (unstable). In the first case we speak of a weak attractor, in the second case 0 is called a weak repellor.

Now consider a Runge-Kutta scheme $\psi(h, x)$ and let R be its stability function. The scheme is called linearly symplectic if the map $x \rightarrow \psi(h, x)$ is symplectic when applied to a linear Hamiltonian system. It is not hard to prove that ψ is linearly symplectic iff $|R(iv)| = 1$ for $v \in \mathbf{R}$. (Remember that a RK scheme ψ is symmetric if $\psi(-h, \psi(h, x)) = x$ holds for all h, x when applied to any system of differential equations, cf. Hairer, Norsett, Wanner (1987). Therefore, a RK scheme is linearly symmetric iff it is symmetric when applied to a linear system. It is easy to see that a scheme is linearly symplectic iff it is linearly symmetric. In our context it seems to be more natural to use the term linearly symplectic). If the scheme is not linearly symplectic then we have $|R(iv)| = 1 + cv^{q+1} + 0(v^{q+2})$, for v in a neighborhood of 0, with $c \neq 0$, $q \geq p$, p being the order of the scheme. Note that all explicit RK schemes are not linearly symplectic. We call the Runge-Kutta scheme expansive if $c > 0$ and contractive if $c < 0$. We have the following result.

Theorem 2 *a) Assume that 0 is a weak attractor. Then for stepsize $h > 0$, sufficiently small, the following holds. (i) If ψ is expansive, then 0 is an unstable fixed point of ψ. (ii) If ψ is either contractive or linearly symplectic, then 0 is an asymptotically stable fixed point of ψ.*
b) Assume that 0 is a weak repellor. Then for stepsize $h > 0$, sufficiently small, the following holds. (i) If ψ is contractive, then 0 is an asymptotically stable fixed point of ψ. (ii) If ψ is expansive or linearly symplectic, then 0 is an unstable fixed point of ψ.

It follows that only linearly symplectic schemes give the correct description in both cases. In case a)(i) and b)(i) the scheme ψ admits a circle-like invariant curve of radius $0(h^{q/2})$, which is attractive in the first case and repelling in the second case. For details, see Lasagni (1990).

b) We consider a singularly perturbed system of ODE's

$$
\begin{aligned}
\dot{x} &= f(x, y) \\
\epsilon \dot{y} &= g(x, y)
\end{aligned}
\tag{2}
$$

where $\epsilon > 0$ is a small parameter. We assume that the second equation of (2) for $\epsilon = 0$ admits a solution $y = s_0(x)$. Under suitable stability assumptions the system (2) admits an invariant manifold $y = s_\epsilon(x)$ for $\epsilon > 0$, sufficiently small, where s_ϵ is close to s_0. We consider two simple numerical schemes for Eq. (2): The Euler Method and the Backward Euler Method.

Theorem 3 *The discrete dynamical systems generated by the Euler and the Backward Euler scheme, respectively, admit an (attractive) invariant manifold (close to $y = s_0(x)$) provided*
(i) *ϵ, h are sufficiently small and $h < c\epsilon$ (where c is some positive constant independent of ϵ, h) in the Euler case*
(ii) *ϵ, h are sufficiently small in the Backward Euler case.*

Note that for the Euler scheme the correct qualitative behaviour is guaranteed only if h is $0(\epsilon)$. On the other hand, in case (ii) there is no dependence of the stepsize h on the smallness parameter ϵ. The reason for this favourable property is that the Backward Euler scheme is A-stable with $|R(\infty)| < 1$ (R still denotes the stability function). We expect that all A-stable RK-methods with $|R(\infty)| < 1$ lead to a result of the type (ii). For details, see Kirchgraber and Nipp (1989).

c) Consider a system of perturbed harmonic oscillators

$$
\dot{z} = f^0(z) + \epsilon f^1(z, \epsilon)
\tag{3}
$$

i.e. for $\epsilon = 0$ (3) reduces to a system of m harmonic oscillators with frequency vector ω, ϵ is a small parameter. In terms of polar coordinates, (3) may be rewritten as follows

$$
\dot{\phi} = \omega + \epsilon R(\phi, a) \qquad \dot{a} = \epsilon T(\phi, a)
\tag{4}
$$

We assume that ω satisfies a diophantine condition and denote the average of T with respect to ϕ by $\overline{T}(a)$; we assume that $a' = \overline{T}(a)$ admits an exponentially asymptotically stable equilibrium A. Then it is well known that (3) admits an attractive invariant m-torus τ_ϵ for $\epsilon > 0$, sufficiently small.

Let us discuss the application of a Runge-Kutta scheme ψ to Eq. (3). A first result is as follows.

Theorem 4 *Assume that ψ is a Runge-Kutta scheme which is not linearly symplectic. There is a neighborhood $U(\tau_0)$ of τ_0 and positive constants C_1, C_2 such that the following statements hold (for all $\epsilon > 0$, $h > 0$, sufficiently small):*
(i) If $\epsilon < C_1 h^q$, then every orbit generated by ψ leaves $U(\tau_0)$
(ii) If $\epsilon > C_2 h^q$, then ψ admits an (attractive) invariant m-torus close to τ_ϵ.

Thus for ϵ small we either have to use a very small step size or else the discrete scheme "overlooks" the torus! The question is whether there are more suitable schemes. The linearly symplectic schemes indeed behave better! In addition to Theorem 4 we have

Theorem 5 *Let ψ be linearly symplectic. ψ then admits an attractive invariant m-torus close to τ_ϵ provided ϵ, h are sufficiently small.*

Thus in case of a linearly symplectic scheme there is no dependence of h on ϵ.

Finally, we consider symplectic Runge-Kutta schemes. ψ is symplectic if the map $x \rightarrow \psi(h, x)$ is symplectic for Hamiltonian systems. Let the $s \times s$-matrix A and the s-vector b define the Runge-Kutta scheme. With $B = diag(b_1, \dots, b_s)$ we have

Theorem 6 ψ *is symplectic if $BA + A^T B - bb^T = 0$.*

This result was found independently by Lasagni (1988), Sanz-Serna (1988), Suris (1988). Note that symplectic RK-schemes are linearly symplectic and therefore Theorem 5 applies. Moreover, for symplectic RK-schemes, Theorem 5 essentially extends to the case of perturbed integrable systems (in the sense of Arnold), for details see Stoffer (1991).

3.2 "Hidden" Invariant Manifolds

In a previous paper, Kirchgraber (1986), it was shown that associated with a strongly stable multi-step method there is a one-step method which essentially generates the same data. The key in the construction of the associated one-step method is the existence of a certain invariant manifold.

Here we briefly discuss a different example, which is similar in spirit. Consider the system $\dot{x} = f(x)$, denote its flow by $\phi(t, x)$ and let $\psi(h, x)$ be a one-step method of order p. Let $\ell(h, x) = \psi(h, x) - \phi(h, x)$ denote the local error. In order to design an algorithm to choose the stepsize we introduce an estimate $\hat{\ell}(h, x)$ of $\ell(h, x)$. For instance, let $\chi(h, x)$ be a one-step method of order $p+1$ applied to $\dot{x} = f$. If we put $\hat{\ell}(h, x) := \psi(h, x) - \chi(h, x)$

we conclude $\hat{\ell}(h, x) = \ell(h, x) + 0(h^{p+2}) = h^{p+1} L(h, x) = h^{p+1} c(x) + h^{p+2} R(h, x)$, for suitable functions L, c, R.

With a point x and a stepsize h, we would like to associate the next point $\bar{x}$ and the next stepsize $\bar{h}$. It is obvious to choose $\bar{x} = \psi(h, x)$. However, which choice of $\bar{h}$ shall we adopt? The expression $|\ell(\bar{h}, \bar{x})|/\bar{h}$ is called the local error per unit step and ideally one would like this quantity to be equal to some tolerance $TOL =: \epsilon^p$. Since the local error per unit step is not available we consider the following sequence of approximations

$$|\ell(\bar{h}, \bar{x})| / \bar{h} \approx |\hat{\ell}(\bar{h}, \bar{x})| / \bar{h} \approx \bar{h}^{p+1} |L(h, x)| / \bar{h} = \bar{h}^p |L(h, x)| \overset{!}{=} TOL = \epsilon^p$$

This leads to the choice $\bar{h} = \epsilon / |c(x) + h R(h, x)|^{1/p} =: H(h, x, \epsilon)$. Thus we have to consider the map $P : (x, h) \rightarrow (\psi(h, x), H(h, x, \epsilon))$. It is easy to prove that for $\epsilon > 0$, sufficiently small, P admits an (attractive) invariant manifold $h = \epsilon g(x, \epsilon)$! The implication is: Asymptotically, the choice of the stepsize is independent of the previous stepsize! For details and an application, see Stoffer and Nipp (1990).

3.3   The Use of Invariant Manifolds to design Algorithms

a)   We consider the simple delay-equation $\dot{x}(t) = \epsilon f(x(t-1))$, where $x \in \mathbf{R}^m$, ϵ is a small parameter. Given an initial function ϕ on $[-1, 0]$ there is a uniquely determined solution $x(t)$. The goal is to compute $x(t)$ for large t efficiently. In the space of initial functions we introduce the map $\mathcal{P}_\epsilon \phi := x(. + 1)|_{[-1, 0]}$. We then want to compute $\mathcal{P}_\epsilon^n \phi$ for n large. The following remark is trivial but decisive: $\mathcal{P}_0$ projects the space of initial functions onto the constants. To take advantage of this fact we introduce the splitting $\phi = z + y$ where $z = \phi(0), y = \phi - z$. The map $\mathcal{P}_\epsilon$ then may be rewritten as follows

$$\mathcal{P}_\epsilon : (z, y) \longrightarrow \left(z + \epsilon \int_{-1}^0 f(z + y(\sigma)) \, d\sigma, -\epsilon \int_\sigma^0 f((z + y(\sigma)) \, d\sigma \right)$$

Of course, in practice we cannot compute these integrals. They are approximated numerically. Think of Riemann sums for simplicity. We then have a finite but high dimensional approximation of $\mathcal{P}_\epsilon$

$$\mathcal{P}_\epsilon : (z, y) \longrightarrow (z + \epsilon e_0(z, y), -\epsilon e(z, y)) \text{ with } e_i = \sum_{j=i}^{n-1} \frac{1}{n} f(z + y_j)$$

where $y = (y_1, \ldots, y_n)$, $e = (e_1, \ldots, e_n)$.

Note that $\mathcal{P}_0$ projects the z-y-space onto the z-plane, the z-plane being invariant pointwise. Thus for $\epsilon > 0$, sufficiently small, $\mathcal{P}_\epsilon$ admits an attractive invariant manifold $M_\epsilon = \{(z, g(z)) \, | \, z \in \mathbf{R}^m\}$. The restriction of $\mathcal{P}_\epsilon$ to M_ϵ leads to a map $p_\epsilon : \mathbf{R}^m \rightarrow \mathbf{R}^m$ which can be easily constructed with high precision because M_ϵ is highly attractive. It remains to compute high powers of p_ϵ. Since p_ϵ is a near identity map there are efficient algorithms available, cf. Kirchgraber and Willers (1986) and Kirchgraber (1988). Finally, it is interesting to study the relation between the finite dimensional and the infinite dimensional problem. It turns out that M_ϵ tends to $\mathcal{M}_\epsilon$, the invariant manifold associated with $\mathcal{P}_\epsilon$, as $n \to \infty$. Similar results hold if the integrals are approximated by more sophisticated numerical schemes. For details, see Kirchgraber (1991).

b)   Numerically speaking, systems of type (2) with ϵ small are a subclass of the class of stiff systems which in general are difficult to treat. The attractive invariant manifold $y = s_\epsilon(x)$ of Eq. (2) may be used to design an efficient numerical integration method. The solutions of Eq. (2) on this manifold are described by the system

$$\dot{x} = f\left(x, s_\epsilon(x)\right) \tag{5}$$

which is of lower dimension compared to Eq. (2) and which is no longer stiff as $\epsilon \to 0$. The approach developed in K. Nipp (1990) is based on determining numerically an expansion of $s_\epsilon(x)$ with respect to ϵ. Then, close to the invariant manifold the reduced system (5) with s_ϵ replaced by its approximation is integrated by a standard numerical scheme.

References

Boxler, P.: A Stochastic Version of Center Manifold Theory, Probab. Th. Rel. Fields (1989), p. 504-545.

Collet, P., Eckmann, J.P.: *Iterated Maps on the Interval as Dynamical Systems*, (1980).

Foias, C., Sell, G.R., Temam, R.: Inertial manifolds for nonlinear evolutionary equations, J. Differential Eq. (1988), p. 309-353.

Hadamard, J.: Sur l'itération et les solutions asymptotiques des équations différentielles, Bull. Soc. Math. France (1901), p. 224-228.

Hairer, E., Norsett, S.P., Wanner, G.: Solving Ordinary Differential Equations I, (1987).

Kirchgraber, U.: On the Stiefel-Baumgarte Stabilization Procedure, J. Appl. Math. and Phys. (ZAMP) (1979), p. 272-291.

Kirchgraber, U.: Multi-Step Methods are Essentially One-Step Methods, Numer.Math. (1986), p. 85-90.

Kirchgraber, U.: An ODE-Solver Based on the Method of Averaging, Numer. Math. (1988), p. 621-652.

Kirchgraber, U. The Use of Invariant Manifolds to Design Efficient Algorithms for Delay Equations, in preparation

Kirchgraber, U., Nipp, K.: Geometric Properties of RK-Methods Applied to Stiff ODE's of Singular Perturbation Type, Research Report No. 89-02, Seminar f. Angewandte Mathematik, ETH-Zürich.

Kirchgraber, U., Stiefel, E.: *Methoden der analytischen Störungsrechnung,* (1978).

Kirchgraber, U., Stoffer, D.: On the Definition of Chaos, Z. für Angew. Math. und Mech. (1989), p. 175-185.

Kirchgraber, U., Stoffer, D.: Chaotic Behaviour in Simple Dynamical Systems, SIAM Review (1990), Vol. 32.

Kirchgraber, U., Willers, G.: An Extrapolation Method for the Efficient Composition of Maps with Applications to Non-Linear Oscillations, Computing (1986), p. 343-354.

Lasagni, F.: Canonical Runge-Kutta Methods, J. Appl. Math. and Phys. (ZAMP) (1988), p. 952-953.

Lasagni, F.: Integration Methods for Hamiltonian Differential Equations, (1990), preliminary report.

Nipp, K.: Numerical integration of stiff ODE's of singular perturbation type, J. Appl. Math. and Phys. (ZAMP) (1990).

Sanz-Serna, J.M.: Runge-Kutta Schemes for Hamiltonian Systems, BIT (1988), p. 877-883.

Stoffer, D.: Transversal homoclinic points and hyperbolic sets for non-autonomous maps I, II, J. of. Appl. Math. and Phys. (ZAMP) (1988), p. 518-549 and p. 783-812.

Stoffer, D.: Averaging for Almost Identical Maps and Weakly Attractive Tori, to appear in Numer. Math. (1991)

Stoffer, D., Nipp, K. Invariant curves for variable step size integrators, to appear in BIT (1990).

Suris, Y.B.: On the Preservation of the Symplectic Structure in the Course of Numerical Integration of Hamiltonian Systems, in S.S. Filippov, M.V. Keldyshi (Eds.), *Numerical Solution of ODE's (1988),* p. 140-160.

International Series of Numerical Mathematics, Vol. 97, © 1991 Birkhäuser Verlag Basel

Combined Analytical – Numerical Analysis of Nonlinear Dynamical Systems

W. Kleczka, E. Kreuzer, and C. Wilmers

TU Hamburg-Harburg, Arbeitsbereich Meerestechnik II
Eißendorfer Straße 42, D-2100 Hamburg 90, F.R.G.

Summary: In this paper we demonstrate a methodology for the analysis of local bifurcations with codimension one of nonlinear dynamic systems using a combination of analytical and numerical methods. Special emphasis is devoted towards the analysis of mechanical systems with harmonic excitation. Instead of dealing with periodic solutions of the continuous system we study an equivalent approximation of the corresponding Poincaré map. An iterative scheme is used to compute the bifurcation points. The stability analysis for these critical values is carried out using center manifold theory in order to decouple the system into a stable and a low-dimensional critical subsystem. This approach will be demonstrated for the surge motion of a moored pontoon.

1 Introduction

The analysis of nonlinear dynamic systems is mostly done using purely numerical methods, since analytical approaches are feasible for low-dimensional problems only. The advent of general-purpose Computer Algebra programs like MACSYMA, MAPLE or REDUCE, however, stretches the limits for analytical calculations far beyond the pen and paper approach.

By combining well-established numerical methods with analytical calculations we may achieve further improvements. Using this combination, a framework has been developed for the determination and analysis of generic bifurcations for ordinary differential equations. Our idea was to combine the strong parts of both numerical and symbolical methods:

- Computer Algebra for the calculation of partial derivatives and for algebraic manipulations.

- Numerical methods for the solution of eigenvalue, boundary and initial value problems.

The first step in this approach is the approximation of the Poincaré map in the neighborhood of a periodic solution using Taylor series expansion. The periodic orbit is computed with a boundary value solver. The analytical formulation of the Poincaré map allows the accurate calculation of bifurcation points with codimension one.

In order to determine the stability of the system at the bifurcation point, the nonlinear terms have to be considered in addition to the linear ones. The application of center manifold theory yields a low-dimensional critical subsystem. The stability of the complete system is then determined by this low-dimensional bifurcation system.

Two programs have been developed for this bifurcation analysis, Figure 1. In order to achieve a semi-automatic procedure, emphasis is placed on the interface between the numerical and symbolical parts of the calculations, i.e. the problem-dependent code for the numerical routines is generated as FORTRAN-code via MAPLE-routines and vice versa.

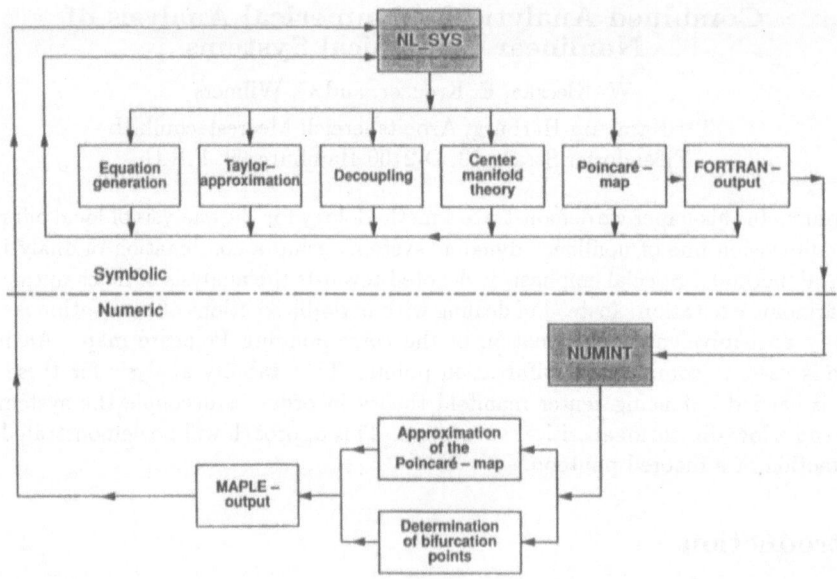

Figure 1: Flowchart of the combined analytical-numerical approach

2 Approximation of the Poincaré map

The analysis of periodic solutions is considerably simplified by using Poincaré maps. Hence, the ordinary differential equation is replaced by a discrete map,

$$\dot{x}(t) = f(x, \mu, t) \qquad \Longrightarrow \qquad x_{n+1} = g(x_n, \mu) \quad , \tag{1}$$

where μ denotes the parameter vector. For harmonically driven oscillators, an appropriate choice for the discretization time is the period T of the excitation.

In general the function $g(x)$ cannot be derived in an analytical form; however, an approximation for $g(x)$ in the neighborhood of a periodic solution $\bar{x}(t)$ can be computed by using Taylor series expansion:

$$g(x) = \left.\frac{\partial g(x)}{\partial x}\right|_{\bar{x}} x + \frac{1}{2}\left.\frac{\partial^2 g(x)}{\partial x^2}\right|_{\bar{x}} \{x, x\} + \frac{1}{6}\left.\frac{\partial^3 g(x)}{\partial x^3}\right|_{\bar{x}} \{x, x, x\} + \dots \quad . \tag{2}$$

The coefficients of the Taylor series can be determined as described in [1] and [2].

The complete symbolic setup of this coupled set of ODEs can be performed via generally applicable MAPLE-procedures. The solution, however, has to be calculated numerically.

In order to study the influence of the control parameter μ, system (1) is expanded by the trivial equation $\dot{\mu} = 0$ and the procedure for generating an approximation of the Poincaré map is applied to this extended system, [2].

Linear stability analysis fails for nonlinear systems with critical eigenvalues, and the stability is determined by the nonlinear parts of the differential equation. Center manifold theory provides an appropriate means to properly account for the influence of the noncritical variables on the stability of the system.

The stability of the complete system (1) can then be analyzed by just considering the *bifurcation system*

$$u_{n+1} = A^c u_n + g^c(u_n, h(u_n))$$

with a polynomial approximation for the center manifold $h(u)$. For simple bifurcations with codimension one equation (3) is one-dimensional and the stability is determined by the first nonvanishing nonlinear term. For the stability analysis of higher dimensional bifurcation systems the computation of the normal form is required.

3 Example

Typical examples for nonlinear dynamic systems in offshore engineering are floating bodies with excitation through current, waves and wind, [3, 4]. We will study the motion of a moored pontoon exposed to regular waves, Figure 2. The surge motion of this pontoon

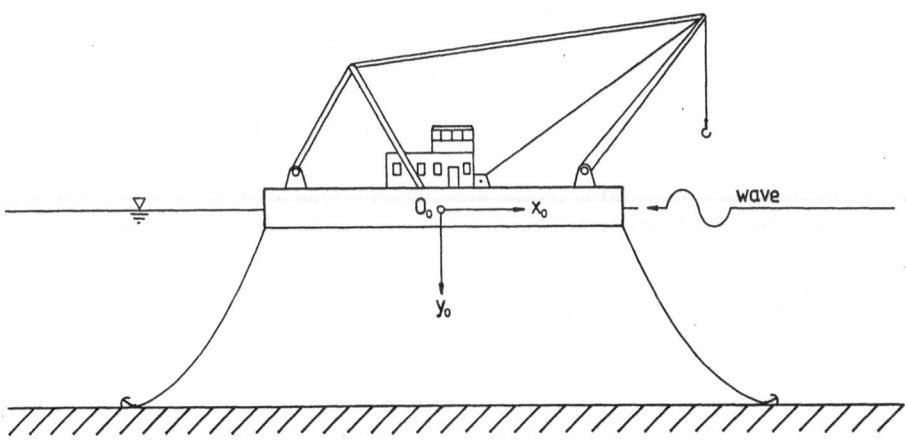

Figure 2: Moored pontoon with excitation

is decoupled if we consider an excitation by regular periodic waves acting in longitudinal direction only and neglect wind and current. The equation of motion for the rigid body is

$$[m + a(\infty)] \ddot{x}_0 + b(\infty)\dot{x}_0 + \frac{1}{2}\rho BT c_D \mid \dot{x}_0 \mid \dot{x}_0 + c_1 x_0 + c_3 x_0^3 = s_0 + F_s(t) \quad . \tag{4}$$

The parameters of the system are the mass m, width B and draught T of the pontoon, added mass $a(\infty)$ and damping coefficient $b(\infty)$ due to hydrodynamic effects and the drag coefficient

c_D. The linear and nonlinear restoring forces from the mooring system are calculated with the coefficients c_1 and c_3.

The hydrodynamic response force $s_0(t)$ is determined from

$$\dot{s}_{n-k} = s_{n+1-k} - A_k s_0 - B_k \dot{x}_0, \tag{5}$$

with $k = 1,\dots,n$, and $s_{n+1}(t) = 0$. The coefficients A_k and B_k are identified to fit the frequency response calculated from potential theory. In our example n is set to 3.

The exciting force F_s due to regular waves consists of a periodic part and a constant part representing the drag,

$$F_s = a\, p_{dyn}\, cos(\omega t) + a^2\, p_{drag} \quad , \tag{6}$$

with the wave height a and the coefficients p_{dyn} and p_{drag} determined from potential theory. Using the coupled equations (4) and (5) the governing differential equation with order 6 for the complete system is

$$\frac{d}{dt}\begin{pmatrix} x_0 \\ \dot{x}_0 \\ s_0 \\ s_1 \\ s_2 \\ s_3 \end{pmatrix} = \begin{pmatrix} \dot{x}_0 \\ \frac{1}{m+a(\infty)}\left[-b(\infty)\dot{x}_0 - \frac{1}{2}\rho BT c_D \mid \dot{x}_0 \mid \dot{x}_0 - c_1\, x_0 - c_3\, x_0^3 + \right. \\ \left. +s_0 + a\, p_{dyn}\, cos(\omega t) + a^2\, p_{drag}\right] \\ s_1 - A_3\, s_0 - B_3\, \dot{x}_0 \\ s_2 - A_2\, s_0 - B_2\, \dot{x}_0 \\ s_3 - A_1\, s_0 - B_1\, \dot{x}_0 \\ -A_0\, s_0 - B_0\, \dot{x}_0 \end{pmatrix}. \tag{7}$$

The wave height a is chosen as control parameter. The various periodic solutions are shown in Figure 3.

Using the methodology proposed in [2] the first period-doubling bifurcation was found after a few iteration steps for a wave amplitude of 2.34 m. Application of center manifold yields the one-dimensional bifurcation system

$$u_{n+2} = 1.00u_n - \underbrace{0.186 * 10^{-5}}_{\cong 0}\, u_n^2 - 0.649 u_n^3 + \mathcal{O}(u_n^4) \quad . \tag{8}$$

The stability is determined by the first nonvanishing nonlinear term. For this case, the negative cubic term implies a supercritical flip-bifurcation; the new two-periodic solution can be seen in the bifurcation diagram, Figure 3.

4 Conclusions

The analysis of nonlinear oscillators with harmonic excitation like rigid bodies floating in regular waves can be considerably simplified by using Computer Algebra packages like MA-PLE. A combination of analytical computations with numerical methods can avoid some of the immanent limitations of Computer Algebra (e.g. calculation of eigenvalues for dimension > 5, expression swell), and allow for a broader analysis than purely numerical investigations.

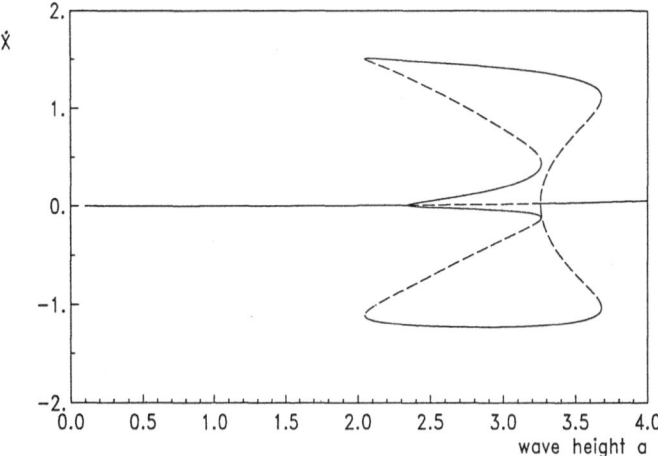

Figure 3: Bifurcation diagram for different wave heigths,
stable solutions: ———   unstable solutions: – – –

A Poincaré map is determined in order to gain an analytical expression for a periodic solution. This point map in analytical form is the basis for the calculation of bifurcation points and the subsequent stability analysis. The bifurcation equations are derived by means of center manifold theory. Again, the required calculations are performed analytically.

The applicability of the presented approach was demonstrated for a six-dimensional system. Standard numerical methods have been used for solving the eigenvalue problem and for computing the periodic orbit.

References

[1] E. Lindtner, A. Steindl, and H. Troger. Generic one-parameter bifurcations in the motion of a simple robot. In: H. D. Mittelmann and D. Roose, (eds.), *Continuation Techniques and Bifurcation Problems*, Basel / ...: Birkhäuser Verlag, 1990.

[2] M. Kleczka, W. Kleczka, and E. Kreuzer. Bifurcation analysis: A combined numerical and analytical approach. In: D. Roose et al., (eds.), *Continuation and Bifurcations: Numerical Techniques and Applications*, Dordrecht: Kluwer Academic Publishers, 1990.

[3] J. Jiang. Untersuchung chaotischer Schiffsdynamik an Beispielen aus der Offshoretechnik. Private communication. Hamburg: Germanischer Lloyd, June 1990.

[4] T. E Schellin, S. D. Sharma, and T. Jiang. Crane ship response to regular waves: Linearized frequency domain analysis and nonlinear time domain simulation. Proceedings of the Eigth Joint International Conference on Offshore Mechanics and Polar Engineering, The Hague, The Netherlands, March 19-23, 1989.

International Series of Numerical Mathematics, Vol. 97, © 1991 Birkhäuser Verlag Basel

Monotony methods and minimal and maximal solutions for nonlinear ordinary differential equations

M.Koleva   N.Kutev

The aim of this paper is to investigate different boundary value problems/b.vps/ for general second order nonlinear ordinary differential equations/o.d.es/ which appear in mechanics,chemistry,biology,geometry etc. i.e.

(1) $u'' = f(x,u,u')$ in (a,b) ; $u(a)=A$, $u(b)=B$

(2) $u'' = f(x,u,u')$ in (a,b) ; $u'(a)-\sigma u(a)=A$, $u'(b)+\tau u(b)=B$

(3) $u'' = f(x,u,u')$ in (a,b) ; $u'(a)-\sigma u(a)=A$, $u(b)=B$

where A,B are arbitrary constants and σ,τ are positive ones,although the methods hold for more general boundary conditions as well.Under some natural conditions for the right-hand side $f(x,z,p)$ existence of a minimal and maximal solution u_{min},u_{max} of (1)-(3) is proved so that for every solution u of (1)-(3) the inequalies $u_{min} \leq u \leq u_{max}$ hold.Moreover u_{min},u_{max} are obtained as limits of monotony sequences of solutions of special o.d.es and in particular of linear o.d.es.

A constructive method based on the properties of monotony compact operators was suggested by Chandra and Davis $[C-D]$ for quasilinear equations

(4) $(p(x)u')' = f(x,u)u'+g(x,u)$ for $x \in (a,b)$

with nonlinearity depending only on the unknown function.As is pointed out in $[C-D]$,more general nonlinearities essentially change the situation.That is the reason why the example of the lubrication problem is investigated in $[C-D]$ only after a suitable change of the variables which reduces the equation to type (4).In the present paper the result in $[C-D]$ is exended to general nonlinear o.d.es under the most general solvability conditions for (1)-(3) and it is shown which of the conditions are necessary and which of them are close to the best possible ones.

Let us recall that a lot of publications are devoted to abstract existence theorems for b.v.ps for (1)-(3).The main methods for the proof of the existence theorems are the topological me-

thods,particularly the degree theory.As a rule the solvability conditions which are usually used are only sufficient and it is not
clear whether they are the best possible ones.For example,the Nagumo condition,i.e. the quadratic growth of $f(x,z,p)$ with respect
to p for $|p| \to \infty$,which is most often used in the existence theory,
is not necessary for the solvability of (1)-(3).As is proved in
the present paper,only in a neigbourhood/ngbh/ of the end points a
and/or b,where Dirichlet data are prescribed,the quadratic growth
of $f(x,z,p)$ with respect to p is necessary for the general solvability of the Dirichlet problem.For example,the equations of curves
with prescribed curvature $K(x,u)$ i.e.

(5) $u'' = K(x,u)(1+u'^2)^{3/2}$ in (a,b) ; $u(a)=A$, $u(b)=B$
are generally solvable under the condition $K(a,A)=K(b,B)=0$,$K \in C^2$
and some additional assumptions for the amplitude of K in spite of
the fact that Nagumo condition fails.

Let us note that in case the minimal and maximal solutions co-
inside,we have an answer to the complicated problem for the uniqueness of (1)-(3).Since u_{min}, u_{max} are a monotony limit of solutions of special o.d.es for which uniqueness theorems hold,the monotony algorithm advanced in the present work enables us to con-
structively/numerically/ prove uniqueness or nonuniqueness as well
as bifurcation phenomena for wide classes of nonlinear o.d.es.
Moreover,it is shown that for every pair of lower and upper solutions u_0, v_0,satisfying the inequality $u_0 \leq v_0$,there exist a minimal
and maximal solution of (1)-(3) between u_0, v_0.Thus if (1)-(3) possess two different pairs lower and upper solutions u_0, v_0 and u^0, v^0
resp.,such that $u_0 \leq v_0 \leq u^0 \leq v^0$ then (1)-(3) have at least two different solutions between u_0 and v^0.In this way the problem of existence of multiple solutions for (1)-(3) is reduced to the existence of two or more pairs of lower and upper solutions.In the present paper,under the conditions which are the best possible ones,
existence of a lower and upper solution $u_0, v_0, u_0 \leq v_0$ is proved,such
that every solution u of (1)-(3) satisfies the inequalities
$u_0 \leq u \leq v_0$.Let us also note that the existence only of a lower solution u_0 of (1)-(3) can be used numerically to check whether there
exists a solution u greater than u_0.Analogously,by means only of

an upper solution v_0 we can numerically prove existence or non-
existence of a solution u of (1)-(3) less than v_0.

In order to formulate the results let us recall the definiti-
ons of lower and upper solutions for (1)-(3).

Definition:The function $u_0 \epsilon C^2(a,b) \cap C^1[a,b]$ is a lower solution
of (1)/(2) or (3)/ if $u_0'' - f(x,u_0,u_0') \geq 0$ in (a,b), $u_0(a) \leq A$ /$u_0'(a)-$
$-\sigma u(a) \geq A$/ and $u_0(b) \leq B$ /$u_0'(b)+\tau u(b) \leq B$ or $u_0(b) \leq B$ resp./.

Analogously the function $v_0 \epsilon C^2(a,b) \cap C[a,b]$ which satisfies the
oposit inequalities is upper solution.

Using the lower solution u_0 for a first approximation we can
build the sequence of successive approximations $\{u_n\}$ corresponding
to the b.v.ps (1)-(3) i.e.

(6) $L_n u_n = u_n'' - f(x,u_{n-1},u_n') = 0$ in (a,b) , n=1,2,...

and u_n satisfies one of the boundary conditions (1)-(3).

Analogously,using the upper solution v_0 for a first approximation we can construct the sequence $\{v_n\}$ of solutions of (6).In the
following theorem existence and convergence of the sequences $\{u_n\}$,
$\{v_n\}$ resp.,to the minimal and maximal solution of (1)-(3) between
u_0 and v_0 will be proved.For this purpose let us note that only in
case of Dirichlet data at the point a and/or b in a sufficiently
small neighbourhood/ngbh/ N_a and/or N_b of the corresponding points
a,b we need the additional assumptions.

(7)$_a$ $|f(x,z,p)| \leq \phi^{1+\beta}(|p|)(x-a)^\beta |p|^{2+\beta}$ for $x \epsilon N_a$,

(7)$_b$ $|f(x,z,p)| \leq \phi^{1+\beta}(|p|)(b-x)^\beta |p|^{2+\beta}$ for $x \epsilon N_b$

and for inf u $\leq z \leq$ sup u ,$|p| \geq L_0$.Here $\beta \geq 0, L_0 \geq 0$ are some nonnegative
constants and the positive nondecreasing function $\phi \epsilon C[0,\infty)$ satisfies the condition

(8) $\int_1^\infty dt/t\phi(t) = \infty$ / for example $\phi(t)=\ln(2+t)$ /.

Theorem 1:Let $f \epsilon C^1([a,b] \times R^2), f_z(x,z,p) \leq 0$ for $(x,z,p) \epsilon [a,b] \times R^2$
and f satisfy condition (7) for inf $u_0 \leq z \leq$ sup v_0 but only at the
point a and/or b where Dirichlet data are prescribed.Suppose $u_0 \leq v_0$
are resp.,lower and upper solution of b.v.ps (1)/(2) or (3)/ and
at least one of the following conditions
(9)$_i$ $|f(x,z,p)| \leq C|p|^2$;

$(9)_{ii}$ $-f_z(x,z,p)-f_x(x,z,p)/p \leq 0$;

$(9)_{iii}$ $-(C/(C+1))^{1/2}f_z(x,z,p)-f_x(x,z,p)/p \leq |pf_p(x,z,p)-f(x,z,p)|/2(C$

$+1)$osc u , and $pf_p(x,z,p)-f(x,z,p)$ has a constant sign

hold for $x \in [a,b]$, inf $u_0 \leq z \leq$ sup v_0, $|p| \geq L_1$ and for some positive con-
stants C, L_1.

Then the sequences $\{u_n\}, \{v_n\}$ exist and $\{u_n\}$, uniformly and monotonically increasing, tends to the minimal solution u_{min} and $\{v_n\}$,
uniformly and monotonically decreasing, tends to the maximal solution u_{max} of (1)/(2) or (3)/. If u is a solution of (1)/(2) or (3)/
between u_0 and v_0 then the inequality $u_0 \leq u_1 \leq ... \leq u_n \leq ... u_{min} \leq u \leq u_{max} \leq$
$\leq ... v_n \leq ... \leq v_1 \leq v_0$ holds.

As is clear from theorem 1, the existence of one or more pairs of
lower and upper solutions is the key to the existence of multiple
solutions for (1)-(3). For many concrete equations which appear in
mechanics, chemistry, biology etc. it is possible to find out exple-
cit lower or upper solutions. Nevertheless, in the following theorem
we prove existence of a lower and upper solution u_0 and v_0 which
are less or greater resp., from every solution of (1)-(3).

Theorem 2: Suppose $f \in C([a,b] \times R^2)$ satisfies condition

     $-f(x,z,p)$sign z $\leq g(x)/h(p)$ for $x \in [a,b]$, $|z| \geq m_0$, $p \in R$

(10)
     $\int_a^b g(x)dx < \int_{-\infty}^{\infty}h(p)dp$

for some positive function $g \in C[a,b]$, $h \in C(R)$ and some nonnegative
constant m_0. Then b.v.p. (1)/(2) or (3)/ has at least one lower and
upper solution $u_0, v_0, u_0 \leq v_0$. If u is a solution of (1)/(2) or (3)/
the inequalities $u_0 \leq u \leq v_0$ hold for every $x \in [a,b]$.

Remark 1: Condition (10) is the best possible one since for the
right-hand side $f(x,z,p)=g(x)/h(p)$ if b.v.p.(1)/(2) or (3)/ has a
classical $C^2(a,b) \cap C^1[a,b]$ solution then (10) must hold. Indeed, in-
tegreating equations (1)-(3) from a to b we obtain inequality (10)

i.e. $\int_a^b g(x)dx = \int_a^b u^{\prime\prime}h(u^\prime)dx = \int_{u^\prime(a)}^{u^\prime(b)}h(p)dp < \int_{-\infty}^{\infty}h(p)dp$

Conditions $(7)_a$ and/or $(7)_b$ are also necessary for the general
solvability of b.v.ps (1),(3) for all data A,B, in case the Dirichlet data at a and/or b are prescribed. In fact, they guarantee boun-

dary gradient a priori estimates for the solutions of (1),(3) at the point a and/or b.More precisely,if one of the following conditions hold i.e.

(11)$_a$ $|f(x,z,p)| \geq \Phi^{1+\beta}(|p|)(x-a)^\beta|p|^{2+\beta}$ for $x \in N_a$,

(11)$_b$ $|f(x,z,p)| \geq \Phi^{1+\beta}(|p|)(b-x)^\beta|p|^{2+\beta}$ for $x \in N_b$

and for $z \in R, |p| \geq L_2$,for some nonnegative constants $\beta \geq 0, L_2 \geq 0$ and for some positive nondecreasing function $\Phi \in C[0,\infty)$ satisfying the condition

(12) $\int_1^\infty dt/\Phi(t)t < \infty$ / for example $\Phi(t) = \ln^{1+\varepsilon}(2+t)$ /

then b.v.ps (1),(3) are not solvable for every data A,B.

   Theorem 3:Suppose $f \in C([a,b] \times R^2)$ satisfies (11)$_a$/or (11)$_b$/ and either

(13) $f(x,z,p) \geq 0$ for $x \in [a,b]$,$z \geq m_1$,$p \leq -L_3$ or
       $f(x,z,p) \leq 0$ for $x \in [a,b]$,$z \leq -m_1$,$p \geq L_3$

holds for some nonnegative constants m_1, L_3.

   Then b.v.p.(1)/or (3)/ has no classical $C^2(a,b) \cap C[a,b]$/ or $C^2(a,b) \cap C^1[a,b]$ / solution for all constants A/or B/.

   As for assumptions (9) they are only sufficient and guarantee global gradient estimates in [a,b] for the solutions of (1)-(3).It is not clear wheather they are necessary for the general solvability of the b.v.ps.In spite of this,these conditions are close to the best possible ones as simple examples show.For example,in case that Nagumo condition (9)$_i$ fails in a ngbh of some interior point $c \in (a,b)$,as in theorem 3,we can show that super a priori estimates for the modulus of continuity of the solutions of (1)-(3) or for their Hölder norm hold with constants independent of the boundary data A,B.Unfortunately,it is not clear how to use these super a priori estimates in order to prove that the Nagumo condition (9)$_i$ is the best possible one for the general solvability of the b.v.ps.

R E F E R E N C E S

[C-D] .Chandra J,,Davis P.:A monotony method for quasilinear boundary value problems.Arch.Ration.Mech.Anal.57,257-266 (1973)
[G-M] .Gaines R.,Mawhin J.:Coincidence degree and nonlinear differential equations.Lect.Notes in Math.,568 (1977).

International Series of Numerical Mathematics, Vol. 97, © 1991 Birkhäuser Verlag Basel

Interior Crisis in an Electrochemical System

K. Krischer, M. Lübke, W. Wolf, M. Eiswirth and G. Ertl

Fritz-Haber-Institut der Max-Planck-Gesellschaft,

Faradayweg 4-6, W-1000 Berlin 33, Germany

1. INTRODUCTION

Temporal oscillations in electrochemical reactions (such as the dissolution of Cu or Fe electrodes in certain electrolytes) have been reported by a number of authors over many years. There has been renewed interest in such systems recently /1/, mainly because a rich variety of complex dynamical behavior (such as mixed-mode oscillations quasiperiodicity and different routes to chaos) could be observed with a good S/N-ratio for these reactions.

The present paper deals with a galvanostatic oscillator, namely the oxidation of hydrogen on a platinum anode, which was found to exhibit (among other complex oscillations /2/) a Feigenbaum route to chaos as well as a chaos-to-chaos transition through an interior crisis of a strange attractor.

2. EXPERIMENTAL

The galvanostatic oxidation of hydrogen was carried out in 1 M $HClO_4$ in the presence of $Cu(ClO_4)_2$ with concentrations between 10^{-6} and 10^{-2} M on a (polycrystalline) Pt wire as anode using a Pt cathode and a Hg/Hg_2SO_4 reference electrode. The working electrode was cleaned in hot oxidizing acid and subsequent adjustment of a high positive potential (1.2 V vs. Hg/Hg_2SO_4) for two minutes. The quality of the electrode was checked with current-voltage diagrams. Using a galvanostat, an amperemeter and a voltmeter, the potential oscillations were recorded digitally for constant current. The current density i was varied as external control parameter. A detailed description of the experimental set-up and the potentiostatic measurements is given elsewhere /2/.

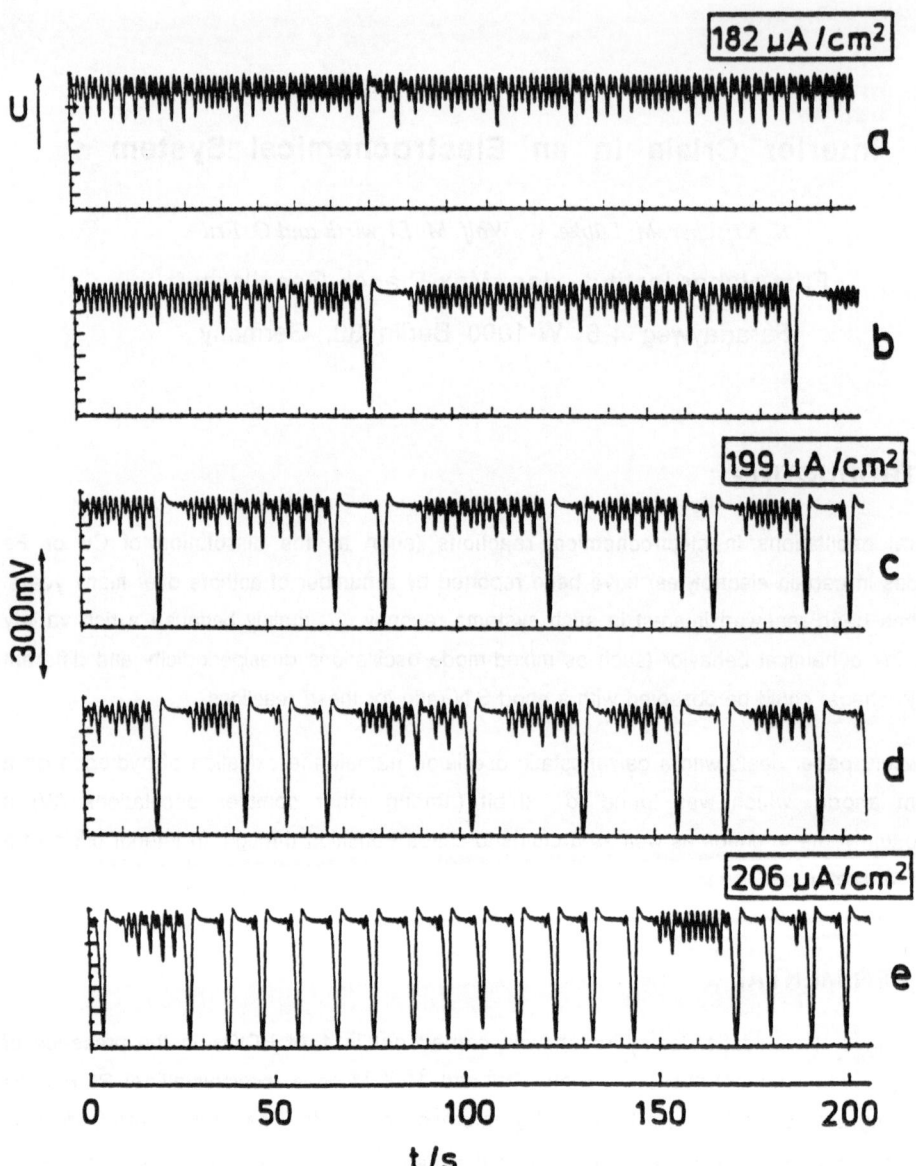

<u>Fig. 1:</u> Galvanostatic oscillations in the H_2 oxidation on a Pt wire (1 M $HClO_4$, 3×10^{-5} M $Cu(ClO_4)_2$). Upon increase of the current density suddenly large amplitudes occur.

Fig. 2: Poincaré maps for the time series of fig. 1.

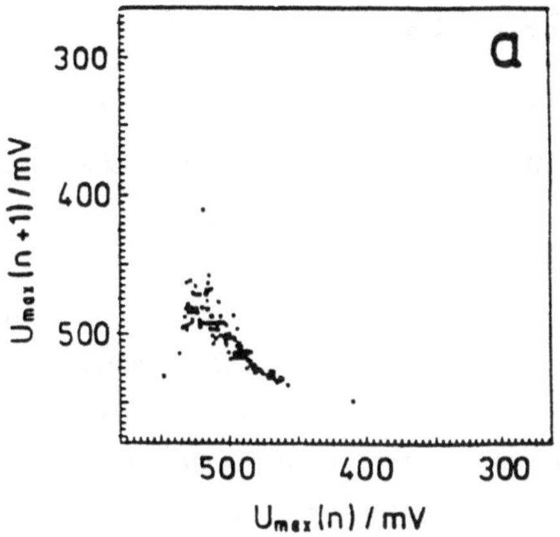

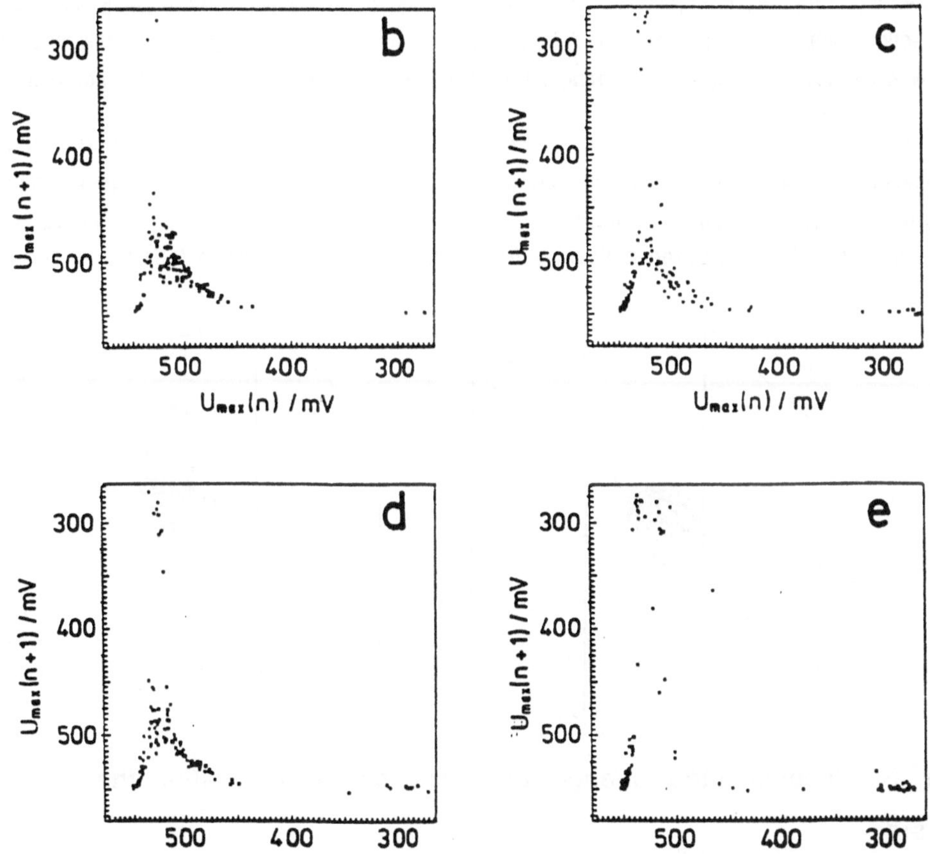

3. RESULTS

Two different kinds of chaotic behavior were found in the system described above. At low current densities (ca. 182 mA/cm^2) chaotic oscillations with small amplitudes around 100 mV occurred (fig. 1a). These stemmed from a Feigenbaum route in which two successive period-doubling bifurcations were observed /2/. Upon further increase of the control parameter, sharp large bursts of about 300 mV appeared, which at first occurred quite rarely (fig. 1b), then with increasing frequency towards higher current densities (Fig. 1 c-e). This behavior also becomes apparent from the Poincaré maps (next-minimum-maps) reproduced in fig. 2. Note that the form of the Poincaré map in fig. 2a is largely retained just after the bifurcation, with only few large amplitudes in addition (fig. 2b), but changes shape further away from it (fig. 2e). Phase space reconstructions (obtained with the time delay method) of attractors before and after large amplitudes developed are shown in fig. 3. After a large excursion reinjection took place right into the (hitherto empty) center of the original smaller attractor, followed by slow spiraling-out indicating the presence of a saddle-focus inside the attractor.

Quantitative characterization of the different strange attractors was also carried out. No significant change of the maximum Lyapounov exponent (calculated with a modified WSSV algorithm /3,4/) was found (around 0.12 bit/s for both types), whereas the information dimension D_1 /5,6/ increased noticeably from 1.95±0.1 for the small to 2.5±0.1 for the big attractor.

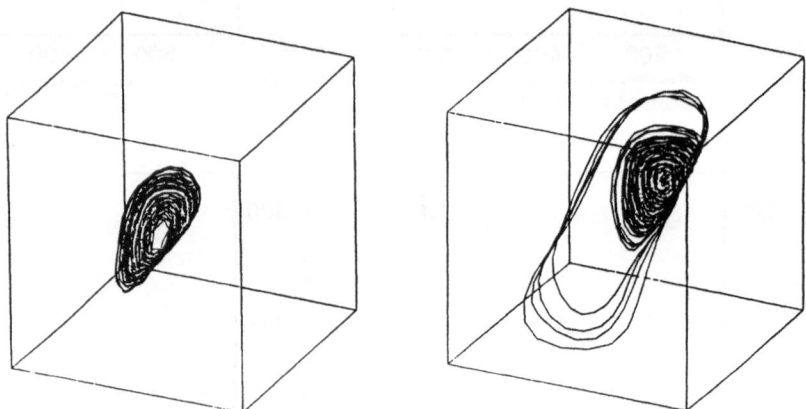

Fig. 3: Reconstructed strange attractors before and after the interior crisis

4. Discussion

In the previous section, experimental data were presented where a sudden (catastrophic) change in size of a strange attractor occurred in such a way that the new larger attractor completely contains the smaller one. Only as the bursts get frequent, is the form of the smaller attractor destroyed. Such a bifurcation is referred to as an interior crisis /7/ and has been described theoretically for maps (e.g. the logistic map /8/) as well as for systems of ordinary differential equations /9,10/. This phenomenon is attributed to a collision of the small-amplitude attractor with the inset of saddle (point or orbit) /7/, leading to a large excursion, and subsequent reinjection. Further increase of the bifurcation parameter then leads to a gradual increase of the number of excursions (cf. figs. 1 and 2). Such a scenario is exactly in line with the measurements reported in section 3, which represent, to our knowledge, the first experimental evidence of an interior crisis in a chemical oscillator.

REFERENCES

/ 1 /   J. L. Hudson and M. R. Bassett, Reviews in Chemical Engineering (in press)

/ 2 /   K. Krischer, Ph.D. thesis, Freie Universität Berlin 1990

/ 3 /   A. Wolf, J. B. Swift, H. L. Swinney and J. A. Vastano, Physica D **16**, 285 (1985)

/ 4 /   Th. Kruel, Diplomarbeit, Universität Würzburg 1987

/ 5 /   W. v. d. Water and P. Schram, Phys. Rev. A **37**, 3118 (1988)

/ 6 /   R. Badii and A. Politi, Phys. Lett. **104** A, 303 (1984)

/ 7 /   J. M. T. Thompson and H. B. Stewart, Nonlinear Dynamics and Chaos, Wiley, Chichester
        1986

/ 8 /   R.W. Leven, B.P. Koch and B. Pompe, Chaos in dissipativen Systemen, Akademie-
        Verlag, Berlin   1989

/ 9 /   Y. Ueda, in R. H. G. Helleman (ed.): Nonlinear Dynamics, New York Academy of
        Sciences, New York 1980

/ 1 0 / J. L. Hudson, O. E. Rössler and H. Killory, Chem. Eng. Commun. **46**, 159 (1986)

International Series of Numerical Mathematics, Vol. 97, © 1991 Birkhäuser Verlag Basel

MULTIPLE BIFURCATION OF FREE-CONVECTION FLOW BETWEEN VERTICAL PARALLEL PLATES

by M. Kropp and F.H. Busse

Institute of Physics, University of Bayreuth, 858 Bayreuth, FRG

1. INTRODUCTION

The fluid flow between two vertical parallel plates one of which is heated and the other is cooled has been the subject of many experimental and theoretical studies over the last 30 years because of its importance in heat transfer engineering [1-7].

In the limit of large aspect ratio Γ (=height/width of the layer) there exist two instabilities of the basic flow: stationary rolls occur at low, oscillatory rolls occur at high Prandtl numbers. The properties of the instabilities have been investigated separately, but the problem of their interaction at moderate Prandtl numbers has not been addressed in the literature. A weakly nonlinear analysis of the interaction problem yields a system of coupled amplitude equations. Secondary bifurcations from the pure mode solutions lead to mixed solutions. Recently, after the work reported here had been completed, we have received a preprint of the work by Fujimura and Mizushima [9] in which the problem of interaction has also been addressed.

2. MATHEMATICAL FORMULATION OF THE PROBLEM

Consider two vertical infinitely extended parallel plates with distance d, which are kept at constant but different temperatures T_1 and T_2, as shown in figure 1. The gap between the plates is filled with a fluid of kinematic viscosity ν, thermal expansivity γ and thermal diffusivity κ. Using d as length scale, d^2/ν as time scale and $\nu^2/(\gamma g d^3)$ as temperature scale, we find the Grashof number $G=(T_2-T_1)\gamma g d^3/\nu^2$ and the Prandtl number $P=\nu/\kappa$ as nondimensional parameters of the problem. The basic state is given by the velocity field $\mathbf{u_0}=\frac{1}{6}G(\frac{1}{4}z-z^3)\mathbf{j}$ and the temperature field $T_0=Gz$. Writing the divergence free velocity field in the form $\mathbf{u}=\mathbf{u_0}+\mathbf{u_s}=\mathbf{u_0}+\bar{\mathbf{u}}_s+\nabla\times\nabla\times\mathbf{k}\phi+\nabla\times\mathbf{k}\psi$ and the temperature field in the form $T=T_0+\vartheta$, we obtain the following equations for the scalar functions ϕ,ψ,ϑ:

$$\left[(\nabla^2-\partial/\partial t)\nabla^2-u_0\cdot\nabla\nabla^2+u_0''\cdot\nabla\right]\Delta_2\phi + j\cdot\nabla\partial\vartheta/\partial z = k\cdot\nabla\times\nabla\times(u_s\cdot\nabla u_s) \qquad (2.1a)$$

$$u_0'\cdot\nabla\times k\Delta_2\phi + \left[\nabla^2-\partial/\partial t-u_0\cdot\nabla\right]\Delta_2\psi - i\cdot\nabla\vartheta = k\cdot\nabla\times(u_s\cdot\nabla u_s) \qquad (2.1b)$$

$$G\Delta_2\phi + \left[P^{-1}\nabla^2-\partial/\partial t-u_0\cdot\nabla\right]\vartheta = u_s\cdot\nabla\vartheta \qquad (2.1c)$$

where primes denote z-derivatives of functions depending only on the z-coordinate and $\Delta_2=\nabla^2-(k\cdot\nabla)^2$.

These equations must be supplemented by an equation for the mean flow $\bar{u}_s$,

$$\left[(k\cdot\nabla)^2-\partial/\partial t\right]\bar{u}_s = \left[\partial/\partial z\overline{(-\Delta_2\phi\ \nabla\cdot(j\partial\phi/\partial z-i\psi))} - \bar{\vartheta} + \eta\right]j \qquad (2.2)$$

where the bar indicates the mean over the x,y-plane. We have anticipated that no mean flow will be generated in the x- direction. The constant η represents a pressure gradient that will be generated in order to satisfy the condition that the averaged mass flux vanishes. The boundary conditions at the isothermal rigid walls are $\phi = \partial\phi/\partial z = \psi = \vartheta = \bar{u}_s = 0$.

Writing equation (2.1) in the form:

$$\left(\mathbb{L} - \partial_t\ \mathbb{M}_1 - G\ \mathbb{M}_2\right)\cdot X = N(X,X) \qquad (2.3)$$

with the linear operators $\mathbb{L}$, $\mathbb{M}_1$ and $\mathbb{M}_2$ and the nonlinear operator N, we introduce an expansion in powers of the amplitude A of convection

$$G = G_0 + AG_1 + A^2G_2 + \ldots$$
$$X = A(X_0 + AX_1 + A^2X_2 + \ldots)$$
$$\omega = \omega_0 + A\omega_1 + A^2\omega_2 + \ldots$$

where X_0 is given by

$$X_0 = X_0(z) \exp(i(\alpha x+\beta y+\omega_0 t)) + \text{conjugate complex}$$

and is normalized to fulfill the condition $<|u_s|^2>=A^2$. In the order A^1 we obtain the linear stability problem. Of particular interest is the minimum G_0 of the Grashof number as a function of α,β and the corresponding value $\pm\omega_0$ of the frequency. Because of the symmetry of $N(X_0,X_0)$, the solvability condition of the second order requires G_1 and ω_1 to vanish, whereas the solvability condition in third order yields finite G_2 and ω_2. Alternatively, we may assume that A is weakly time dependent and obtain the amplitude equation with complex coefficients

$$\dot{A} = \lambda(G-G_0)A + \alpha|A|^2A \qquad (2.4)$$

Close to the codimension-two-point (G_{co}, P_{co}) in the Grashof-, Prandtl number plane this approach can be generalized by including the coupling of the stationary and the oscillatory solutions through the nonlinear terms. The coupled amplitude equations can thus be written in the form

$$\dot{A}_i = \left(\lambda_i(G-G_{co}) + \mu_i(P-P_{co}) + \sum_{j=1}^{m} \alpha_{ij}|A_j|^2 \right) A_i \ , \quad i=1,2,3 \qquad (2.5)$$

where $A_1(T)$ and $A_2(T)$ are the amplitudes of the oscillatory solutions with $+\omega_0$ and $-\omega_0$ and $A_3(T)$ is the amplitude of the stationary solution. The fixed point solutions of equations (2.5) and their stability with respect to small perturbations within the framework of equations (2.5) can be calculated analytically.

3. WEAKLY NONLINEAR RESULTS AND STABILITY PROPERTIES

The codimension-two-point under consideration is given by $G_{co}=7873$ and $P_{co}=12.45$. The wavevector of the stationary solution is $\alpha_c^s=0$, $\beta_c^s=2.7$, the wavevector and frequency of the oscillatory solution are given by $\alpha_c^o=0$, $\beta_c^o=0.7$ and $\omega_0=40$ [3]. The bifurcating solutions described by equations (2.5) are shown in figures 2 and 3. The properties of physical interest are:

(i)   Both the stationary and the oscillatory instabilities set in supercritically.

(ii) The standing wave pattern $|A_1^s|=|A_2^s|$ is always stable with respect to travelling waves $A_1=0$ and $A_2^t \neq 0$ or vice versa. Because of the weak coupling the equilibrium amplitude $|A_1^s|$ exceeds the amplitude $|A_1^t|$ only by a small amount.

(iii) A secondary bifurcation in the form of a stable mixed mode with $A_3 \neq 0$, $A_1 \neq 0$ and $A_2 \neq 0$ occurs at a Grashof number close to the Grashof number for the onset of the pure stationary solution.

(iv) Another secondary bifurcation occurs in the form of an unstable mixed mode corresponding to the combination of the stationary solution and the travelling waves.

Properties (i) and (ii) hold for all Prandtl numbers. These results agree with those of Fujimura and Mizushima [9] and disagree with those of Riley and Wynne [8], who expect a subcritical Hopf bifurcation.

The stable mixed mode (iii) is a superposition of stationary rolls and two antisymmetric waves travelling in opposite directions.

In the case of the unstable mixed mode (iv) the Amplitude A_3 is weakly time dependent describing a drift of the rolls in a direction opposite to that of the travelling wave. This property is caused by the asymmetry associated with the travelling wave solutions which tend to be attached to one or the other walls of the system.

The stationary solution is only slightly affected by the existence of the oscillatory solutions, whereas the latter are much influenced by the stationary mode.

REFERENCES

[1] Bergholz, R. F., J. Fluid Mech. 84 (1978), 743-768

[2] Chait, A. & Korpella, S. A., J. Fluid Mech. 200 (1989), 189-216

[3] Chen, Y.-M. & Pearlstein, A. J., J. Fluid Mech. 189 (1989), 513-541

[4] Elder, J. W., J. Fluid Mech. 23 (1965), 77-98

[5] Gershuni, G. Z. & Zhukovitskii, E. M., Appl. Math. Mech. 33 (1969), 830-835

[6] Nagata, M. & Busse, F. H., J. Fluid Mech. 135 (1983), 1-26

[7] Seki, N., Fukusaka, S. & Inaba, H., J. Fluid Mech. 84 (1978), 695-704

[8] Riley, D. S. & Wynne, M. C., Proc. Roy. Soc. Lond. A 420 (1988), 419-443

[9] Fujimura, K. & Mizushima, J., to appear in Proc. IUTAM Symp. on Nonlinear Hydrodynamic Stability and Transition (1990)

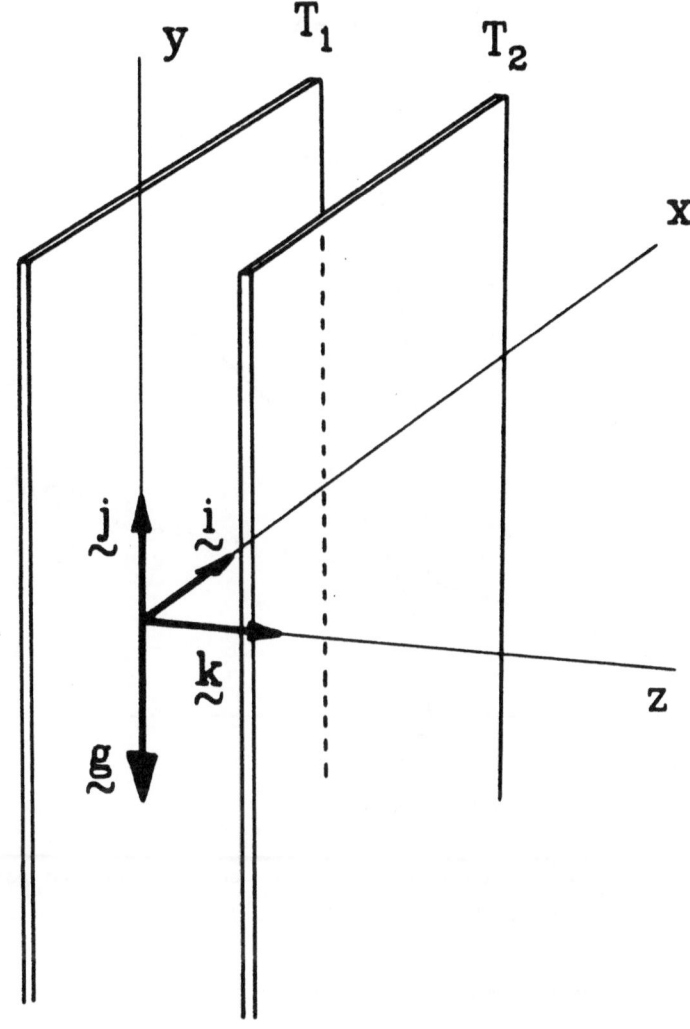

Figure 1   Geometrical configuration of the problem of convection in
                            the gap between two differentially heated parallel plates.

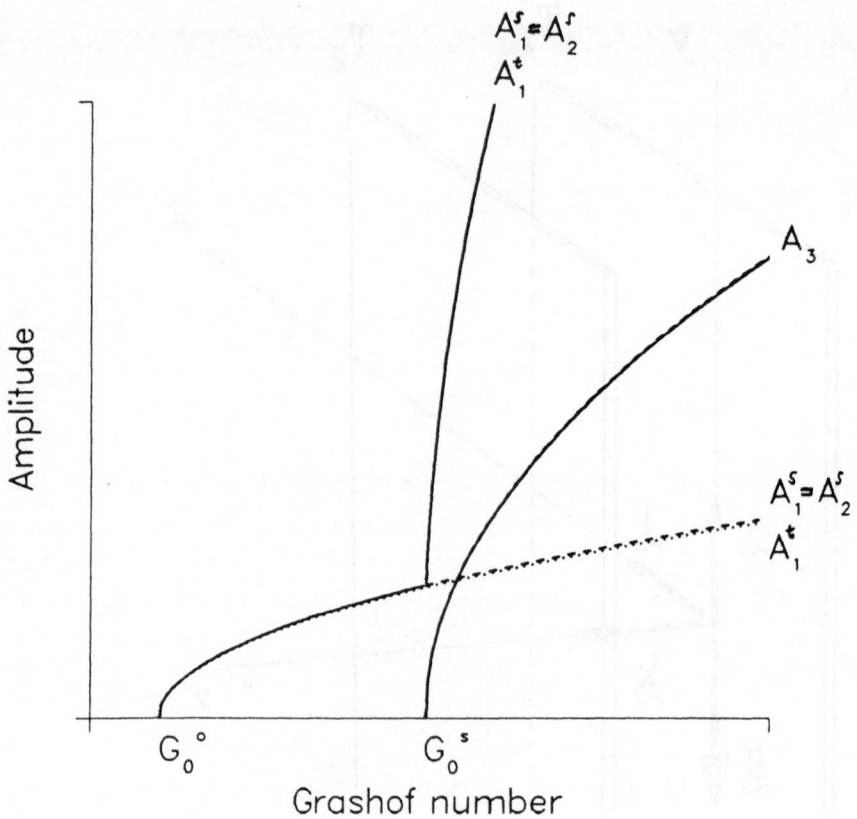

Figure 2   Bifurcation diagram near the codimension-two-point, when
the critical Grashof number of the oscillatory solution is
lower than that of the stationary solution. All solutions
involving $A_3 \neq 0$ nearly coincide. The bifurcation points for
the pure stationary solution and the mixed solutions also
coincide within the accuracy of the drawing. Stable
solutions are indicated by solid lines, unstable pure modes
by dashed lines and unstable modes containing the
travelling oscillatory mode by dotted lines.

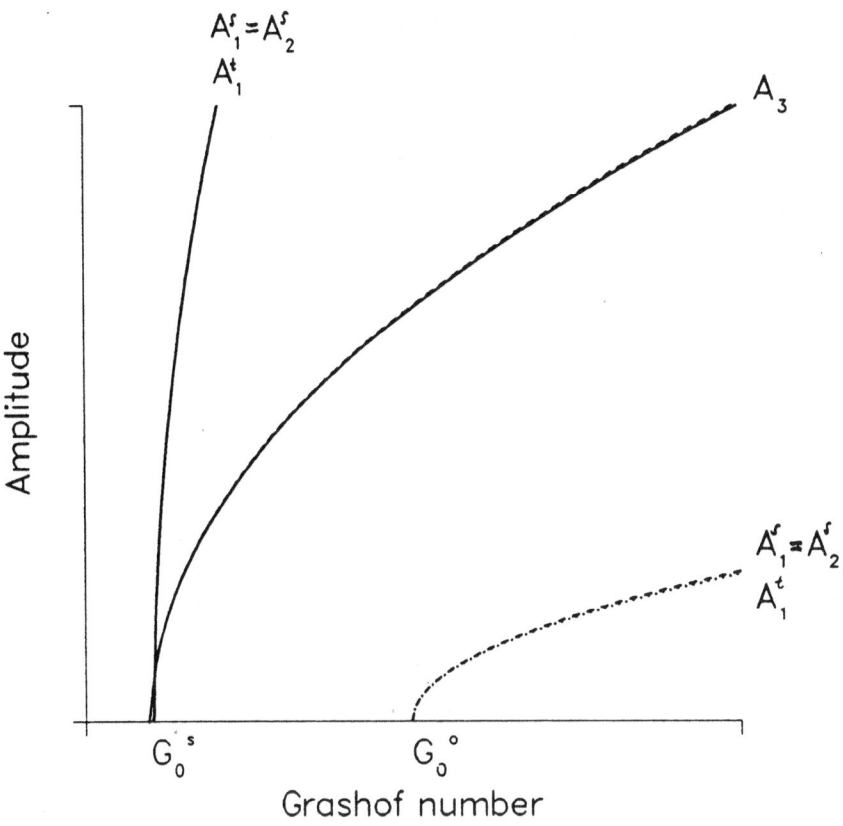

Figure 3   Same as figure 2 in the case when the the critical Grashof
                  number of the oscillatory solution exceeds that of the
                  stationary solution.

International Series of Numerical Mathematics, Vol. 97, © 1991 Birkhäuser Verlag Basel

Description of Chaotic Motion by an Invariant Distribution at the Example of the Driven Duffing Oszillator

A.Kunert, F.Pfeiffer

Technical University Munich, Institut B for Mechanics

Arcisstraße 21, D-8000 München 2

Summary

For chaotic motion the calculation of single trajectories under slightly different conditions results in completely different time series although the motion is bounded on the same strange attractor. Due to the exponential divergence of initially neighboring trajectories the state of the system becomes unpredictable in the long time behavior inspite of deterministic equations. According to this inherent property of chaotic systems the calculation of the system's state for a certain time point contains little information about the variety of possible states. A more adequate description is given by a probability distribution for the state space variables within the area of attraction because it represents all possible trajectories. For autonomous chaotic systems this distribution is invariant. To approximate these probability distributions an additional stochastic excitation of small intensity is taken into account. This not only reduces problems concerning numerical convergence of the solution but is also more realistic because a finite level of noise is present everywhere in reality. Due to the additional noise excitation the calculation of the probability density can be achieved directly by the solution of a Fokker-Planck-Equation, if the excitation is given by white noise. For decreasing noise intensities the result tends to the distribution for the stochastically unperturbed system.

Global Description of Motion

To show that a global description of chaotic motion by invariant distributions is useful, the widely examined Duffing oscillator is investigated. In an autonomous formulation the equations are given by:

$$\begin{aligned}
\dot{x} &= y \\
\dot{y} &= -dy + x - x^3 + a\cos\tau \\
\dot{\tau} &= 1 \ .
\end{aligned} \qquad (1)$$

The solution can be found by integration of a single trajectory for given initial conditions in the three-dimensional state space. The equations include only two parameters: the excitation amplitude a and the damping ratio d. For vanishing excitation there are two stable equilibria at $x = \pm 1$ and an unstable equilibrium at $x = 0$ due to the negative linear stiffness. Parameter combinations (a, d) for periodic, subharmonic, chaotic and coexsisting chaotic/periodic motion can be found [3]. In Fig. 1a a typical Poincaré section for chaotic motion is shown ($a = 0.3, d = 0.185$). An observation of single time series turns out to be disadvantageous in the case of chaotic motion because even smallest variations of the initial conditions result in completely different time histories, although the trajectories belong to the same attractor. This corresponds to a basic property of chaotic systems: the divergence of neighboring trajectories, which results in a positive Ljapunov exponent. So, for an investigation of the basic properties of chaotic motion it is more convenient to calculate a Poincaré section. Even for different trajectories the same structure will be achieved. Nevertheless, additional information is required to fully appreciate the chaotic motion as different areas of attraction have different probabilities. This probabilistic point of view leads to a global description of motion and resolves the problem of unpredictability of single trajectories. It deals with probabilities for different state space elements. The probability density $\varrho(x, y, \tau)$ states with which probability certain areas of the state space are crossed. For autonomous systems this distribution is invariant because of the ergodic property of chaotic motion. It is also independent on the initial conditions if only a single strange attractor exists. Instead of time-consuming integration and averaging a direct calculation of the distribution would be comfortable. In general it is quite difficult to calculate the probability distribution directly because of the fractal structure of the attractor. Numerical problems are caused by the steep gradients of the distribution, especially if iterative solution techniques are used. These problems can be reduced by an additive noise excitation with diffusive effect. So, for a direct, approximative calculation of the probability density of the state space variables additional noise excitation of small intensity is taken into account. This not only reduces problems concerning numerical convergence of the solution but is also more realistic because a finite level of noise is present everywhere in reality. If the random excitation is chosen to be Gaussian white noise, the solution for the stochastic process is given exactly by the associated Fokker-Planck-Equation

$$\frac{\partial \varrho(x, y, \tau)}{\partial \tau} = -\frac{\partial}{\partial x}(y\varrho) + \frac{\partial}{\partial y}((dy - x + x^3 - a\cos \tau)\varrho) + \frac{1}{2}S_o\frac{\partial^2 \varrho}{\partial y^2}. \tag{2}$$

In this partial differential equation the additional noise excitation introduced to the dynamical system is described by the diffussive term proportional to S_o. The noise excitation is easy to handle, no additional numerical effort is necessary to produce random data. For small noise intensities the result tends to the distribution for the stochastically unperturbed system. Therefore, the solution obtained from the Fokker-Planck-Equation is an approximation for the distribution of the dynamical system (1). The influence of the additional random excitation can be seen easily for the stationary solution of eq.(2) with $a = 0$. In this case the solution is given analytically by

$$\varrho(x, y) = N \, \exp(-\frac{d}{S_o}(y^2 - x^2 + x^4/2)). \tag{3}$$

The value of N is determined by the normalization condition

$$\int_{-\infty}^{\infty} \int_{-\infty}^{\infty} \varrho(x,y) dx dy = 1. \tag{4}$$

Extrema of the probability density (3) at $(x = \pm 1, y = 0)$ and the saddle point at $(x = 0, y = 0)$ correspond to stable and unstable equilibria of the Duffing Oscillator. The parameter ratio d/S_o influences the degree of exponential decay of probability for greater displacements and velocities. For decreasing noise intensity S_o and constant dissipation the probability distribution degenerates to dirac distributions at the stable equilibria.

Numerical solution of the Fokker-Planck-Equation

In the following the solution of eq. (2) is found numerically by a finite difference procedure. Instead of the stationary solution for the density in three-dimensional state space (x, y, τ), an instationary solution is calculated in the discretized (x, y) space. For a numerical solution the continuous density $\varrho(x, y, \tau)$ is approximated by discrete densities $d_{j,l}^n$ at gridpoints x_j, y_l and timelevel τ_n. A single timestep is devided into two substeps $\tau_n \rightarrow \tau_{n+1/2}$ and $\tau_{n+1/2} \rightarrow \tau_{n+1}$. In each substep one of the difference operators L_x and L_y respectively is solved implicitly.

$$d_{j,l}^{n+1/2} = d_{j,l}^n + \frac{1}{2}\Delta\tau \left(L_x d_{j,l}^{n+1/2} + L_y d_{j,l}^n \right) \tag{5}$$

$$d_{j,l}^{n+1} = d_{j,l}^{n+1/2} + \frac{1}{2}\Delta\tau \left(L_x d_{j,l}^{n+1/2} + L_y d_{j,l}^{n+1} \right) \tag{6}$$

This socalled ADI (alternating direction implicit) procedure results in a set of tridiagonal sets of equations which can be solved efficiently. For the given example of Fig. 1a the invariant distribution is approximated by the solution of the Fokker-Planck-Equation with $S_o = 0.003$. Fig. 1b shows the contour lines of the probability density at $\tau = 0$. For a better comparibility in Fig. 2a,2b the cross sections of the probability distribution at $y = 0.255$ are shown. Steep gradients merge in the stochastic motion but basic structures of the distribution are preserved. If the parameters are changed to (a=0.3 , d=0.15), a coexisting periodic solution appears indicated by the encircled single point in the Poincaré section of Fig. 3. In the stochastically unperturbed case of eq. (1) it depends on the initial conditions, which one of the attractors is achieved by calculation of a single trajectory. In the presence of noise the intensity of the stochastic excitation is significant. If the noise intensity is high enough an interchange of probability between coexisting attractors could become possible depending also on the sensitivity of the dynamical system. Fig. 4 gives the numerical solution of the Fokker-Planck-Equation for the probability density at $\tau = 0$ with $S_o = 0.005$. It shows that the periodic solution becomes much more probable than the chaotic one. Using a probability density as initial condition lying completely in one basin of attraction, the solution converges in every case to a distribution containing both attractors. So, a probability flux between the two attractors exists due to the white noise excitation, and the solution becomes independent from the initial condition. If the basins of attraction and the attractors themselves are seperated more cleary like in the case of coexisting periodic solutions the relation of the probability for different attractors is preserved.

Conclusion and future work

A description by an invariant distribution has shown to be adequate for chaotic systems because it results in an invariant probability measure instead of unpredictable single trajectories. An approximative solution can be achieved directly by the solution of the Fokker-Planck-Equation. The advantage of the description by a probability density is maintained even in the case of coexisting periodic solutions. It also gives results on the stability of periodic solutions under random excitation. Improvements and new numerical procedures for the solution the Fokker-Planck-Equation are under investigation in order to use smaller noise intensities and to handle steeper gradients. We like to thank the VOLKSWAGENSTIFTUNG which granted this research work under project I/62 888.

Literature

[1] K.Karagiannis: Analyse stoßbehafteter Schwingungssysteme mit Anwendung auf Rasselschwingungen beim Zahnradgetriebe. VDI Fortschrittberichte, Reihe 11 Schwingungstechnik, Nr. 125, 1989.

[2] A. Kunert; F. Pfeiffer: 'Stochastic Model for Rattling in Gear-Boxes'. Nonlinear Dynamics in Engineering Systems, IUTAM Symposium Stuttgart Germany 1989, W. Schiehlen (ed),Proceedings Springer Verlag, Berlin- Heidelberg, 1990

[3] E. Kreuzer: 'Numerische Untersuchung nichtlinearer dynamischer Systeme'. Springer Verlag, Berlin - Heidelberg - New York, 1987

[4] F.Pfeiffer:'Seltsame Attraktoren in Zahnradgetrieben'. Ingenieur-Archiv 58,S.113-125, 1988

[5] F. Pfeiffer; A. Kunert: 'Rattling Models from Deterministic to Stochastic Processes' . to be published in Nonlinear Dynamics.

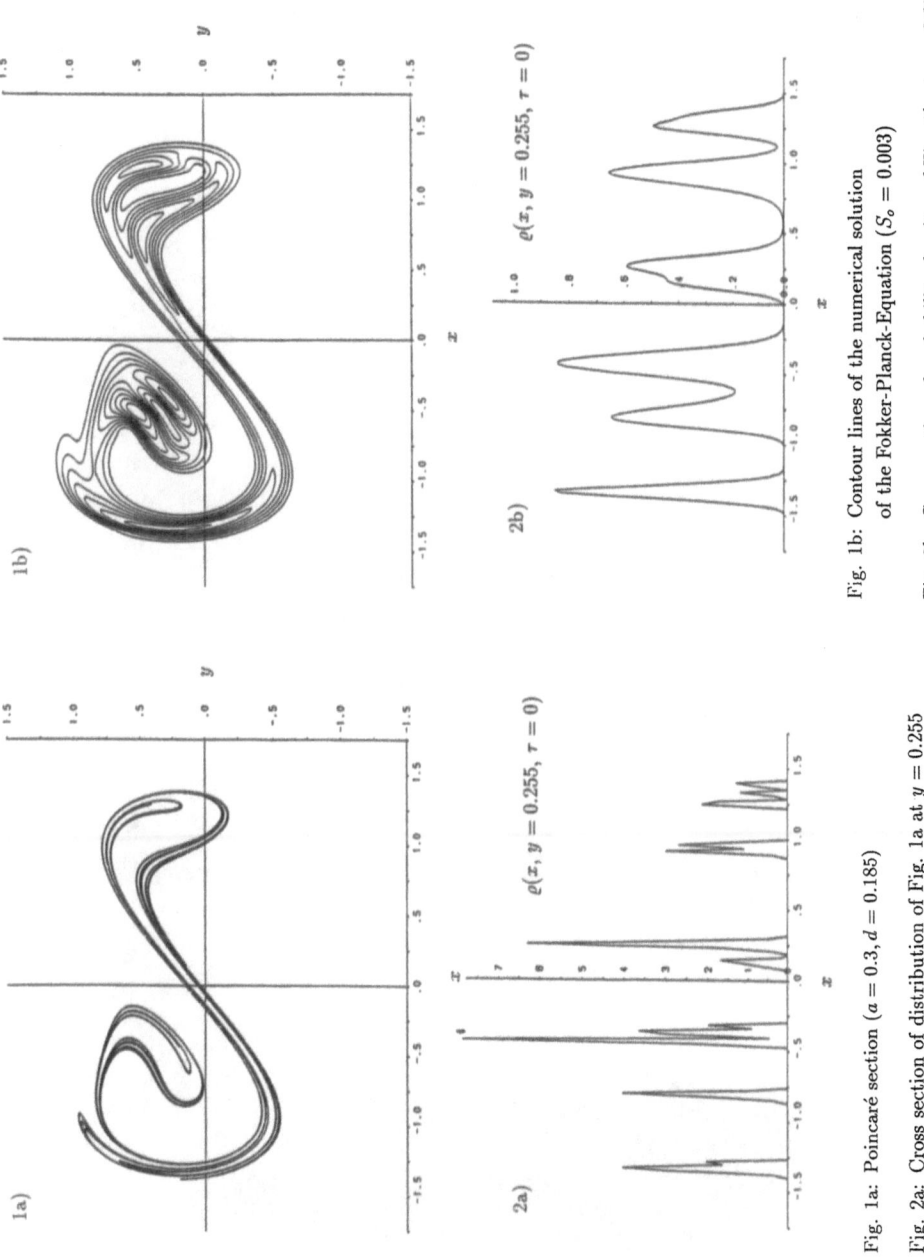

Fig. 1a: Poincaré section ($a = 0.3, d = 0.185$)

Fig. 2a: Cross section of distribution of Fig. 1a at $y = 0.255$

Fig. 1b: Contour lines of the numerical solution
of the Fokker-Planck-Equation ($S_o = 0.003$)

Fig. 2b: Cross section of probability density of Fig. 1b at $y = 0.255$

A. Kunert and F. Pfeiffer

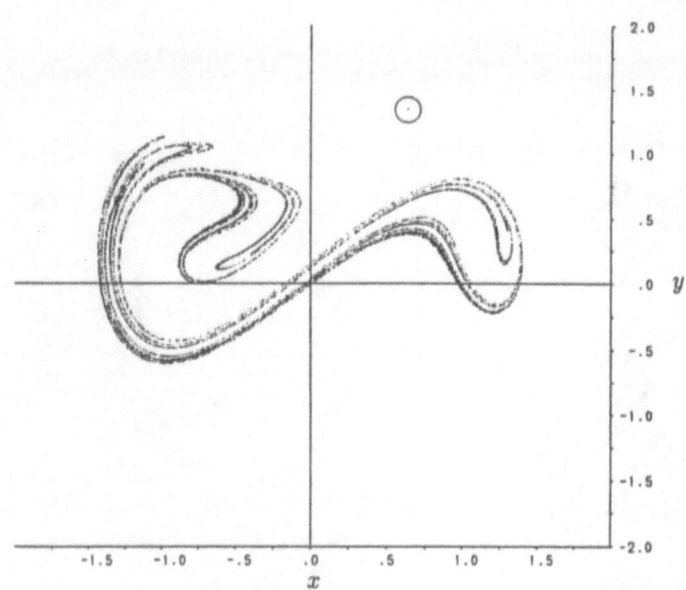

Fig. 3: Poincaré section for coexisting chaotic
and periodic attractors (a = 0.3 , d = 0.15)

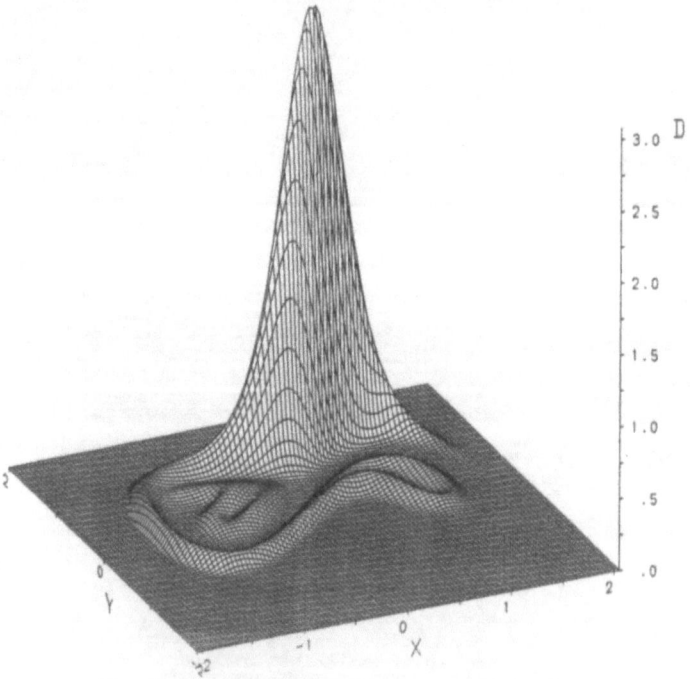

Fig. 4: Probability density at $\tau \bmod 2\pi = 0$;
(a = 0.3 , d= 0.15, S_o = 0.005)

International Series of Numerical Mathematics, Vol. 97, © 1991 Birkhäuser Verlag Basel

AUGMENTED SYSTEMS FOR GENERALIZED TURNING POINTS

Peter Kunkel

Fachbereich Mathematik

Universität Oldenburg

Postfach 2503

D-2900 Oldenburg

Abstract. We consider augmented systems for singular points with a given rank drop which shall satisfy some further conditions. Regularity results are given and applied to a family of problems which include Bogdanov-Takens points.

1 Introduction

In studying parametrized systems of nonlinear equations

$$(1.1) \qquad f(x,\tau) = 0, \quad f\colon \mathbf{R}^n \times \mathbf{R}^p \to \mathbf{R}^n$$

or related dynamical problems, interest has grown in evaluating singular points, i.e. solutions (x^*, τ^*) of (1.1) with

$$(1.2) \qquad \operatorname{rank} L = n - k, \quad L = f_x(x^*, \tau^*), \quad k \geq 1.$$

In addition, they are often required to satisfy some further conditions. Presumably, the best-known singular points of this kind are the so-called Bogdanov-Takens points where we have $k = 1$, i.e. simple turning points, and a second zero eigenvalue is required. See, e.g., [2,3,8] for analytical and numerical details.

A standard technique to compute singular points numerically is by means of augmented systems where (1.1) is embedded into a larger system such that the desired point (x^*, τ^*) becomes part of a regular solution of the new system. In the case of the Bogdanov-Takens points, such a system can be found in [8].

In [1], more complicated singular points fitting into the above general problem type are reported to occur in a Brusselator model but no suitable augmented system was given. It is the aim of the present paper to show the applicability of a family of augmented systems first introduced in [5] for the numerical treatment of this problem type.

2 Generalized Turning Points

There are two basic concepts in analyzing singular points, namely Lyapunov-Schmidt reduction and contact equivalence, see, e.g., [4]. For simplicity of notation, let all functions be C^∞ in the corresponding neighborhoods where they are used.

The Lyapunov-Schmidt reduction is applied to eliminate the regular part of (1.1) in a neighborhood of (x^*, τ^*). To introduce the necessary notation, we sketch this technique in the following. Let

$$(2.1) \qquad U^T L V = \begin{pmatrix} \Sigma & 0 \\ 0 & 0 \end{pmatrix}, \qquad \Sigma \text{ regular}, \quad U, V \text{ orthogonal}$$

be a singular value decomposition of L. We split the orthogonal matrices according to the rank drop k of L into

$$(2.2) \qquad U = (\bar{z}, z), \quad V = (\bar{t}, t).$$

Consequently, the columns of z and t form orthogonal bases of corange L and kernel L, respectively. Defining the orthogonal projectors $P = tt^T$ and $Q = zz^T$, we can split (1.1) into

$$(2.3) \qquad (I - Q)f(v + w, \tau) = 0, \quad Qf(v + w, \tau) = 0$$

with $x = v + w$, $v = Px$. By construction, the first part of (2.3) is uniquely solvable in a neighborhood of (x^*, τ^*) by $w = w^*(v, \tau)$ due to the implicit function theorem and we remain with the so-called reduced function

$$(2.4) \qquad g(\xi, \tau) = z^T f(t\xi + w^*(t\xi, \tau), \tau), \quad g \colon \mathbf{R}^k \times \mathbf{R}^p \to \mathbf{R}^k.$$

By contact equivalence, these reduced functions are divided into classes due to their local behavior at (ξ^*, τ^*) where $\xi^* = t^T x^*$ by the following equivalence relation. Two functions g and h are called contact equivalent at (ξ^*, τ^*) if there is a neighborhood of (ξ^*, τ^*) such that

$$(2.5.a) \qquad h(\xi, \tau) = T(\xi, \tau)g(\varrho(\xi, \tau), \lambda(\tau))$$

with

$$(2.5.b) \qquad \begin{aligned} &T \colon \mathbf{R}^k \times \mathbf{R}^p \to \mathbf{R}^{k,k}, & &T(\xi^*, \tau^*) \text{ regular}, \\ &\varrho \colon \mathbf{R}^k \times \mathbf{R}^p \to \mathbf{R}^k \times \mathbf{R}^p, & &\varrho_\xi(\xi^*, \tau^*) \text{ regular}, \quad \varrho(\xi^*, \tau^*) = (\xi^*, \tau^*), \\ &\lambda \colon \mathbf{R}^p \to \mathbf{R}^p, & &\lambda_\tau(\tau^*) \text{ regular}, \qquad \lambda(\tau^*) = \tau^*. \end{aligned}$$

We then have the important property that two different reduced functions of the same function are contact equivalent. Thus, we have a well-defined equivalence class $[g]$ connected with a singular point (x^*, τ^*) of f.

As first step towards the treatment of our general problem, we define the notion of a generalized turning point.

Definition 1. A singular point (x^*, τ^*) is called generalized turning point if and only if it satisfies

$$(2.6) \qquad\qquad z^T f_x(x^*, \tau^*) t = 0$$

and

$$(2.7) \qquad B = \begin{pmatrix} 0 & z^T f_\tau(x^*, \tau^*) \\ z^T f_{xx}(x^*, \tau^*)(t_1, t) & z^T f_{x\tau}(x^*, \tau^*)(t_1) \\ \vdots & \vdots \\ z^T f_{xx}(x^*, \tau^*)(t_k, t) & z^T f_{x\tau}(x^*, \tau^*)(t_k) \end{pmatrix} \quad \text{regular,} \quad t = (t_1, \ldots, t_k).$$

∎

Note that (2.6) restates the rank drop of k in L and that in the case $k = 1$ these conditions simplify to the known definition of a simple turning point. The following theorem shows that the above definition fits into the concept of contact equivalence.

Theorem 2. The properties (2.6) and (2.7) imply $h_\xi(\xi^*, \tau^*) = 0$ and

$$(2.8) \qquad B_h = \begin{pmatrix} 0 & h_\tau(\xi^*, \tau^*) \\ h_{\xi\xi}(\xi^*, \tau^*)(e_1, \cdot) & h_{\xi\tau}(\xi^*, \tau^*)(e_1) \\ \vdots & \vdots \\ h_{\xi\xi}(\xi^*, \tau^*)(e_k, \cdot) & h_{\xi\tau}(\xi^*, \tau^*)(e_k) \end{pmatrix} \quad \text{regular}$$

where $e_1, \ldots, e_k$ is the canonical basis of $\mathbf{R}^k$ for all h being contact equivalent to a reduced function g of f.

Proof. Since g is a reduced function of f, we have

$$g_\xi(\xi^*, \tau^*)(a_1) = z^T f_x(x^*, \tau^*)(t a_1),$$
$$g_{\xi\xi}(\xi^*, \tau^*)(a_1, a_2) = z^T f_{xx}(x^*, \tau^*)(t a_1, t a_2),$$
$$g_{\xi\tau}(\xi^*, \tau^*)(a_1)(b_1) = z^T f_{x\tau}(x^*, \tau^*)(t a_1)(b_1) +$$
$$+ z^T f_{xx}(x^*, \tau^*)(t a_1, -L^+ f_\tau(x^*, \tau^*)(b_1)).$$

Therefore (2.6) is equivalent to $g_\xi(\xi^*, \tau^*) = 0$. Furthermore, by elementary row and column operations we have rank $B_g = $ rank B.

Now, let h be contact equivalent to g. By the relations

$$h_\xi(\xi^*, \tau^*)(a_1) = T(\xi^*, \tau^*) g_\xi(\xi^*, \tau^*)(\varrho_\xi(\xi^*, \tau^*)(a_1)),$$
$$h_\tau(\xi^*, \tau^*)(b_1) = T(\xi^*, \tau^*) g_\tau(\xi^*, \tau^*)(\lambda_\tau(\tau^*)(b_1)),$$
$$h_{\xi\xi}(\xi^*, \tau^*)(a_1, a_2) = T(\xi^*, \tau^*) g_{\xi\xi}(\xi^*, \tau^*)(\varrho_\xi(\xi^*, \tau^*)(a_1), \varrho_\xi(\xi^*, \tau^*)(a_2)),$$
$$h_{\xi\tau}(\xi^*, \tau^*)(a_1)(b_1) = T(\xi^*, \tau^*) g_{\xi\tau}(\xi^*, \tau^*)(\varrho_\xi(\xi^*, \tau^*)(a_1))(\lambda_\tau(\tau^*)(b_1)) +$$
$$+ T(\xi^*, \tau^*) g_{\xi\xi}(\xi^*, \tau^*)(\varrho_\xi(\xi^*, \tau^*)(a_1), \varrho_\tau(\xi^*, \tau^*)(b_1)) +$$
$$+ T_\xi(\xi^*, \tau^*)(a_1) g_\tau(\xi^*, \tau^*)(\lambda_\tau(\tau^*)(b_1)),$$

we immediately have $h_\xi(\xi^*, \tau^*) = 0$. Moreover, inserting these relations into B_h we can transform B_h into B_g by a sequence of elementary row and column operations thus yielding rank B_h = rank B_g. With this, the proof is finished.

∎

3 Augmented Systems

For the numerical treatment of generalized turning points, we introduce the augmented systems

(3.1)
$$\begin{aligned}
f_x(x,\tau)t_i - z\beta_i = 0, \quad t_i^T t_{i'} = \delta_{ii'}, \quad 1 \le i' \le i \le k, \\
f_x(x,\tau)^T z_j - t\gamma_j = 0, \quad z_j^T z_{j'} = \delta_{jj'}, \quad 1 \le j' \le j \le k, \\
f(x,\tau) = 0, \\
b(x,\tau,t,z) = 0
\end{aligned}$$

which are due to [5,7]. Here b must map into $\mathbf{R}^p$. Because of Definition 1, we want b to consist of the k^2 scalar equations (2.6). Thus, we must require $p = k^2$ which is indeed the codimension of the manifold of (n,n)-matrices with rank deficiency k in $\mathbf{R}^{n,n}$. Note that this b also guarantees that $\beta_i = 0$ and $\gamma_j = 0$ at a solution. The next theorem states regularity for the augmented systems (3.1).

Theorem 3. A generalized turning point (x^*, τ^*) is part of a regular solution of (3.1) with b consisting of (2.6), i.e. it is part of a solution of (3.1) where the corresponding Jacobian has full row rank.

Proof. The only nontrivial part is to show regularity. But this holds due to a more general result in [7] which in this case says that regularity of the solution is equivalent to (2.7).

∎

For every further scalar condition we want to impose on a generalized turning point, we need a further parameter. Let us therefore assume that $b = (b_1, b_2)$ in (3.1) with b_1 consisting of (2.6) and b_2 mapping into $\mathbf{R}^{\bar{p}}$, i.e. we have $\bar{p}$ additional conditions. Then the following theorem holds.

Theorem 4. Let (x^*, τ^*, t^*, z^*) be part of a solution of (3.1) with $p = k^2 + \bar{p}$ and $b = (b_1, b_2)$ as above. In addition, let b_2 satisfy

(3.2)
$$\frac{\partial b_2}{\partial t_i}(x^*, \tau^*, t^*, z^*)(t_{i'}^*) = 0, \quad \frac{\partial b_2}{\partial z_j}(x^*, \tau^*, t^*, z^*)(z_{j'}^*) = 0,$$
$$i, i' = 1, \ldots, k, \quad j, j' = 1, \ldots, k,$$

i.e. b_2 is independent of the special computed bases t^* and z^*. Accordingly to b, let $\tau = (\tau_1, \tau_2)$ with $\tau_2 \in \mathbf{R}^{\bar{p}}$. Assume that in a neighborhood of τ_2^*, the above solution extends to a smooth manifold of generalized turning points given by $(x(\tau_2), \tau_1(\tau_2), t(\tau_2), z(\tau_2))$. Then, this solution is part of a regular solution of (3.1) if and only if

$$(3.3) \qquad \det\Big(\frac{db_2}{d\tau_2}(x(\tau_2), \tau_1(\tau_2), \tau_2, t(\tau_2), z(\tau_2))\big|_{\tau_2=\tau_2^*}\Big) \neq 0,$$

i.e. if and only if b_2 has a simple root on the above defined manifold.

Proof. The above result is an immediate consequence of a generalization of Schur's complement to the elimination with respect to a matrix with full row rank instead of one which is regular. For validity of this generalization, assumption (3.2) is needed. Due to space limitations, we must refer to [7] for details.

∎

4 Applications

For the given general problem, the above result immediately yields a suitable augmented system and states the corresponding regularity condition for its solutions. We want to demonstrate this by applying it to the examples mentioned in the introduction.

A Bogdanov-Takens point is a simple turning point with a second zero eigenvalue. Generically, we then have a nontrivial Jordan block for the zero eigenvalue. This can be obtained by requiring $z^T t = 0$, i.e. we have $k = 1, \bar{p} = 1$, and

$$(4.1) \qquad b_2(x, \tau, t, z) = z^T t.$$

Theorem 4 states that b_2 must have a simple root on the manifold of simple turning points. Note that compared with the augmented system of [8] we have the same regularity condition but the system given there has a totally different structure.

In [1], they found Bogdanov-Takens points with an additional rank drop of the Jacobian L. Thus, we have $k = 2$. To obtain a nontrivial Jordan block as before, we require the determinant of the (2,2)-matrix $z^T t$ to be zero, i.e.

$$(4.2) \qquad b_2(x, \tau, t, z) = \det(z^T t)$$

which in this form can immediately be generalized to an arbitrary rank drop k.

5 Concluding Remarks

The above results can easily be extended to functions $f: \mathbf{R}^{n+l} \times \mathbf{R}^p \to \mathbf{R}^n$ with $l, p \geq 0$. Due to the techniques in [6,7], the augmented systems (3.1) can be solved by a superlinearly convergent modified Gauß-Newton method requiring in leading order one decomposition of $f_x(x, \tau)$ per iteration step which is the same as for the computation of regular solutions of (1.1).

6 References

[1] DE DIER, B., ROOSE, D.: *Computation of branches of codimension two in a three-parameter space*, Techn. Rep. TW99, Katholieke Universiteit Leuven (1987).

[2] FIEDLER, B.: *Global Hopf bifurcation of two-parameter flows*, Arch. Rat. Mech. Anal. 94, 59–81 (1986).

[3] FIEDLER, B., KUNKEL, P.: *A quick multiparameter test for periodic solutions*, in "Bifurcation: Analysis, Algorithms, Applications", Küpper, Seydel, Troger (eds.), Birkhäuser, Basel (1987).

[4] GOLUBITSKY, M., SCHAEFFER, D.: *Singularities and Groups in Bifurcation Theory I*, Springer-Verlag, New York (1984).

[5] KUNKEL, P.: *Quadratically convergent methods for the computation of unfolded singularities*, SIAM J. Numer. Anal. 25, 1392–1408 (1988).

[6] KUNKEL, P.: *Efficient computation of singular points*, IMA J. Numer. Anal. 9, 421–433 (1989).

[7] KUNKEL, P.: *A unified approach to the numerical treatment of singular points*, submitted (1990).

[8] ROOSE, D.: *Numerical computation of origins for Hopf bifurcation points in a two-parameter problem*, in "Bifurcation: Analysis, Algorithms, Applications", Küpper, Seydel, Troger (eds.), Birkhäuser, Basel (1987).

International Series of Numerical Mathematics, Vol. 97, © 1991 Birkhäuser Verlag Basel

NUMERICAL ANALYSIS OF THE ORIENTABILITY OF HOMOCLINIC TRAJECTORIES

Yu.A. Kuznetsov

Research Computing Centre of the USSR Academy of Sciences,

Pushchino, Moscow Region, 142292, USSR

Abstract. Two numerical methods for the analysis of the orientability of homoclinic trajectories (i.e. the orientability of invariant manifolds of the corresponding saddle) are presented. The methods are developed for the case when the saddle has only one positive eigenvalue. As an example, one of the methods is applied to Lorenz equations.

1. Introduction

Many applications require the computation of parameter values corresponding to the appearance of homoclinic trajectories in ODEs and the analysis of the behavior of close trajectories for small parameter perturbations.

Let us consider the following system of ordinary differential equations:

$$\dot{x} = A(\alpha)x + F(x; \alpha), \tag{1}$$

where $x \in \mathbb{R}^n$, $\alpha \in \mathbb{R}^1$, F is a smooth function with $\partial F(0; \alpha)/\partial x = 0$. We suppose that the matrix $A(\alpha)$ has only one eigenvalue $\lambda(\alpha)$ with positive real part for all α, while all other eigenvalues have negative real part. In this case system (1) has a saddle type equilibrium $x = 0$ with two invariant manifolds: a stable $(n-1)$-dimensional manifold $W^s(0)$ composed by all incoming trajectories and an unstable one-dimensional manifold $W^u(0)$ composed by the two outgoing trajectories $\Gamma^1(0)$ and $\Gamma^2(0)$. For some parameter value $\alpha = \alpha^*$ the trajectory $\Gamma^1(0)$ may return to the saddle, thus forming a *homoclinic trajectory* or *loop*. Obviously, this trajectory belongs also to the stable invariant manifold.

We suppose also that for $\alpha = \alpha^*$ the eigenvalue $\mu(\alpha^*)$ with maximal negative real part is unique (and hence real). This eigenvalue is called *leading* [5]. Almost all trajectories on $W^s(0)$ tend to 0 along the eigenvector corresponding to the leading eigenvalue. All other trajectories form the so called *nonleading* manifold $W^{s0}(0)$ of dimension n - 2. The stable manifold $W^s(0)$ is subdivided by the nonleading manifold into two parts. We assume that for $\alpha = \alpha^*$ the loop is generic: this means that $\Gamma^1(0)$ does not belong to the nonleading manifold of the saddle.

The topology of the stable invariant manifold $W^s(0)$ around the loop $\Gamma^1(0)$ plays a key role in the analysis of the homoclinic bifurcation. Let us fix a small tubular neighborhood of the loop. In the generic case, the closure of the part of the manifold $W^s(0)$ containing the loop is a nonsmooth manifold M^0 which is homeomorphic to a simple or a twisted $(n - 1)$-dimensional belt in $\mathbb{R}^n$ [5]. The nonleading manifold $W^{s0}(0)$ belongs to M^0. Therefore, M^0 is either orientable (non-twisted) or nonorientable (twisted). In the first case we say that the loop is *orientable* while in the second case we say it is *nonorientable*.

For $\lambda(\alpha^*) + \mu(\alpha^*) > 0$ the sing of the perturbation of α giving rise to a *saddle limit cycle* $L(\alpha)$ in system (1) is determined by the orientability of the loop [5] (see **Fig.1**). For $\lambda(\alpha^*) + \mu(\alpha^*) = 0$ the system (1) exhibits a codimension two bifurcation: a *resonant side-switching* if the loop is orientable and a *resonant homoclinic doubling* in the opposite case [2].

Numerical methods for testing of the loop orientability have been briefly outlined in [4]. In the present paper we assume that the homoclinic trajectory is known (see, for example,[3],[4]) and concentrate on methods for determining its orientability.

Let us transform system (1) for $\alpha = \alpha^*$ to the following form:

$$\dot{y} = \lambda y + G\ (y,u,v)$$

$$\dot{u} = \mu u + H\ (y,u,v) \tag{2}$$

$$\dot{v} = Bv + Q\ (y,u,v),$$

where $\lambda = \lambda(\alpha^*)$, $\mu = \mu(\alpha^*)$, $y,u \in \mathbb{R}^1$, $v \in \mathbb{R}^{n-2}$ and B is $(n - 2)\times(n - 2)$ matrix. For system (2) the y- and u-axes are the eigenvectors of the Jacobian matrix corresponding to eigenvalues λ and μ respectively. Without loss of generality, assume that the homoclinic trajectory outgoes from the origin in the positive y-direction and returns to it along the positive u-semiaxis.

2. Direct method

Let us consider the following two $(n - 1)$-dimensional hyperplanes: $\Pi^1 = \{(y,u,v)$: $y = \varepsilon\}$ and $\Pi^2 = \{(y,u,v) : u = \delta\}$, where ε and δ are small and positive. Π^1 and Π^2 intersect transversally with the loop $\Gamma^1(0)$ near the saddle. Then introduce new coordinates (u^1,v^1) in Π^1 and (y^2,v^2) in Π^2 in such a way that the intersections of $W^s(0)$ with these planes are $u^1 = 0$ and $y^2 = 0$ respectively, and the homoclinic trajectory goes through the origin of these new coordinate systems. Assume also that the new coordinates have the same positive directions as the original ones. System (2) defines a map $\Lambda: \Pi^1 \to \Pi^2$, or $y^2 = \Phi\ (u^1,v^1)$, $v^2 = \Psi\ (u^1,v^1)$.

The direct method consists of evaluating

$$\Delta = sign\ (\Phi\ (\gamma,0)),$$

for γ sufficiently small and positive by numerical solution of system (2). The homoclinic trajectory is orientable if $\Delta = 1$ and nonorientable if $\Delta = -1$. While determining the sign of $\Phi\ (\gamma,0)$ numerically one may choose ε and δ small and use the original coordinates (u,v) in Π^1 and (y,v) in Π^2 and linear approximations for the invariant manifolds near the saddle.

The numerical solution of (2) near the loop gives rise in many cases to computational difficulties due to strong divergence of the trajectories. The following technique has proved to be useful to evaluate the sign of $\Phi\ (\gamma,0)$. It is based upon ideas used in the analysis of the stability of motion, for example, in the computation of Lyapunov exponents [1]. Let $x^*(t) = (y^*(t),u^*(t),v^*(t))$ be the homoclinic trajectory and $t = 0$ and $t = T$ the times of the intersection of the loop with Π^1 and Π^2. The linearized equations of (2) along the homoclinic trajectory are

$$\dot{\xi} = \lambda\xi + \partial G(x^*(t))/\partial y\ \xi + \partial G(x^*(t))/\partial u\ \eta + \partial G(x^*(t))/\partial v\ \zeta$$
$$\dot{\eta} = \mu\eta + \partial H(x^*(t))/\partial y\ \xi + \partial H(x^*(t))/\partial u\ \eta + \partial H(x^*(t))/\partial v\ \zeta \qquad (4)$$
$$\dot{\zeta} = B\zeta + \partial Q(x^*(t))/\partial y\ \xi + \partial Q(x^*(t))/\partial u\ \eta + \partial Q(x^*(t))/\partial v\ \zeta,$$

where $\xi,\ \eta \in \mathbb{R}^1$ and $\zeta \in \mathbb{R}^{n-2}$. Rewrite system (4) as

$$\dot{z} = C(t)z, \qquad (5)$$

where $z = (\xi,\ \eta,\ \zeta)$. If $z(0)$ is a vector with unit norm and $z(t)$ is a corresponding solution of (5), then $w(t) = z(t)/\|z(t)\|$ satisfies the following nonlinear system:

$$\dot{w} = C(t)w - <C(t)w,\ w>w, \qquad (6)$$

where $w = (\xi^1,\ \eta^1,\ \zeta^1)$ and $<,>$ denote the standard scalar product in $\mathbb{R}^n$. Thus, we have avoided the exponential growth of the solutions of (5).

By solving system (6) with the initial condition $(0,\ 1,\ 0) \in \Pi^1$ from $t = 0$ to $t = T$ we get $\Delta = sign\ (\xi^1(T))$.

3. Tangent plane method

At each point $x^*(t)$ of the homoclinic trajectory the tangent plane to the stable invariant manifold $W^s(0)$ is well defined. If this tangent plane is denoted by $\Pi^T(t)$ then $\Pi^T(+\infty) = \{ (y,u,v) : y = 0 \}$ with the normal $n^+ = (1, 0, 0)$. Starting from $\Pi^T(+\infty)$ we can determine the tangent plane $\Pi^T(t)$ *backward in time* and get $\Pi^T(-\infty)$. The corresponding normal vector will be $n^- = (0, \Delta, 0)$, where $\Delta = \pm 1$. Thus, orientability can be determined by computing such a vector..

In practice, we can start from a point $x^*(T^0)$ with T^0 sufficiently large (near the saddle) and take $\Pi^T(+\infty)$ as initial plane. We have to compute only n - 2 independent vectors lying on $\Pi^T(t)$, because one vector on the plane is known: $z(t) = \dot{x}^*(t)$. To compute tangent vectors avoiding numerical difficulties one can solve system (6) introduced in the previous section. An additional normalization may be required to get an orthogonal basis of dimension (n - 1) on $\Pi^T(t)$. Having this basis for all t we can continue the normal vector backward in time and stop the computations for a large negative time thus approximating n^-.

4. Application to Lorenz system

Let us consider the well known Lorenz system:

$$\dot{X} = -\sigma X + \sigma Y$$

$$\dot{Y} = rX - Y - XZ \qquad\qquad (7)$$

$$\dot{Z} = -bZ + XY.$$

For $\sigma = 10.0$, $b = 8/3$ and $r^* = 13.926...$ system (9) has a saddle point at the origin with a two-dimensional stable and a one-dimensional unstable manifold. These manifolds intersect one each other along two symmetric homoclinic trajectories. The Z-axis is a leading direction and condition $\lambda + \mu > 0$ is satisfied.

We have applied the tangent plane method to determine the orientability of the homoclinic trajectory. Since this system is three-dimensional, we have to compute only one tangent vector starting, for example, with a normal vector tangent to the nonleading manifold at the origin. The results are presented in **Fig.2** where a family of vectors tangent to $W^s(0)$ along the homoclinic trajectory is shown. The final vector is essentially the same as the initial one, which means that the loop is *orientable*.

5. Acknowledgement

The author would like to thank Prof. S. Rinaldi from the Polytechnical University of Milan for careful reading of the manuscript and useful discussions.

6. References

[1] G. Benettin, L. Galgani, J.-M. Strelcyn. *Kolmogorov entrophy and numerical experiments.* Phys. Review A, 1976, v. 14, 6, pp. 2338-2345.

[2] S.-N. Chow, B. Deng and B.Fiedler. *Homoclinic bifurcation at resonant eigenvalues.* Konrad-Zuse-Zentrum für Infirmationstechnik Berlin. Preprint SC 88-10, 1988, 75 p.

[3] B.D. Hassard. *Computation of invariant manifolds.* In: Ph.Holmes (ed.), "New Approaches to Nonlinear Problems in Dynamics. SIAM, Philadelphia, 1980, p. 27-42.

[4] Yu.A. Kuznetsov. *Computation of invariant manifold bifurcations.* In: D.Roose, Bart de Dier and A.Spence (eds.), "Continuation and Bifurcations: Numerical Techniques and Applications". Kluwer Academic Publishers, Netherlands, 1990, p. 183-195.

[5] L.P. Shil'nikov. *On the generation of a periodic motion from trajectories doubly asymptotic to an equilibrium state of saddle type.* Math. USSR Sbornik, 1968, 6, pp. 427-437.

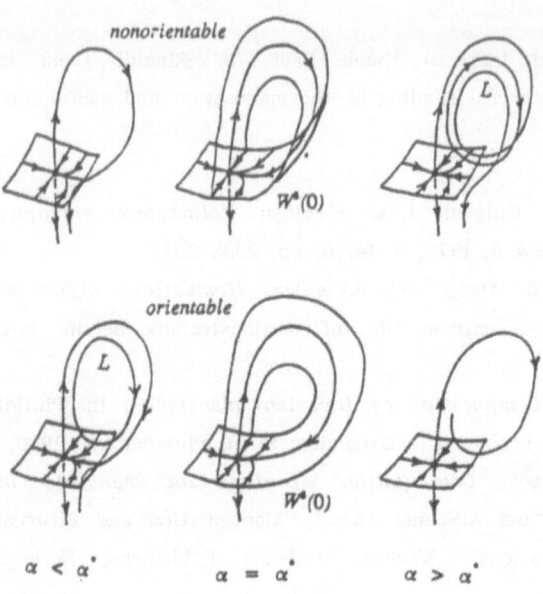

nonorientable

L

$W^s(0)$

orientable

L

$W^s(0)$

$\alpha < \alpha^*$ $\alpha = \alpha^*$ $\alpha > \alpha^*$

Fig.1. Bifurcations of nonorientable and orientable loops in $\mathbb{R}^3$.

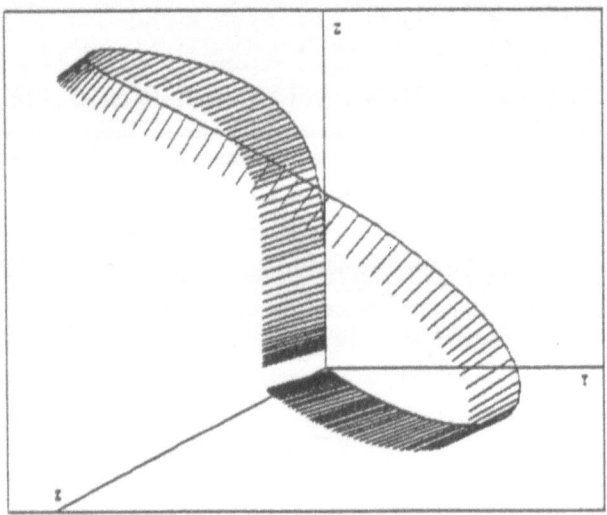

Fig.2. An orientable loop in the Lorenz system.

International Series of Numerical Mathematics, Vol. 97, © 1991 Birkhäuser Verlag Basel

QUALITATIVE AND QUANTITATIVE BEHAVIOUR OF
NONLINEARLY ELASTIC RINGS UNDER HYDROSTATIC PRESSURE

Franz Karl Labisch

1. Introduction

This paper is concerned with the equilibrium configurations of a circular nonlinearly elastic ring under hydrostatic pressure. The pressure acts orthogonal to the deformed middle line, which is assumed to deform by flexure and extension and to describe the shape of the deformed ring. The qualitative behaviour of such rings has been studied by ANTMAN [1], ANTMAN & DUNN [2] and LIBAI & SIMMONDS [3]. In the present paper a qualitative and quantitative analysis of the solution set of the underlying boundary value problem (BVP) is carried out for the case of a particular constitutive law. The applied mathematical model is geometrically exact.

2. The Boundary Value Problem

The BVP for the problem under study is published in [1] and [2]. It consists of the equilibrium equations

$$\left[\frac{\hat{M}'\ (\xi,\ \mu)}{1\ +\ \xi} \right]' \ -\ (1\ +\ \mu)\ \hat{N}\ -\ p(1\ +\ \xi)\ =\ 0,\ \ (1\ +\ \xi)\ \hat{N}'\ +\ (1\ +\ \mu)\ \hat{M}'\ =\ 0\ , \quad (2.1)$$

the geometric relations

$$\Theta'(s)\ =\ 1\ +\ \mu(s),\ \ X'(s)\ =\ (1\ +\ \xi(s))\ \cos\ \Theta(s)$$

$$Y'(s)\ =\ (1\ +\ \xi(s))\ \sin\ \Theta(s)\ , \quad (2.2)$$

the periodicity conditions

$$\mu(s\ +\ 2\pi)\ =\ \mu(s),\ \ \int_{0}^{2\pi}\mu(s)ds\ =\ 0,\ \ \ \ \xi(s\ +\ 2\pi)\ =\ \xi(s) \quad (2.3)$$

and the boundary conditions

$$X(0)\ =\ 0,\ \ Y(0)\ =\ -(1\ +\ a)\ . \quad (2.4)$$

The unit circle and the arc length on this circle are choosen as the reference configuration and the material variable, respectively. Derivatives with respect to s are denoted by primes. In the deformed configuration denote $\Theta(s)$, $S(s)$, $M(s)$, $N(s)$, $\mu(s) = \Theta'(s) - 1$ and $\xi(s) = S'(s) - 1$ the tangent angle, the arc length, the couple resultant, the resultant force component in the direction of the tangent line, the flexural strain and the extensional strain, respectively. p is the pressure intensity per unit deformed length. For a homogeneous material depend N and M on s only through $\xi(s)$ and $\mu(s)$, so that $N(s) = \hat{N}(\xi(s), \mu(s))$, $M(s) = \hat{M}(\xi(s), \mu(s))$. The morphology of the solution set depends on the functions $\hat{N}$ and $\hat{M}$. The admissible constitutive laws must be physically reasonable. Therefore a number of conditions imposed on $\hat{N}$ and $\hat{M}$ is cited in [2]. For the particular constitutive law

$$\hat{M}(\xi, \mu) = \beta\mu, \quad \hat{N}(\xi, \mu) = \frac{m \; \xi^3}{1 + \xi} \; ; \; m > 0, \; \beta > 0 \tag{2.5}$$

considered here, the equilibrium equations (2.1), take the form

$$-\mu''(1+\xi)+\xi'\mu' +m_b(1+\mu)(1+\xi)\xi^3+p_b(1+\xi)^3 = 0 \tag{2.6}$$

$$m_b(3 +2\xi) \; \xi^2\xi' + (1 + \xi) \; (1 + \mu) \; \mu' = 0 \; , \tag{2.7}$$

where $m_b = \dfrac{m}{\beta}$, $p_b = \dfrac{m}{\beta}$.

3. Linearized Morphology Analysis

For each p > 0 there exists a unique deformed circular equilibrium configuration. The set of deformed circular configurations corresponding to p $\in$ [0, ∞) is called trivial solution path. The trivial solution path and all bifurcation points on this path can be found by a linearized analysis. The admissible strains

$$\xi = a + \sum_{k=2}^{\infty} (a_k \cos ks + b_k \sin ks), \quad \mu = \sum_{k=2}^{\infty} (c_k \cos ks + d_k \sin ks) \tag{3.1}$$

are inserted into (2.6), (2.7). Only linear terms in a_k, b_k, c_k and d_k are retained. A coefficient comparison leads for k = 0 to the equation for the trivial solution path

$$p_b R^2 + m_b a^3 = 0, \; R = 1 + a \tag{3.2}$$

and for k = 2, 3, ... to the homogeneous linear algebraic system

$$m_b a^2(3 + 2a) \begin{bmatrix} a_k \\ b_k \end{bmatrix} + R \begin{bmatrix} c_k \\ d_k \end{bmatrix} = 0 \tag{3.3}$$

$$(m_b 3a^2 + 2R \; p_b) \begin{bmatrix} a_k \\ b_k \end{bmatrix} + (k^2 + m_b a^3) \begin{bmatrix} c_k \\ d_k \end{bmatrix} = 0 \; . \tag{3.4}$$

This system has a nonzero solution only if

(3.5)

$$\det \begin{bmatrix} 3 + 2a & 1 \\ 3 + a & k^2 + m_b a^3 \end{bmatrix} = m_b(2a^4 + 3a^3) + (2k^2 - 1)a + 3(k^2 - 1) = 0 .$$

For m_b = const exists a sequence of bifurcation points, see Table 1. The locus of bifurcation points appearing for the control pair (p_b, m_b) is shown in Fig. 3. Bifurcation points do not appear for $p_b = 0$ and $k = 1$.

4. Nonlinear Morphology Analysis

The construction of the solution branches emanating from the bifurcation points requires a nonlinear analysis. Due to the nodal properties preserved at the solution branches (see e. g. [1], [2]) and the foregoing linear analysis the admissible strains are for a given k restricted to the form

$$\xi = a + a_k \cos ks + b_k \sin ks, \quad \mu = c_k \cos ks + d_k \sin ks . \quad (4.1)$$

Insertion of (4.1) into (2.6), (2.7) leads when some trigonometric identities are applied via a coefficient comparison to a nonlinear algebraic system of the form

$$F_1 = m_b Ra^3 + p_b R^3 + r^2 f_1(a, r^2, \lambda; m_b, p_b, k) = 0$$

(4.2)

$$F_2 = r f_2(a, r^2, \lambda; m_b) = 0, \quad F_3 = r f_3(a, r^2, \lambda; m_b, p_b, k) = 0 ,$$

where $c_k = \lambda a_k, \quad d_k = \lambda b_k, \quad r^2 = a_k^2 + b_k^2 .$

The solution is determined up to $r^2 = a_k^2 + b_k^2$. Solution branches corresponding to $k = 2$ and $k = 3$ are for some range of p_b shown in Fig. 1. The set of singular solution points consists of the subset of bifurcation points given by

$$r = 0, \quad f_2 = 0, \quad f_3 = 0 \quad (4.3)$$

and the subset of singular solution points given by the extended system

$$F_1 = 0, \quad f_2 = 0, \quad f_3 = 0, \quad \frac{\partial (F_1, f_2, f_3)}{\partial (a, r^2, \lambda)} = 0 . \quad (4.4)$$

The set of limit points appearing on the solution branch corresponding to $k = 2$ is depicted in Fig. 2. The singular solution points at which the solution branches end are not shown here. The asymptotic approximation of such a point can be deduced for $k = 2$ from Fig. 1c. Deformed equilibrium configurations of

246 F.K. Labisch

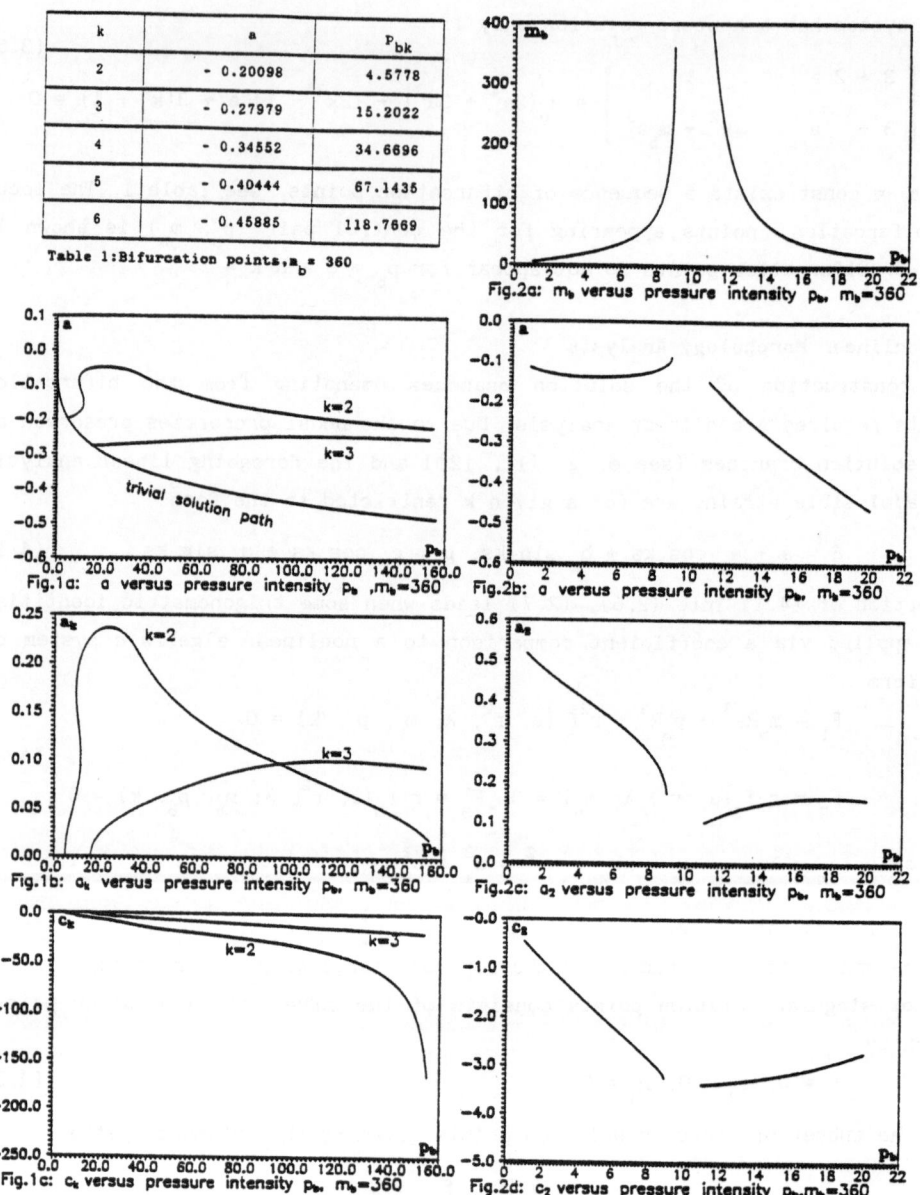

k	a	P_{bk}
2	- 0.20098	4.5778
3	- 0.27979	15.2022
4	- 0.34552	34.6696
5	- 0.40444	67.1435
6	- 0.45885	118.7669

Table 1:Bifurcation points, m_b = 360

Fig.2a: m_b versus pressure intensity p_b, m_b=360

Fig.1a: a versus pressure intensity p_b ,m_b=360

Fig.2b: a versus pressure intensity p_b, m_b=360

Fig.1b: a_k versus pressure intensity p_b, m_b=360

Fig.2c: a_2 versus pressure intensity p_b, m_b=360

Fig.1c: c_k versus pressure intensity p_b, m_b=360

Fig.2d: c_2 versus pressure intensity p_b,m_b=360

the ring are for k = 2 and k = 3 shown in Fig. 4. The 2 last configurations for k = 2 and k = 3 are enlarged. It is assumed that parts of the ring can cross. A discussion of the influence of $1 + \mu = 0$ can therefore be avoided. If a_k and b_k are changed so that $a_k^2 + b_k^2 = r^2$ = const, then the deformed ring seems to be rotated in the picture plane.

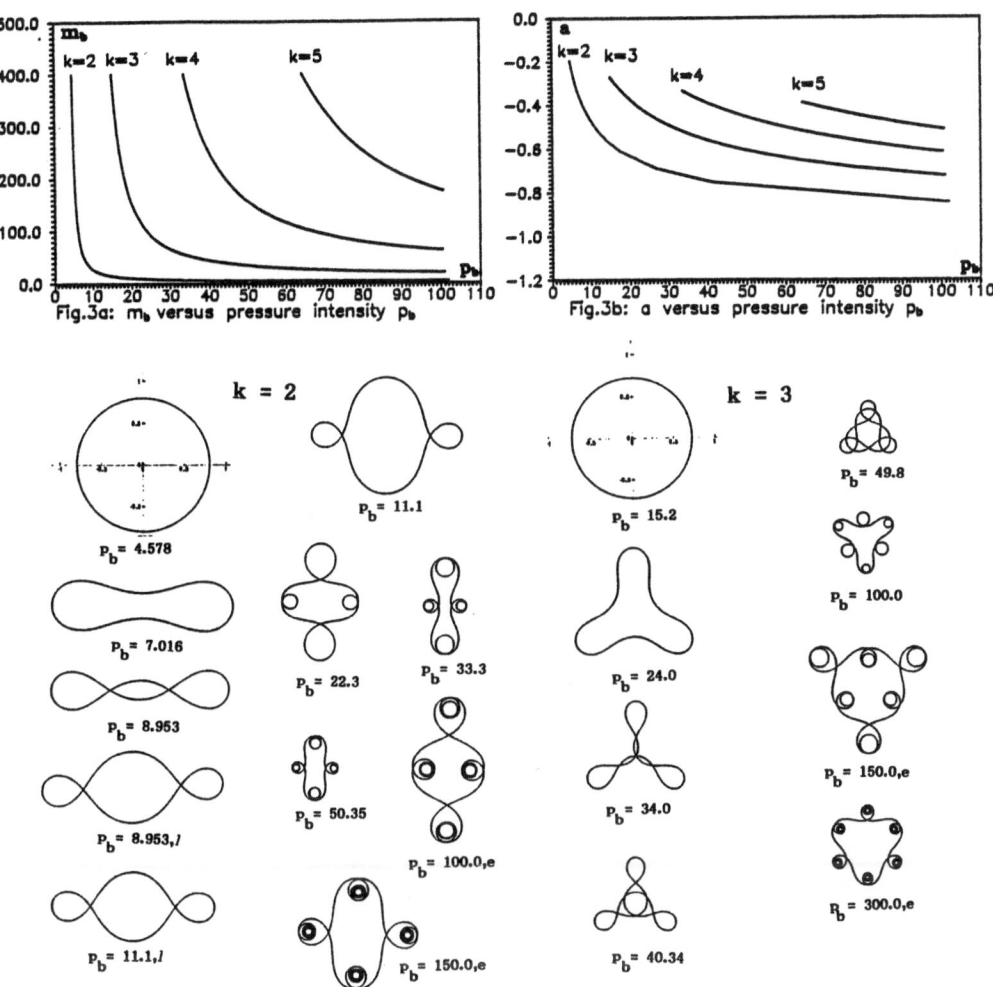

Fig.3a: m_b versus pressure intensity p_b

Fig.3b: a versus pressure intensity p_b

Fig. 4: Deformed ring configurations; l - limit point, e - enlarged.

5. References

[1] Antman, S. S. (1970) The shape of buckled nonlinearly elastic rings, Z. Angew. Math. Phys. **21**, 422 - 438

[2] Antman, S. S. and Dunn, J. E. (1980) Qualitative behaviour of buckled nonlinearly elastic arches, Journal of Elasticity **10**, 225 - 239

[3] Libai, A. and Simmonds, J. G. (1983) Highly nonlinearly cylindrical deformations of rings and shells, J. Non-Linear Mechanics **18**, 181 -197

F. K. Labisch, Lehrstuhl für Allgemeine Mechanik, Ruhr-Universität Bochum

International Series of Numerical Mathematics, Vol. 97, © 1991 Birkhäuser Verlag Basel

COMPUTATION OF BASINS OF ATTRACTION
FOR THREE COEXISTING ATTRACTORS

C.-H. LAMARQUE and J.-M. MALASOMA

**Ecole Nationale des Travaux Publics de l'Etat
1 rue Maurice Audin 69518 Cedex-FRANCE.**

Introduction

We consider a one degree-of-freedom mechanical system with cubic nonlinearity and parametric excitation:

$$\ddot{x} + a\,\dot{x} - 0.5\,(\,1. - 2\,f\cos(\omega t) - x^2\,)\,x = f\cos(\omega t) \qquad (1)$$

with $a = 0.2$, $f = 0.9$. We show that this system modelling a shallow-arch exhibits a new behaviour in the area [1.8;2.2] of the parameter of bifurcation ω : two strange attractors can coexists with a limit cycle.

I - Study of the global behaviour with Interpolated Cell Mapping Method

Finding all the coexisting attractors and limit cycles and drawing their basins of attraction are extremely time-consuming tasks. Therefore, the analysis of the global behaviour of a nonlinear system is indeed a major challenge. To address this difficulty, we used an approach introduced by Tongue [8], [9], [10]. This technique, called Interpolated Cell Mapping, replaces the Poincaré map for the system by a locally bilinear approximation.

More precisely, trajectories are computed for a rectangular grid of initial conditions in a bounded region of the phase space. The time duration of the trajectories will be the forcing period, and the terminal position of each trajectory is recorded. After that, extending each trajectory only involves a repeated iteration of the interpolation scheme. Interpolated Cell Mapping technique, clearly reduces the computational cost, and was successfully used to investigate the fractal dimension of basin boundaries [8].

The basins of attraction were found to be in the region $-5. \leq x \leq 5.$, $-10. \leq \dot{x} \leq 10.$, by generating a 201x401 array of initial conditions and integrating equation (1) for one forcing period, for each of these starting points. This equation was solved by using a fourth order Runge-Kutta algorithm. Then, Interpolated Cell Mapping was used to produce the basins of attraction to a resolution of 800x800.

II Numerical results.

II - 1 The first cascade of period doubling.

Szemplinska-Stupnicka et al. [7] investigated the response of this system by integrating equation (1) with initial condition $x(0) = 1.02$ and $\dot{x}(0) = 0$. When the parametric frequency ω is decreased, taking the values 2.15, 2.09, 2.06 and 2.05, they found limit cycles of period 2, 4, 8 and a chaotic attractor. The limit cycles encircling the right fixed point, and the system experienced a cascade of period-doubling bifurcations leading to chaos.

Using the Interpolated Cell Mapping method, we were able to find all the others coexisting attractors in the region $-5. \leq x \leq 5.$ and $-10. \leq \dot{x} \leq 10.$ .There are two limit cycles, one of same period as the period of the parametric excitation, lying between the left and the central fixed points, and a large period-two limit cycle encircling all three fixed points.The first one was found by Szemplinska-Stupnicka et al. [7] at $\omega = 2.00$ and the second one at $\omega = 1.95$. But these two limit cycles are also present in the studied region $2.05 \leq \omega \leq 2.15$, and with the limit cycle that goes through the cascade of period-doubling there are no other attractors.

The resulting plots of the basins of attraction of the three attractors is shown in figure 1 at the value $\omega = 2.15$ and in figure 2 at the value $\omega = 2.05$. In these figures, the basin of the limit cycle of period one is colored in white. Those of the limit cycle of period 2, that did not bifurcate, is in red. The basin of the third limit cycle that experience the cascade of bifurcations is in blue in figure 1 and the basin of the chaotic attractor is also in blue in figure 2. When ω decreases from the value 2.039 to the value 1.96 the chaotic attractor disappears and only the two limit cycles remain. Figure 3 shows the two basins of attraction, with the same color conventions.

II-2 A second cascade of period-doubling leading to chaos.

Szemplinska-Stupnicka et al.[7] found a chaotic attractor at the value $\omega = 1.90$ and at $\omega = 1.70$. A global analysis shows that besides this attractor coexist the large period two limit cycle. A new stable attractor appears at $\omega = 1.913$ that encircles the three equilibrium points. This is a period four limit cycle. This periodic solution goes through a cascade of period-doubling bifurcations, as ω is decreased. We have studied this sequence of bifurcations in some detail, observing subharmonic solutions up to period 128.

Figure 4 shows some of these limit cycles in the period doubling sequence. Figure 4a shows the period 4 orbit at $\omega = 1.91$, and figure 4b shows the period 8 orbit at $\omega = 1.90$. These period doublings occur very rapidly with decrease of ω as shown by the period 16 orbit in figure 4c at $\omega = 1.898$, period 32 orbit in figure 4d at $\omega = 1.8978$; figure 4e exhibits period 64 orbit at $\omega = 1.8977$; after all chaos obtained at $\omega = 1.8945$ is drawn in figure 4f.

Let ω_n be the threshold for a bifurcation from a solution of period 2^n to a solution of period 2^{n+1}. Since the values,where the bifurcations take place, become ever closer to each other, it becomes also ever more tedious to compute them. Table 1 shows the first fith found values . From these values, the Feigenbaum numbers [2], [3] i.e. the ratios of the ω-interval of period 2^{n+1} orbit, to the ω-interval of period 2^{n+2} orbit :

$$\delta_n = (\omega_n - \omega_{n+1}) / (\omega_{n+1} - \omega_{n+2})$$

may be computed. We find $\delta_2 = 4.39 \pm 0.05$, $\delta_3 = 4.66 \pm 0.09$, $\delta_4 = 4.6 \pm 0.3$ in fair agreement with the universal Feigenbaum number $\delta = 4.6692...$ It therefore appears possible that the sequence of δ_n converges to δ, and that this cascade fit quite precisely the Feigenbaum's scenario of route to chaos.

Upon examination of the sequence of numbers ω_n in table 1 one is struck by the apparent convergence. Assuming geometric convergence, we may write $\omega_\infty - \omega_n = a / \delta^n$ where a is a constant and ω_∞ the limit of the sequence. Using the values of ω_n from table 1, one can predict that the bifurcation values accumulates at the value $\omega_\infty = 1.89766$, where there is a non-periodic orbit.

Table 1. Bifurcation values of the cascade of period doublings

n	ω_n
2	1.902435 ± 0.000005
3	1.898735 ± 0.000005
4	1.897892 ± 0.000001
5	1.897711 ± 0.000001
6	1.897672 ± 0.000001

II-3 Two coexisting chaotic attractors.

Indeed, for a parametric frequency value less than ω_∞, we found a new chaotic attractor coexisting with the first one.Figure 5a shows a Poincaré section of this chaotic attractor at the value ω =1.8945. This chaotic attractor shows the caracteristic band structure that occurs on the chaotic side of period-doubling sequences. Four pieces can be seen in this figure, and figures 5b to 5c are blowups of each piece of this attractor. The blowups shown in figure 5f was computed from 150000 points by plotting only those points that lie within the box indicated in figure 6e.This chaotic attractor has a simple self similar microscopic structure. At each scale of resolution, many sheets are discernible to the eye.

The cascade of period-doubling bifurcations presented in the previous section may be considered to be sufficient evidence of chaotic region for ω values greater than ω_∞. However, even more definitive discriminators of the nature of the attractor, are Lyapunov exponents [1], [5]. One feature of chaos is sensitive dependence on initial conditions. Strange attractor have the property that nearby trajectories have exponential separation locally while confined to a compact subset of the phase space globally. This means that, what distinguishes strange attractors from non-chaotic attractors is the existence of at least one positive Lyapunov exponent. We found the values $\lambda_1 = 0.06$ and $\lambda_2 = -0.35$ that prove the chaotic nature of this attractor. In figure 6 Lyapunov exponents were plotted as a function of time. Finaly figure 7 shows the three basins of attraction at the value $\omega = 1.8945$. The basin corresponding to the limit cycle is in grey, those of the first known chaotic attractor is in black and the basin of the last one chaotic attractor is colored in white.

III Conclusions

Using the Interpolated Cell Mapping technique, we found all the coexisting attractors and drawn their basin of attraction of the system (1), in the region defined by $-5 \leq x \leq 5$ and $-10 \leq \dot{x} \leq 10$. Secondly, we report numerical results of a cascade of period-doubling bifurcations leading to chaos. We show that these results are consistent with the universal features of the noninvertible one-dimensional maps.

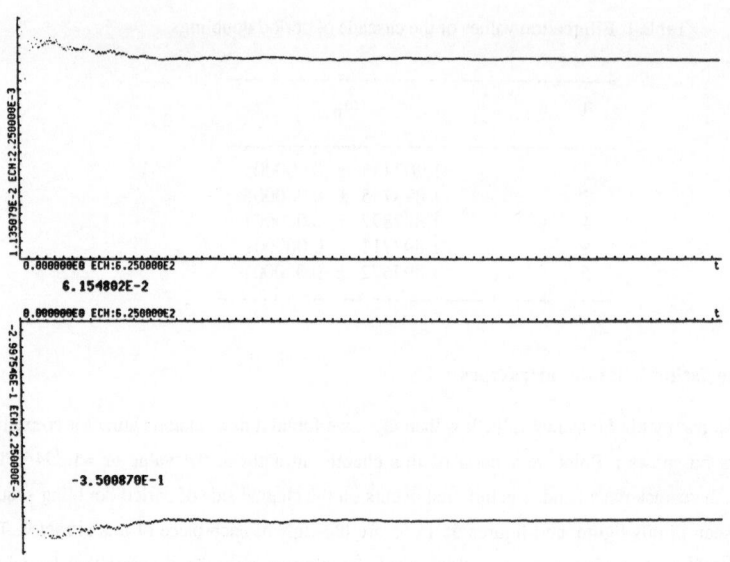

Figure 6. Lyapunov exponents versus time

References

[1] G. Benettin, L. Galgani, A. Giorgilli, J.M. Strelcyn, Lyapunov characteristic exponents for smooth dynamical systems; a method for computing all of them.Part I : theory (1980), Meccanica, 15, 9-20. Part II : numerical applications (1980), Meccanica, 15, 21-30.

[2] M. J. Feigenbaum, Quantitative universality for a class of nonlinear transformations (1978), J. Stat. Phys. 19, 25-52.

[3] M. J. Feigenbaum, The universal metric properties of nonlinear transformations (1979), J. Stat. Phys. 21, 669-706.

[4] V.I. Oseledec, A multiplicative ergodic theorem. Lyapunov characteristic numbers for dynamical systems (1968), Trans. Moscow Math. Soc. , 19, 197-231.

[5] C. Froeschlé, The Lyapunov exponents and applications (1984), Journal de mécanique théorique et appliquée, Numéro spécial, 101-132.

[6] R.H. Plaut and J.C. Hsieh, Oscillations and instability of a shallow arch under two-frequency excitation (1985), J. Sound and Vibration, 102, 189-201.

[7] W. Szemplinska-Stupnicka, R.H. Plaut, J.-C. Hsieh, Period doubling and chaos in unsymmetric structures under parametric excitation, J. Appl. Mech. (1989), 56, 947-952.

[8] B. H. Tongue, On obtaining global nonlinear system characteristic through Interpolated Cell Mapping, Physica 28D (1987),401-408.

[9] B. H. Tongue and K. Gu, Interpolated Cell Mapping of nonlinear systems. J. Appl. Mech. (1988), 55, 461-466.

[10] B. H. Tongue and K. Gu, A theoretical basis for Interpolated Cell Mapping, SIAM J. Appl. Math. (1988), 48, 1206-1214.

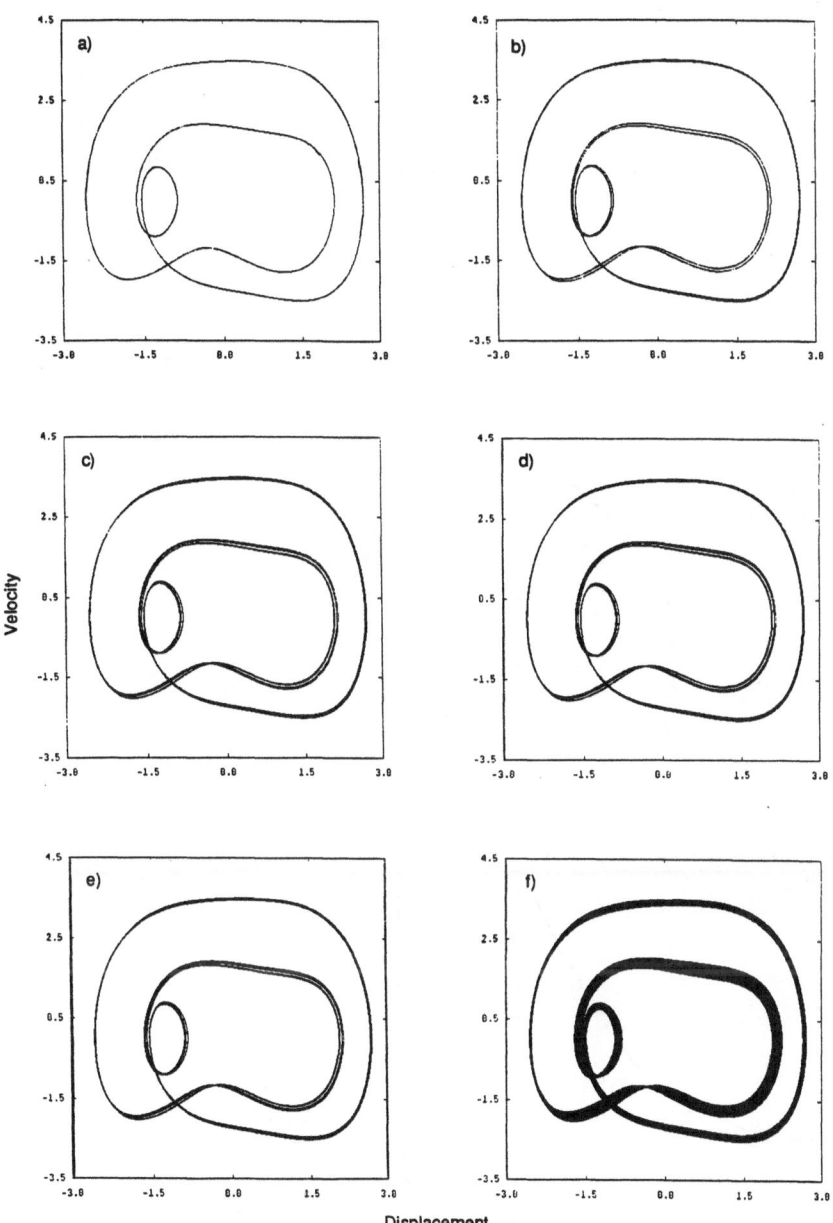

Figure 4. Period doubling sequence

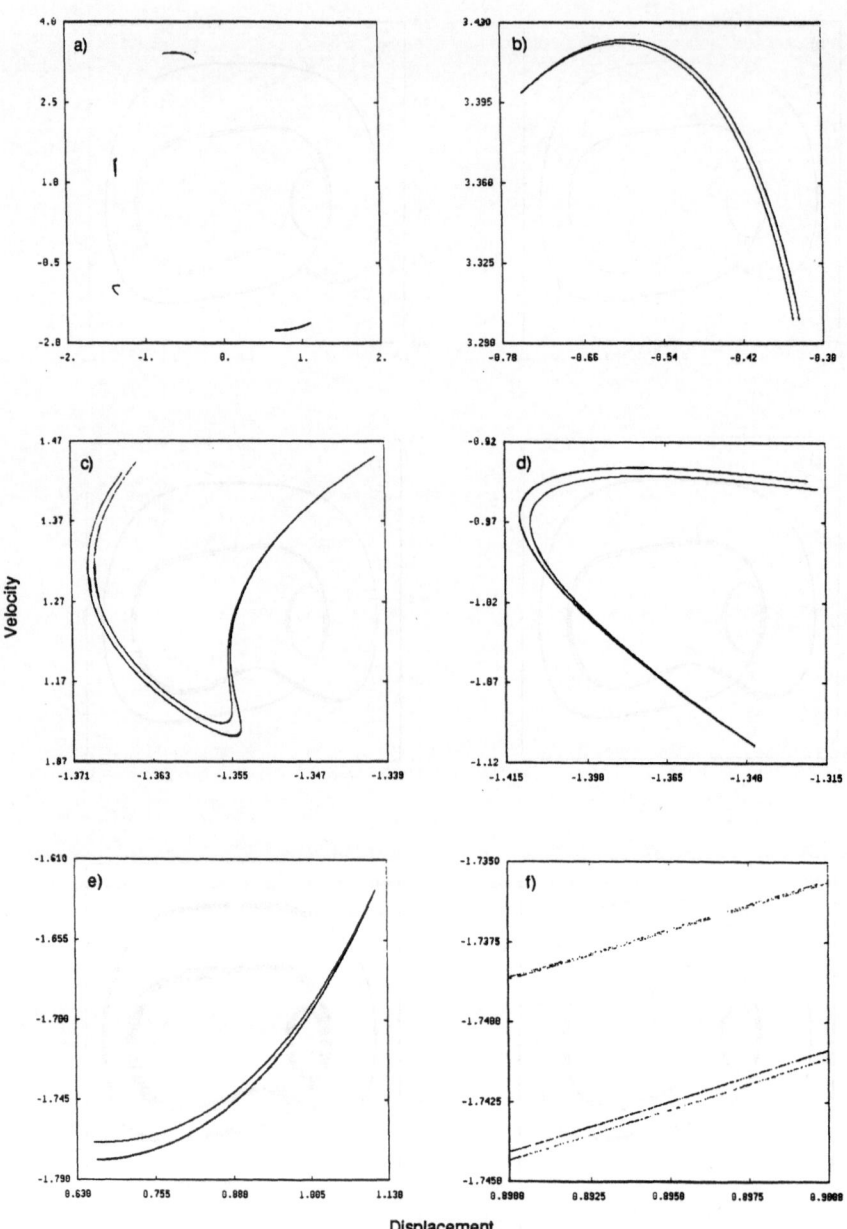

Figure 5. Poincaré sections of the chaotic attractor.

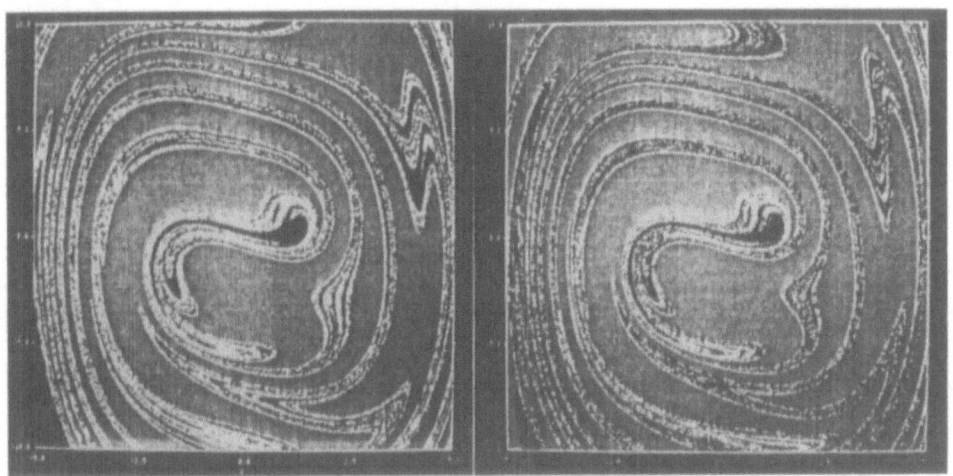

Figure 1. Basins of attraction at $\omega = 2.15$ Figure 2. Basins of attraction at $\omega = 2.05$

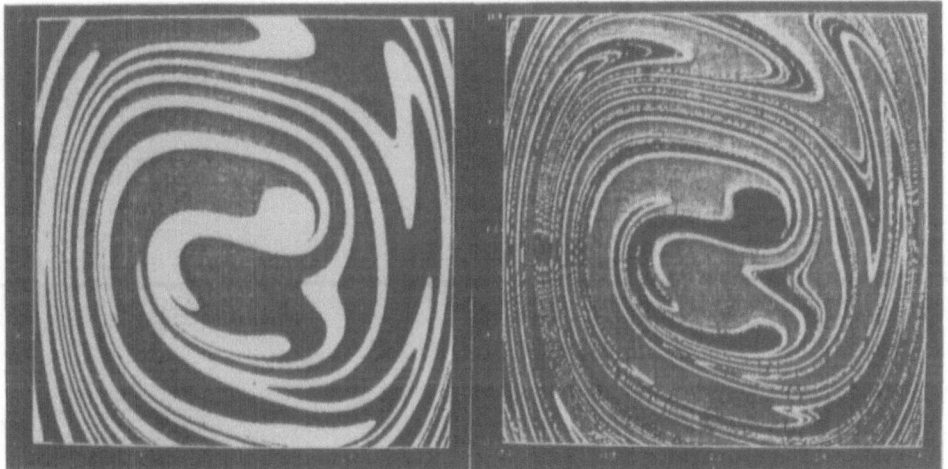

Figure 3. Basins of attraction at $\omega = 2.039$ Figure 7. Basins of attraction at $\omega = 1.8945$

International Series of Numerical Mathematics, Vol. 97, © 1991 Birkhäuser Verlag Basel

CONTROLLABILITY OF LORENZ EQUATION

Luce R. and Kernévez J.P.; UTC, BP 649, Compiègne, France.

ABSTRACT

The Lorenz equations are studied from the point of view of the possibilty to drive the system from a state to another one by acting on the Rayleigh number as a (control) function of time. It is shown that it is numerically possible, and the obtained control and trajectories have a good behaviour. However if we stop controlling at some time during the transfer, the system may evolve in a chaotic way.

I STATEMENT OF THE PROBLEM.

This work is motivated by J.L. Lions conjecture that the Navier-Stokes system is approximately controllable[2].

Our ultimate goal is to check whether Rayleigh-Bénard rolls can be controlled, that is whether we can make them appear or disappear within a given time T. And it is well known [3] that a first approximation of the qualitative behavior of Rayleigh-Benard system is given by the Lorenz equations:

$$(1.1) \qquad \begin{cases} \dot{x} = Pr\,(\,y - x) \\ \dot{y} = x(r - z) - y \\ \dot{z} = x\,y - b\,z \end{cases}$$

where the parameters Pr, the Prandl number, b, the aspect ratio, and r the Rayleigh number are given (Pr $= 10$, $b = \frac{8}{3}$). Defining $Y = \begin{pmatrix} x \\ y \\ z \end{pmatrix}$ these equations can be rewritten in the form

$$(1.2) \qquad \dot{Y} = f(Y,\, r).$$

- **Definition 1.1** (of exact controllability of a dynamical system (1.2)).- A dynamical system (1.2) is said to be exactly controllable if whatever the states Y_0 and $Y_d \in \mathbf{R}^3$, and time T > 0, it is possible to find a control function $r = r(t)$, belonging to some set $\mathcal{U}$ of functions, driving the system from Y_0 at time t $= 0$ to Y_d at time t $=$ T, i.e. such that one has (1.2) together with

$$(1.3) \qquad Y(0) = Y_0 \text{ and } Y(T) = Y_d.$$

Remark 1.1.-The Lorenz system (1.1) is certainly not exactly controllable since if $Y_0 = 0$ then $\dot{Y}(0) = 0$ whatever the control fonction r, so that the system stays at state $Y = 0$..As a consequence in the following Y_0 will be taken $\neq 0$.

The problem we address to in this paper is to drive the system from a state Y_0 to another one $\underline{Y_d}$.

Figure 1 shows the principle of our approach and the kind of result we obtain:

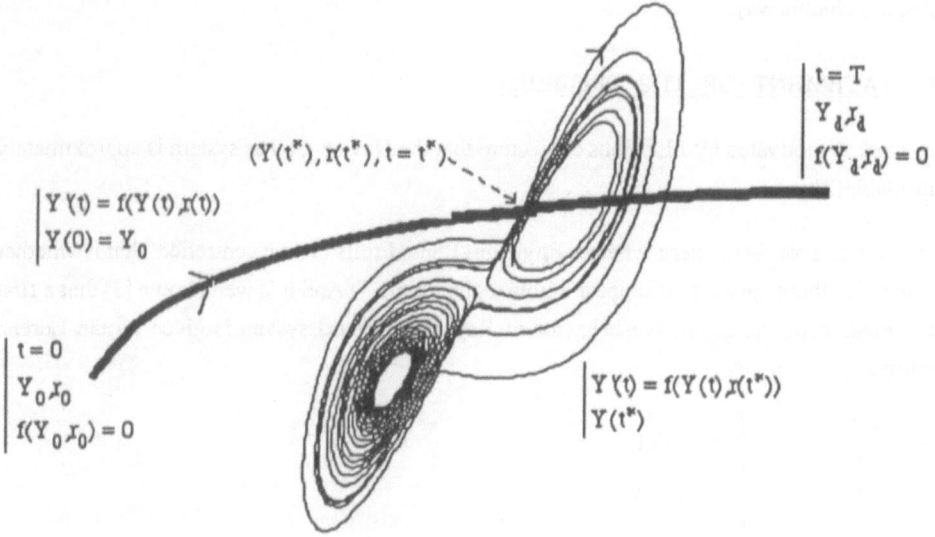

$$\left|\begin{array}{l} Y(t) = f(Y(t),r(t)) \\ Y(0) = Y_0 \end{array}\right.$$

$(Y(t^*), r(t^*), t = t^*)$

$$\left|\begin{array}{l} t = T \\ Y_d, r_d \\ f(Y_d, r_d) = 0 \end{array}\right.$$

$$\left|\begin{array}{l} t = 0 \\ Y_0, r_0 \\ f(Y_0, r_0) = 0 \end{array}\right.$$

$$\left|\begin{array}{l} Y(t) = f(Y(t), r(t^*)) \\ Y(t^*) \end{array}\right.$$

Fig n°1 Example

The control r(t) transfers the system from point Y_0 an equilibrium at time $t = 0$ to another one, Y_d, at time $t = T$. If at time $t = t^*$, r(t) is frozen at $r(t^*)$ then the system undergoes chaotic motion[5].

In Figure 1 the points Y_0 and Y_d are equilibrium points respectively to the values r_0 and r_d of r. Y_0 (resp Y_d) correspond to a state with (resp without) Rayleigh-Benard rolls.

Problem: Drive the system from a state-parameter pair (Y_0,r_0) at time $t = 0$ to the state parameter pair (Y_d,r_d) at time $t = T$.

Orientation: we are going to solve optimal control problems, namely look for controls r(t) enabling us to approach as close as possible the desired state. One can show [4] that if the state Y_d is attainable then the optimal control thus obtained enables us to exactly attain it.

II   THE OPTIMAL CONTROL PROBLEM.

II.1 Transfer from a state to another one.

Let $Y(.; r)$ denote the solution of the Cauchy problem

$$(2.1) \quad \begin{cases} \dot{Y} = f(Y, r) \\ Y(0) = Y_0 \end{cases} \qquad Y = Y(t; r) \qquad 0 \leqslant t \leqslant T.$$

where $r = r(t)$
We consider the following cost function:

$$(2.2) \qquad J(r) = \frac{1}{2} \| Y(T; r) - Y_d \|^2_{R^3}$$

and the problem

$$(2.3) \qquad \min_{r \in \mathcal{U}} J(r),$$

where $\mathcal{U}$ denotes the set of control functions r. It may have no solution. Therefore we consider the regularized cost function

$$(2.4) \qquad J_\varepsilon(r) = \frac{1}{2} \| Y(T; r) - Y_d \|^2_{R^3} + \frac{\varepsilon}{2} \| r \|^2_{\mathcal{U}}$$

and the following minimization problem

$$(2.5) \qquad \min_{r \in \mathcal{U}} J_\varepsilon(r)$$

It is show in [3] that if the state Y_d is attainable then we can pass to the limit as ε tends to 0.

Remark 2.1.-By the previous procedure we can drive the system from a stationary state Y_0, associated to the value r_0, to a point $Y_d \in R^3$ located on a trajectory associated to the value r_d, this trajectory being a periodic orbit or not.

II.2 Numerical   methods.   We applied a gradient method, namely the conjugate gradient method, and obtained the gradient by using an adjoint state $p : r$ being given, $J'(r)$ for J defined by (2.2) is defined by

$$(2.6) \quad \begin{cases} \dot{Y} = f(Y, r), & Y(0) = Y_0 \\ -\dot{p} = f_y^*(Y, r)p, & p(T) = Y(T) - Y_d \\ J'(r) = f_r^* p. \end{cases}$$

We used a Crank-Nicholson discretization of the state equations and the cost function (mid-point rule). The control has been looked for of the form

$$(2.7) \qquad r(t) = r_d + \frac{(t-T)}{T}(r_0 - r_d) + \sum_{i=0}^{Nb} c_i \sin(i \frac{\pi}{T} t)$$

This decomposition of $r(t)$ presents the advantage to verify $\forall c_i$ $r(0) = r_0$ et $r(T) = r_d$. All the results presented here have been obtained for $T = 1$ and for $\varepsilon = 0$. The results obtained by minimization have been checked by using more performing algorithms (methods of type Backward Difference Formulas).

We defined the <u>sensitivity to the coefficient c_i of the control</u> by $S_i(t) = \frac{\partial Y}{\partial c_i}$.

$S(t)$ satisfies the following differential equation:

$$(2.8) \qquad \dot{S} = \nabla_Y f(Y,r) S + \frac{\partial f(Y,r)}{\partial c_i}$$

We compared the controllability of the Lorenz system to that of the following two systems:

$$(2.9) \qquad \begin{cases} \dot{x} = a (y - x) + xy \\ \dot{y} = r x - y - x^2 \end{cases} \qquad + C.I.$$

$$(2.10) \qquad \begin{cases} \dot{x} = -a x - b y - x^3 \\ \dot{y} = -b x - c y - r y^3 \end{cases} \qquad + C.I.$$

II.3 Numerical   results.

II.3.1.-Results obtained for the system (1.1): Numerically we always found a control $t \text{---} > r(t)$ driving the system from a state Y_0 on a solution for $r = r_0$ to a state Y_d on a solution for $r = r_d$. Moreover the sensitivity of the system to the control was a smooth and bounded function of time.

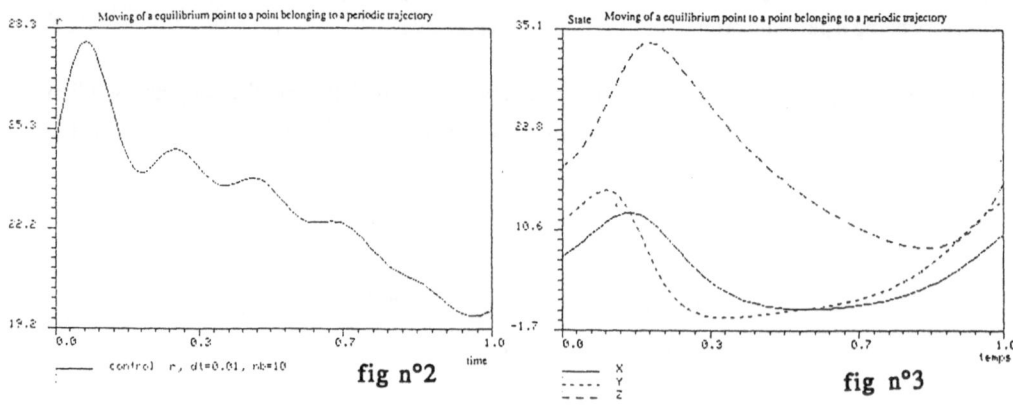

fig n°2   fig n°3

The figure n°2 shows the control obtained to drive the system from a equilibrium state to a state belonging to periodic state (fig n°4). The figure n°3 shows the drawing of the state Y(t,r(t)).

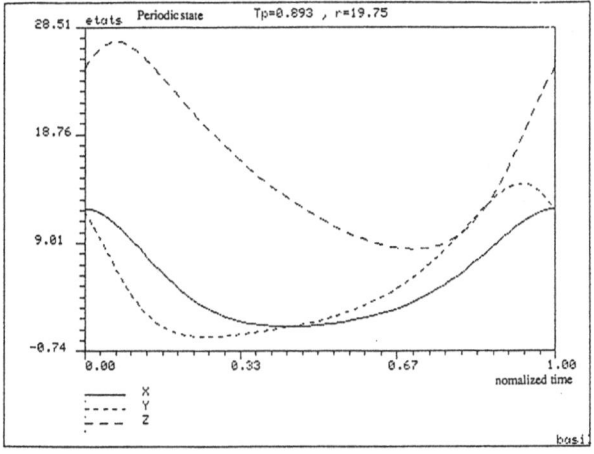

Fig n°4

II.3.2 Results obtained for the system (2.9): Numerically we always found a control r(t) enabling us to pass from an stationary state to another one.

II.3.3 Results obtained for the system (2.10): The system (2.10) is very regularizing and it was not possible to take it from an equilibrium point to another one relatively far from it. .The minimization of (2.2) failed.

References.

[1] Isidori, A., On Linear Control Systems, Second Edition, Springer Verlag.

[2] Lions, J.L., Are there connections between turbulence and controllability? INRIA Meeting in Perpignan, 1990.

[3] Lorenz E.N., "Deterministic non-periodic flow,"J. Atmospherie Sci 20,130 and 448, 1963

[4] Luce, Etude de quelques problèmes mal posés, Thesis, Compiègne 1990

[5] Sparrow, C., The Lorenz Equations: Bifurcations, Chaos, and Strange Attractors, Applied Mathematical Sciences 41, Springer-Verlag.

International Series of Numerical Mathematics, Vol. 97
Bifurcation and Chaos: Analysis, Algorithms, Applications
R. Seydel, F. W. Schneider, T. Küpper, H. Troger (Eds)
© 1991 Birkhäuser Verlag Basel

Spatially Periodic Forcing of Spatially Periodic Oscillators

Mario Markus and Carsten Schäfer

Max-Planck-Institut für Ernährungsphysiologie, Rheinlanddamm 201, D-4600 Dortmund 1,
FRG

1. Introduction

In the past, time-dependent periodic forcing of time-dependent periodic oscillators has
been investigated in detail (see e.g. [1-5]). Here, we investigate the influence of spatially
periodic forcing on periodic waves. We do this for the particular case of excitable media
(for definition and analyses, see [6-9]), choosing the method of cellular automata [10] as an
efficient alternative to the integration of partial differential equations. For the benefit of
low computational cost, cellular automata sacrifice insight into the detailed physical mechanisms. However, if properly designed, they may capture the essential features of complex systems in agreement with experiments, as has been shown for the stimulation of turbulences in fluid dynamics [10,11] and of waves in excitable media [12-14]. For a recent review of the application of cellular automata on chemical systems, see [15]. Being composed
of identical components, each simple, but capable of complex behaviour, cellular automata
are ideally suited for simulations on highly parallel computers.

The simulations presented in this work may be verified experimentally or may form the basis for future experimental design using the light sensitive (ruthenium catalyzed) Belousov-
Zhabotinskii (BZ) reagent [16]. In fact, optical projection of spatially periodic light patterns on periodic waves in this medium provides an experimental setup subject to comfortable variation of the forcing control parameters (amplitude, wavelength and phase of
the perturbation). An experimental project of this kind is underway.

Besides providing hints for experimentation, this work also intends to set up easy-to-handle
prototypes for the simulation of complex patterns in time and space, including turbulences,
by using simple rules. Such prototypes have been established with much success for purely
time-dependent systems, e.g. the logistic map [17], Duffings equation [1,2] or the forced
Sel'kov-Higgins model [3]. In fact, these simple models give us an idea of the variety of

possible dynamic modes. Furthermore, there are some aspects, such as period-doubling cascades, U-sequences and Arnold tongues, which arise in all the time-dependent systems that share some fundamental properties, independently of their degree of complexity. In the same sense, we hope that our particular computer experiments (especially those in one-dimensional systems) will inspire mathematicians to prove general statements. On the other hand, implications of general theorems may easily be visualized using our algorithms.

2.   One-dimensional Systems
2.1   The Model

We consider a ring of N equal cells as illustrated in Fig. 1. Each cell corresponds to an index i, i = 1, 2, ..., N. The neighbourhood of the cell i is defined by the cells i-R, ..., i-1, i+1, ..., i+R for a given natural number R. Each cell has a state S_i (t), which may be receptive, excited or refractory (for the meaning of these states, see [6-9], [12-14], [19]).

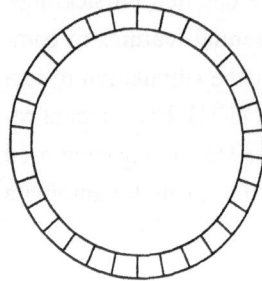

Fig. 1: Geometry of the one-dimensional cellular automaton

In our simulations we start by considering an unforced periodic system. The states S_i are named as follows: S_i = 0 (receptive state), S_i = n+1 (excited state) and S_i = n, n-1, ..., 2,1 (refractory states). The states S_i (t+1) are determined from the S_i (t) as follows:

α) S_i (t+1) = n+1 if S_i (t) = 0 and $\nu \geq$ m,
β) S_i (t+1) = S_i (t)-1 if S_i (t) > 0,
γ) S_i (t+1) = 0 if S_i (t) = 0 and $\nu <$ m.

ν is the number of excited cells within the neighbourhood of i. We use here an approximation in which excitation of refractory states is neglected [12-14]. The threshold m is set to 1 in the present calculations. We are thus left with only two model parameters: R and n. The unforced system consists of a spatially periodic configuration of states (wavelength λ_0 = $(n+2)R$) travelling around the ring. N is chosen as a multiple of λ_0 . Within each wavelength λ_0 , R cells in the state j are followed on their right side by R cells in the state $j+1$ (j = 0, 1, 2, ..., n). For example, if R = 2, n = 3 and N = $2\lambda_0$, the arrangement is 00112233440011223344 (closing to a ring). Such a configuration is stable. We call σ_0 the sequence of 0's. It should be kept in mind that in future works one may consider sequences where σ_0 is longer than R.

After setting up the unforced system in the way just described, we simulate the spatial forcing by switching on a spatially periodic time of relaxation from the excited to the receptive state. We call this time $\tilde{r}_i$ ($\tilde{r}_i$ is a real number, i = 1, 2, ..., N). This is a minimal modelling of a spatial periodicity of light intensity acting on the Ru-catalyzed BZ reagent, as it has been shown that light affects mainly the relaxation kinetics [20]. Our model thus mimics an experiment in which the Ru-catalyzed reaction takes place in a ring, possibly by fixing the catalyst in a gel, cutting a ring-shaped piece out of the gel, and inmersing this piece in a solution containing the reactants.

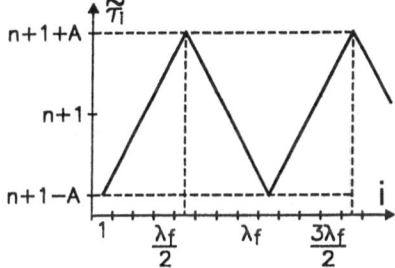

Fig. 2: Typical dependence of the relaxation time on the cell index i for spatially periodic forcing.

The unforced system is described by the particular case $\tilde{r}_i$ = n+1 = const. At the moment of switching on forcing, we let $\tilde{r}_i$ vary periodically between n+1-A and n+1+A, as exemplified in Fig. 2. Since we assume discrete time, we work with the next integer to $\tilde{r}_i$ (ordinate of Fig. 2), which we call r_i. A is the forcing amplitude and λ_f the forcing wavelength. In order to accomodate both an integer number of unforced wavelengths λ_0 and an integer number of forcing wavelengths λ_f, keeping the ring as small as possible, we set N equal to the smallest common multiple of λ_0 and λ_f. We consider no phase shift between the unforced and the forcing wave, meaning e.g. that cell 1 in Fig. 2 coincides with the first cell on the left in σ_0 . In order to have symmetric forcing conditions, as in Fig. 2, we consider only even values of λ_f . In the forced regime, S_i is assumed to be a real instead of an integer number. A relaxation time r_i means that r_i steps of size $(n+1)/r_i$ are necessary to change a cell i from the excited state ($S_i = n+1$) to the receptive state ($S_i = 0$). Thus, rule ß) changes to: $S_i (t+1) = S_i (t) - (n+1)/r_i$ if $S_i (t) > 0$. Rules α) and γ) remain the same. The model parameters in the forced regime are R, n, A and λ_f . If not stated otherwise, simulations were performed by doing 2000 iterations to allow transients to die away and 5000 iterations afterwards to observe the dynamic behaviour. We call a wave chaotic, if there is at least one cell, where no temporal periodicity can be tested within the 5000 iterations.

2.2   Results

For the simple case R = 1 we always obtained periodicity or waves vanishing to zero. For larger R, we also obtained chaos. Fig. 3 shows different modes obtained on a plane defined by the forcing amplitude A and the forcing wavelength λ_f . This type of representation is the spatial counterpart of plots for time-dependent systems on the plane defined by the forcing amplitude and the forcing frequency (see e.g. [3,5]). In Fig. 3 periodicity appears at very low and very high λ_f as well as at low A, wave vanishing appears at high λ_f , and there is a large chaotic region on the plane at high A and intermediate values of λ_f . Figs. 4 through 8 show the time development of the states S_i , given as grey levels, on the ring or magnified parts of the ring. The excited state is shown in white, the receptive state in black, and the refractory states in grey, decreasing S_i being increasingly dark.

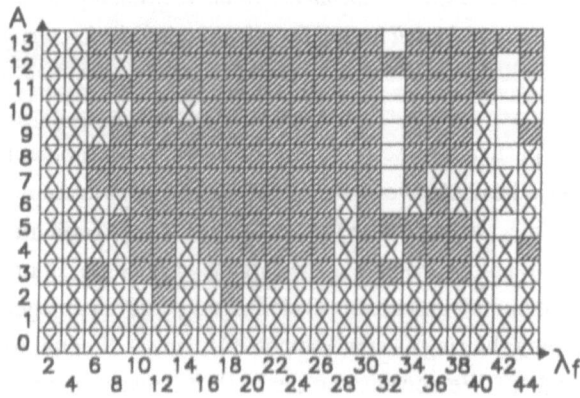

Fig. 3: *Dynamic behaviour of the one-dimensional system, depending on forcing amplitude A and wavelength λ_f. n = R = 14. Crosses: periodic. Dashed: chaotic. Void: vanishing to zero.*

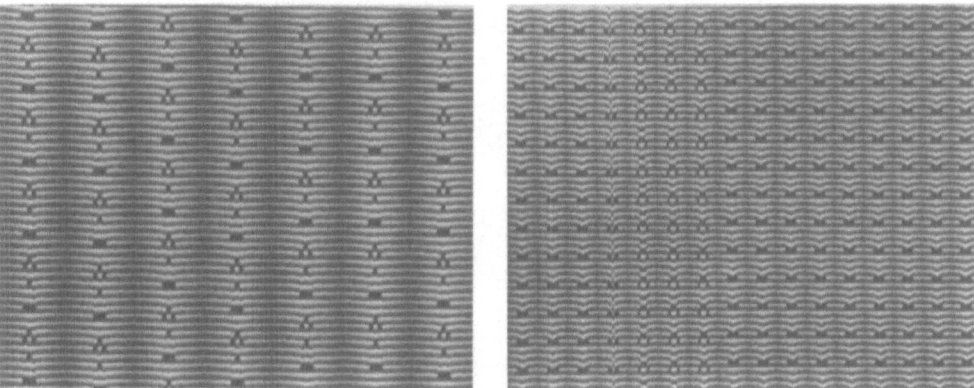

Fig. 4: Periodic responses of the forced one-dimensional system. n = R = 14. Horizontal: space. Vertically downwards: time. Left: A = 4, λ_f = 32. Right: A = 3, λ_f = 14.

Fig. 4 displays the periodic dynamics on two different points of Fig. 3. Chaotic dynamics on two other points of Fig. 3 are shown in the lower part of Fig. 5 and in Fig. 6. A further example of a chaotic wave is given in Fig. 7, which reveals a similar arcade-type structure as Fig. 6. We found this type of structure to be typical for large A/n together with large R/n.

The left, respectively right, part of Fig. 8 shows magnified portions of the upper, respectively lower, part of Fig. 5. Both chaotic patterns display similar spatio-temporal intermittency: periodic structures interrupted by chaotic "faults", the periodic intervals being longer in the upper part of Fig. 5, left part of Fig. 8. Also, the upper part of Fig. 6 seems to reveal intermittent behaviour: large periodic trains of the order of several λ_f are separated by "faults". However, inspection of the magnification shown in the lower part of Fig. 6, where comparison of the cell states is possible with more precision, reveals chaotic behaviour also on a smaller scale.

We found that a mechanism inducing chaos in this type of media is the collision of a periodic wave with a cell which we call "prereceptive", meaning that it is refractory shortly before becoming receptive. A collision of a wave with a prereceptive cell may cause a wave-doubling: a wave travelling in opposite direction appears in addition to the original wave. Both waves may collide with other prereceptive cells as they travel around the ring yielding new wave-doublings, and so on, as illustrated in Fig. 9. A wave-doubling cascade along the ring ends up in chaos.

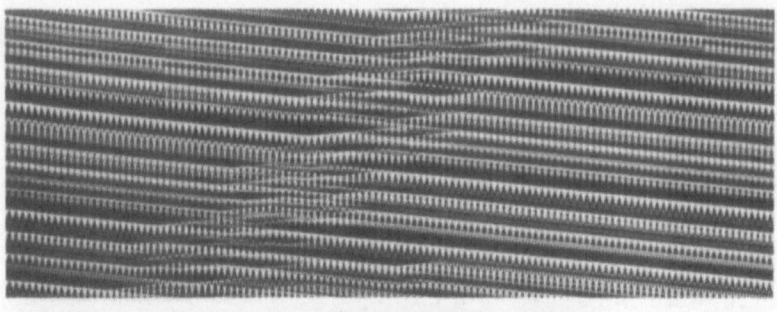

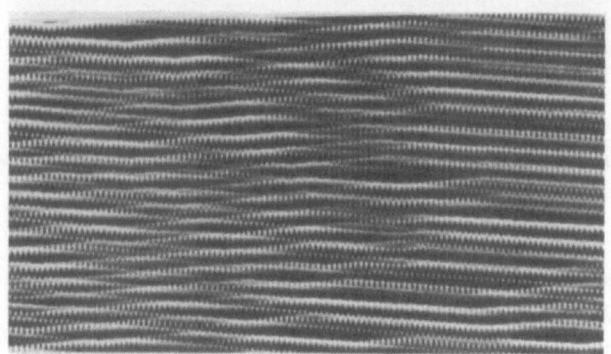

Fig 5: Chaotic responses of the forced one-dimensional system. Horizontal: space. Vertically downwards: time. Upper: $n = 20$, $R = 11$, $A = 5$, $\lambda_f = 10$. Lower: $n = R = 14$, $A = 3$, $\lambda_f = 6$.

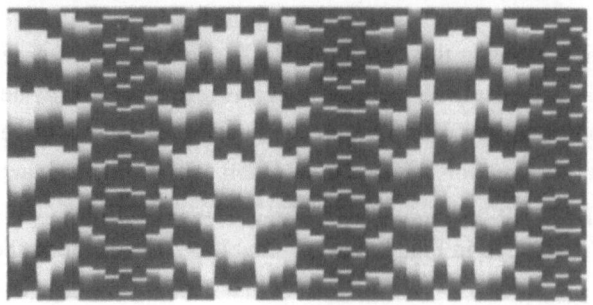

Fig. 6: Chaotic response of the forced one-dimensional system. $n = R = 14$, $A = 8$, $\lambda_f = 16$. Horizontal: space. Vertically downwards: time. Upper: whole ring. Lower: magnified portion.

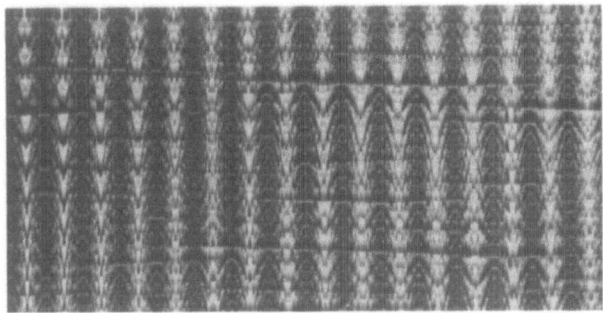

Fig. 7: Chaotic response of the forced one-dimensional system. $n = R = 30$, $A = 20$, $\lambda_f = 60$.

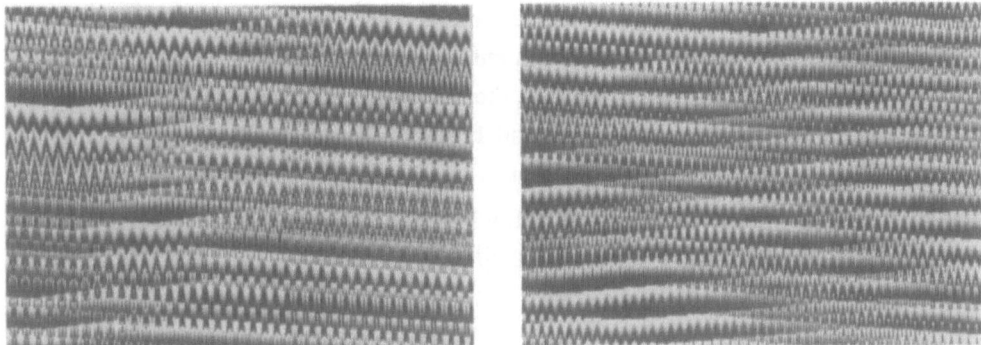

Fig. 8: Chaotic responses of the forced one-dimensional system. The left, resp. right parts of this figure are magnified portions of the upper, resp. lower parts of Fig. 5. The dynamic behaviour here is "intermittent": periodic wave-trains are separated by chaotic "faults".

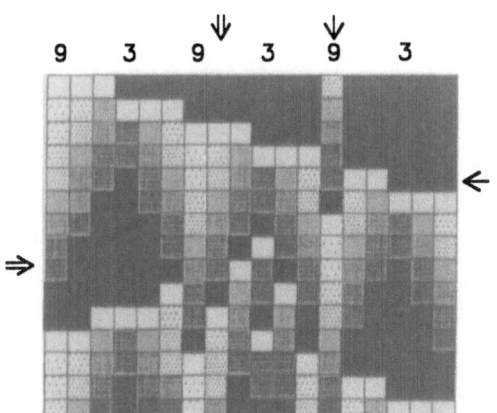

Fig. 9: Mechanism inducing chaos through wave-doublings due to collisions of waves with prereceptive cells. Horizontal: space. Vertically downwards: time. Numbers above: maximum (9) and minimum (3) relaxation time τ_i; White: excited. Shading increases as the receptive state (black) is approached.

In order to get a better insight into the chaotic dynamics, we represented the time dependence of the state S_i only for the cells corresponding to fixed phases φ of the forcing wave.
The possible phases φ are 1, 2, 3, ... λ_f (see abscissa of Fig. 2). We do this in Figs. 10
through 12. Fig. 10 corresponds to the upper part of Fig. 5, Fig. 11 to the lower part of Fig.
5, and Fig. 12 to Fig. 7. Figs. 10 through 12 show that chaos may have different features,
depending on φ, the dynamic behaviour thus differing from cell to cell. An extreme case
can be seen in Fig. 10, where the beviour is nearly periodic for cells with $\varphi = 4$ and chaotic
for neighbouring cells (e.g. $\varphi = 5$). Another extreme case can be seen in Fig. 12
(periodicity for $\varphi = 16$ and chaos in the neighbouring cells). In Fig. 11, different types of
chaotic patterns are found for different φ: mainly "slanting" structures for $\varphi = 1$ and 2, and
mainly "horizontal" structures for $\varphi = 3$ and 4. The differences in the ranges of grey levels
for different φ in Figs. 10 through 12 are related to the values of r_i (see Fig. 2): at $r_i =$
$n+1-A$ the range of refractory states is minimum, leading in extreme cases only to black
and white pictures, whereas at $r_i = n+1+A$ the range of refractory states and thus of grey
levels is maximum. Inspection of Figs. 10 and 11 shows that the method of separating the
phases φ helps to discern dynamic properties which are not visible without this phase
separation (compare with the corresponding intermittent pictures in the upper and lower
parts of Fig. 5 and their magnifications in Fig. 8).

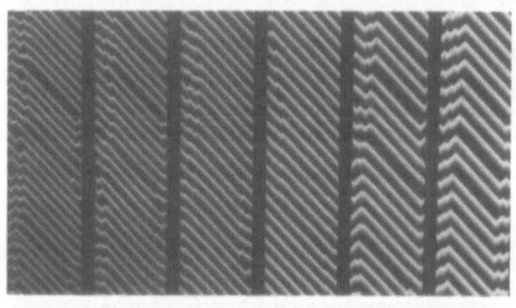

Fig. 10:   *Dynamics at different
forcing phases φ for
the case shown in the
upper part of Fig. 5.
From left to right: $\varphi =$
1, 2, 3, 4, 5, 6*

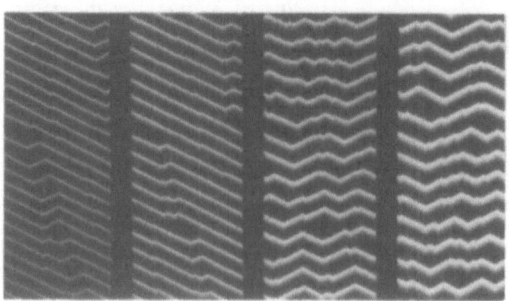

Fig. 11:   *Dynamics at different
forcing phases φ for
the case shown in the
lower part of Fig. 5.
From left to right: $\varphi =$
1, 2, 3, 4*

Fig. 12: Dynamics at different forcing phases φ for the case shown in Fig. 7. From left to right: φ = 1, 4, 7, 10, 13, 16, 19, 22, 25, 28, 31

3. Two-dimensional Systems
3.1 The Model

We consider a plane with M x M square cells and use the same geometry as in refs. [12-14]. One point is placed randomly in each cell at t = 0 and this distribution remains the same throughout the simulations. The neighbourhood of a point is defined by a circle C of radius R and centered at that point. In contrast to one-dimensional systems (see section 2) we had found [12] that the requirements of randomness and of a circular neighbourhood are necessary for isotropic wave propagation. We must also include "local averaging" in the two-dimensional model (averaging the states of all points within C after each iteration, except for the states 0, 1 and n+1), else we get the wrong behaviour at the core of spirals. Thus, we iterate with the same automaton rules as for one-dimensional systems (section 2.1) except for the random distribution of points (indices i), the circular neighbourhood, and local averaging. The natural numbers S_i in the unforced case are calculated by taking the next integer to the average over C. We set here m_0 = 1, p = S_{max} = 0 (see [12-14]). We showed [12,14] that aperiodicities can develop under a periodic variation of the relaxation time, as in Fig. 2, in one spatial direction. Also, we reported that these aperiodicities can be characterized by a positive maximum Lyapunov exponent λ_{max}, and thus have the features of deterministic chaos. A positive λ_{max} means that small perturbations grow exponentially in the average. Here, we investigate the dynamics of pertubation growth including "later" times in which the perturbations are not anymore "small" in the sense that the growth rate is not proportional to the perturbation and thus the growth is not exponential. As initial perturbation we set all points within the neighbourhood of the center Q of a quadratic medium to the excited state at $t = t_o$. We calculated two temporal developments of the

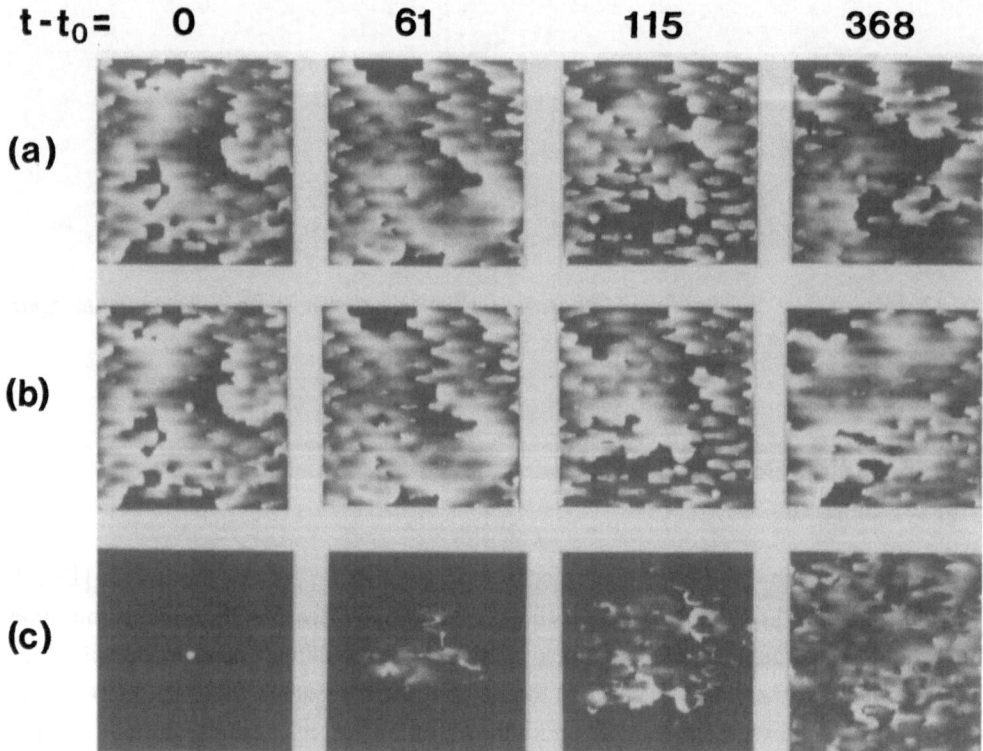

Fig. 13:   Chaotic dynamics in a two-dimensional system at different times $t - t_o$.
              300 x 300 cells, R = 6, n = 12, A = 8, λ_f = 20, m_o = 1, p = S_{max} = 0
              (a): Unperturbed system. (b): System perturbed by exciting all points within a cir-
              cle with radius R (white) at the picture center at $t = t_o$. (c): difference between
              (a) and (b), showing how the perturbation propagates ("butterfly effect").

system: one with and one without this perturbation. The two resulting time series of spatial patterns were then subtracted one from the other leading to a series of perturbation patterns $P(t - t_0)$, $t - t_0 = 1, 2, ... $. We then determined the mean distance travelled by the perturbation, which we call $< d_p >$, as follows. The patterns $P(t - t_0)$ were divided into 60 pairwise adjacent triangles sharing one vertex (Q) and having equal angles (6°) at Q. In each of these triangles, we determined the maximum of the distance between Q and the automaton points where $P(t-t_0)$ is different from zero. For any given $t - t_0$, we then determined the average of this maximum over all 60 triangles and over 8 patterns differing in t_0.

In our model there are two fundamental propagation mechanisms: the excitation process and the local averaging. The excitation process (without local averaging) travels at a velocity equal to R cells/iteration if curvature effects (see [14]) are left out of consideration; we call d_e the travelled distance of this process. A calculation with pure local averaging, i.e. leaving away all other algorithmic steps, was started with a circle having a radius equal to 6 cells, the points within the circles being in the state 6 and those outside in the state 2. Averaging using real numbers was then performed and the half maximum width d_a of the spatial distribution of cells was computed as a function of time.

3.2   Results

Fig. 13 shows a chaotic spatial pattern at four consecutive times without (Fig. 13a) and with (Fig. 13b) a small circular perturbation at $t = t_0$ around the center Q of the medium. Fig. 13c shows the differences between the patterns (a) and (b). As time proceeds, (a) and (b) become different in larger and larger areas around Q, until at $t - t_0 = 368$ iterations the patterns are different all over the medium. The mean distance $< d_p >$ travelled by the perturbation is given in Fig. 14 as a function of time. Fig. 14 also shows the time dependence of the distance d_a travelled by a wave governed solely by local averaging, as well as the time dependence of the distance d_e travelled by excitation. The latter is governed by the law $d_e = R(t - t_0)$ and thus leads to a straight line with slope 1 on the log-log plot in Fig. 14. On the other hand, d_a leads to a slope of approx. 0.5 in this plot, thus indicating that local averaging is a diffusive process described by an equation of the type $d_a \sim \sqrt{D_a t}$. It is remarkable that the propagation of a perturbation in a chaotic regime ($< d_p >$ in Fig. 14) is also described by a slope ≈ 0.5. Thus, we may write $< dp > \sim \sqrt{D_p t}$. However, the diffusion coefficient D_p is larger than D_a ($D_p \approx 3 D_a$). For one-dimensional rings (see section 2), similar results as those shown in Fig. 14 are obtained, D_p being roughly $4 D_a$.

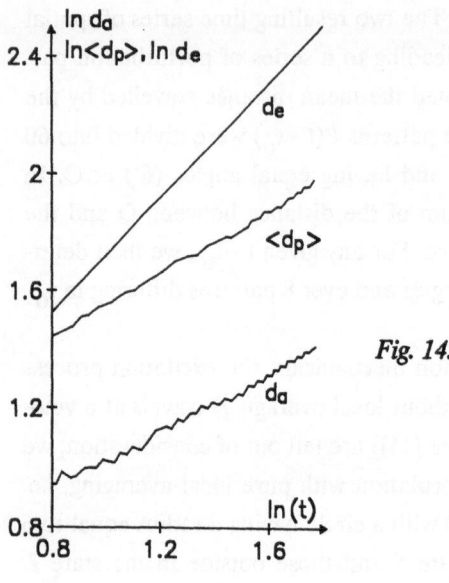

Fig. 14: *ln-ln-plot showing the time dependence of the distance d_e travelled by excitation, of the mean distance $< d_p >$ travelled by a perturbation (see Fig. 13c), and of the distance d_a travelled by pure local averaging.*

Acknowledgements

We thank Prof. Benno Hess for fruitful discussions and the Commission of the European Communities for financial support. We also thank Manfred Krafczyk and Wolf-Dieter Sponheimer for efficient programming, Gesine Schulte for the photographic work and Petra Rathke for editing the manuscript.

References

1.   J.M.T. Thompson and H.B. Stewart, "Nonlinear Dynamics and Chaos", J. Wiley, New York (1986)

2.   F.C. Moon, "Chaotic Vibrations", J. Wiley, New York (1987)

3.   K. Tomita, Phys. Reps. **86**, 113-167 (1982)

4.   M. Markus and B. Hess, Proc. Natl. Acad. Sci. USA **81**, 4394-4398 (1984)

5.   M. Markus and B. Hess, in: "Control of Metabolic Processes" (ed. by A. Cornish-Bowden and M.L. Cardenas) Plenum, New York (1990), pp. 303-313

6.   V.S. Zykov, "Simulations of Wave Processes in Excitable Media", Manchester University Press, Manchester (1988)

7.   A.V. Holden, M. Markus and H.G. Othmer (eds.), " Nonlinear Wave Processes in Excitable Media", Plenum, New York, (in press)

8.   J.P. Keener and J.J. Tyson, Physica **D21**, 307-324 (1986)

9.   J.J. Tyson and J.P. Keener, Physica **D32**, 327-361 (1988)

10.   S. Wolfram, "Theory and Applications of Cellular Automata", World Scientific, Singapore (1986)

11.   G.D. Doolen, S.Orszag and U. Frisch (eds.), "Lattice Gas Methods of Partial Differential Equations", Addison Wesley, Amsterdam (1990)

12.   M. Markus and B. Hess, Nature **347**, 56-58 (1990)

13.   M. Markus, Biomed. Biochim. Acta **49**, 681-696 (1990)

14.   M. Markus and B. Hess, in: "Dissipative Structures in Transport Processes and Combustion" (ed. by D. Meinköhn) Springer-Verlag, Heidelberg (1990), pp. 197-214

15.   R. Kapral, J. Math. Chem. (in press)

16.   L. Kuhnert, Nature **319**, 393-394 (1986)

17.   R.M. May, Nature **261**, 459-467 (1976)

18.   M. Markus, Computers in Physics, Sept./Oct., 481-493 (1990)

19.   B.F. Madore and W.L. Freedman, Science **222**, 615-616 (1983)

20.   Ch. Zülicke and L. Shimansky-Geier, Z. Phys. Chem. **271**, 357-368 (1990)

International Series of Numerical Mathematics, Vol. 97, © 1991 Birkhäuser Verlag Basel   277

Solution Branches at Corank-2 Bifurcation Points with Symmetry

Mei, Zhen

*Fachbereich Mathematik, Universität Marburg, 3550-Marburg/Lahn, FRG and
Department of Mathematics, Xi'an Jiaotong University, PRC*

1. Introduction

Let X be a Hilbert space, Γ be a finite group acting on X with an orthogonal representation. We consider bifurcations of the parameter-dependent equation

$$G(u, \lambda) = 0, \tag{1.1}$$

where $G : X \times \mathbf{R} \mapsto X$ is a Γ-*equivariant* smooth mapping, i.e.

$$G(\gamma u, \lambda) = \gamma G(u, \lambda) \quad \text{for all } \gamma \in \Gamma, \ (u, \lambda) \in X \times \mathbf{R}.$$

At a bifurcaton point (u_0, λ_0), we assume $D_u G_0 := D_u G(u_0, \lambda_0)$ is a Fredholm operator with index 0 and $D_\lambda G_0 := D_\lambda G(u_0, \lambda_0)$ is in range of $D_u G_0$. If the nullspace $N(D_u G_0)$ is m-dimensional ($m \geq 2$), (u_0, λ_0) is called a *corank-m bifurcation point* of (1.1). For a subgroup Σ of Γ, its *fixed point subspace* is defined by

$$X^\Sigma := \{u \in X \mid \sigma u = u, \forall \sigma \in \Sigma\}.$$

For simplicity, we denote $\Sigma_0 := \{\gamma \in \Gamma \mid \gamma u_0 = u_0\}$ the *isotropy group* of u_0 and $G^\Sigma := G|_{X^\Sigma \times \mathbf{R}}$ the restriction of G to $X^\Sigma \times \mathbf{R}$. Due to the Γ-equivariance of G, G maps $X^\Sigma \times \mathbf{R}$ and a solution curve $(u(t), \lambda(t))$ passing through (u_0, λ_0) has necessarily certain symmetry. Namely, there is a subgroup $\Sigma < \Sigma_0$ such that $u(t) \in X^\Sigma$. $(u(t), \lambda(t))$ is also called a Σ-branch of (1.1) into X^Σ, see e.g. Dellnitz/Werner [4]. To determine Σ-branches of (1.1) one may consider the *reduced problem*

$$G^\Sigma(u, \lambda) = 0 \quad \text{for } (u, \lambda) \in X^\Sigma \times \mathbf{R}. \tag{1.2}$$

In the context of bifurcation problems, symmetry and group-theoretic methods are often used to decompose singularities of (1.1) at bifurcation points. In particular, the equivariant branching lemma and its modifications show that if there is a subgroup $\Sigma < \Sigma_0$ such that $\dim N(D_u G_0^\Sigma) = 0$ (resp. 1), then (u_0, λ_0) becomes a nonsingular (resp. simple bifurcation) point of (1.2) and there is a unique (resp. exactly two) solution curve of (1.2) passing through (u_0, λ_0), see e.g. Vanderbauwhede [8], Golubitsky/Stewart/Schaeffer [5], Dellnitz/Werner [4], Allgower/Böhmer/Mei [1]. These solution curves can be traced by the continuation methods and its modifications, see e.g. Allgower/Georg [2], Seydel [7].

In this paper we consider the case that there is a *maximal subgroup* Σ of Σ_0 such that

$$\dim N(D_u G_0^\Sigma) = 2. \tag{1.3}$$

Since (1.3) is not generic, there might be some hidden symmetries in the bifurcating solutions which should be considered. However, it is well known that finding out hidden symmetries is sometimes very difficult. We would like to study a direct method for (1.3).

2. Solution Branches at a Corank-2 Bifurcation Point

We assume (1.3) in the sequel and choose a basis of $N(D_u G_0^\Sigma)$ as $\{\phi_1, \phi_2\}$. For X^Σ we have

$$X^\Sigma = N(D_u G_0^\Sigma) \oplus R((D_u G_0^\Sigma)^*) = N((D_u G_0^\Sigma)^*) \oplus R(D_u G_0^\Sigma), \qquad (2.1)$$

where $\oplus$ denotes an orthogonal sum, $(D_u G_0^\Sigma)^*$ is the adjoint operator of $D_u G_0^\Sigma$. A solution curve $(u(t), \lambda(t))$ of (1.2) with $(u(0), \lambda(0)) = (u_0, \lambda_0)$ allows a unique decomposition

$$\begin{cases} u(t) = u_0 + t\alpha_1(t)\phi_1 + t\alpha_2(t)\phi_2 + t\beta(t)v_0 + t^2 v(t), \\ \lambda(t) = \lambda_0 + t\beta(t), \end{cases}$$

where $\alpha_i(t), \beta(t) \in \mathbf{R}$, $i = 1, 2$, $v(t) \in R((D_u G_0^\Sigma)^*)$ and $v_0 \in X^{\Sigma_0} \cap R((D_u G_0^\Sigma)^*)$ satisfying

$$D_u G_0 v_0 + D_\lambda G_0 = 0.$$

Here we consider $\beta(0) \neq 0$ and take the normalization $\beta(t) \equiv 1$ (locally):

$$\begin{cases} u(t) = u_0 + t\alpha_1(t)\phi_1 + t\alpha_2(t)\phi_2 + tv_0 + t^2 v(t), \\ \lambda(t) = \lambda_0 + t. \end{cases} \qquad (2.2)$$

To determine $v(t)$, $\alpha(t) := (\alpha_1(t), \alpha_2(t))$, we use an enlarged system in $X^\Sigma \times \mathbf{R}^2$:

$$F(v, \alpha, t) := \begin{pmatrix} G^\Sigma(u_0 + t\alpha_1\phi_1 + t\alpha_2\phi_2 + tv_0 + t^2 v, \lambda_0 + t)/t^2 \\ (\phi_1, v) \\ (\phi_2, v) \end{pmatrix} = 0. \qquad (2.3)$$

At $t = 0$, F is defined by its limit. Obviously, the system (2.3) is equivalent to (1.2) on the solution curves in (2.2). Namely, if $(v(t), \alpha(t), t)$ satisfies (2.3), then the corresponding (2.2) is a solution curve of (1.2) and vice versa.

Using the Σ_0-invariance of $N(D_u G_0^\Sigma)$, we define a group action on $X^\Sigma \times \mathbf{R}^2$ by

$$\sigma \begin{pmatrix} v \\ \alpha \end{pmatrix} = \begin{pmatrix} \sigma v \\ \sigma \alpha \end{pmatrix} := \begin{pmatrix} \sigma v \\ A(\sigma)\alpha \end{pmatrix} \quad \text{for all } \sigma \in \Sigma_0, \qquad (2.4)$$

where $A(\sigma) \in \mathbf{R}^{2 \times 2}$ is the induced orthogonal representation of Σ_0 in $\mathbf{R}^2$:

$$\sigma(\phi_1, \phi_2) = (\sigma\phi_1, \sigma\phi_2) = (\phi_1, \phi_2)A(\sigma) \quad \text{for all } \sigma \in \Sigma_0. \qquad (2.5)$$

Lemma 2.1: *The mapping F is Σ_0-equivariant, i.e.*

$$F(\sigma(v, \alpha), t) = \sigma F(v, \alpha, t) \quad \forall \sigma \in \Sigma_0, \ (v, \alpha, t) \in X^\Sigma \times \mathbf{R}^2 \times \mathbf{R}.$$

The existence of solution curves of (2.3) follows from its nonsingular solutions at $t = 0$ and applications of the implicit function theorem. Consequently, we derive from

$$F(v, \alpha, 0) := \begin{pmatrix} D_u G_0^\Sigma v + \frac{1}{2}D^2 G_0^\Sigma(\alpha_1\phi_1 + t\alpha_2\phi_2 + v_0, 1)^2 \\ (\phi_1, v) \\ (\phi_2, v) \end{pmatrix} = 0 \qquad (2.6)$$

and (2.1) the *reduced bifurcation equations* for α (cf. also Decker/Keller [3], Mei [6]):

$$g(\alpha) = \begin{pmatrix} g_1(\alpha) \\ g_2(\alpha) \end{pmatrix} := \begin{pmatrix} (\phi_1^*, D^2 G_0^\Sigma (\alpha_1 \phi_1 + t\alpha_2 \phi_2 + v_0, 1)^2) \\ (\phi_2^*, D^2 G_0^\Sigma (\alpha_1 \phi_1 + t\alpha_2 \phi_2 + v_0, 1)^2) \end{pmatrix} = 0, \tag{2.7}$$

where $\{\phi_1^*, \phi_2^*\}$ is a basis of $N((DG_0^\Sigma)^*)$ and (cf. Golubitsky/Stewart/Schaeffer [5])

$$\sigma(\phi_1^*, \phi_2^*) = (\phi_1^*, \phi_2^*) A(\sigma) \quad \text{for all } \sigma \in \Sigma_0. \tag{2.8}$$

Theorem 2.2: *The mapping g is Σ_0-equivariant, i.e.*

$$g(\sigma \alpha) = \sigma g(\alpha) \quad \text{for all } \sigma \in \Sigma_0, \, \alpha \in \mathbf{R}^2. \tag{2.9}$$

Proof: From (2.4), (2.5) one sees

$$\begin{aligned} g_i(\sigma \alpha) &= (\phi_i^*, D^2 G_0^\Sigma ((a_{11}\alpha_1 + a_{21}\alpha)\phi_1 + (a_{12}\alpha_1 + a_{22}\alpha_2)\phi_2 + v_0, 1)^2) \\ &= (\phi_i^*, D^2 G_0^\Sigma (\sigma(\alpha_1 \phi_1 + \alpha_2 \phi_2 + v_0), 1)^2) \\ &= (\sigma^{-1}\phi_i^*, D^2 G_0^\Sigma (\alpha_1 \phi_1 + \alpha_2 \phi_2 + v_0, 1)^2). \end{aligned}$$

Hence, the conclusion (2.9) follows from (2.7), (2.8) and $A(\sigma^{-1}) = A(\sigma)^T$. ∎

Since (2.7) is a simple system of quadratical polynomials, in many cases it can be solved directly. If an isolated solution α^0 of (2.7) is known, (2.1) yields a unique nonsingular solution $(v^0, \alpha^0, 0)$ of (2.6). Applying the implicit function theorem to F at $(v^0, \alpha^0, 0)$ we obtain a unique solution curve $(v(t), \alpha(t), t)$ of (2.3) passing through $(v^0, \alpha^0, 0)$ at $t = 0$. The corresponding solution curve of (1.2) is given in (2.2). We conclude this process in an algorithm which is also used in the numerical approximations.

Algorithm 2.3: *Path following of solution branches of (1.2) beyond (u_0, λ_0).*
Step 1) *Computing (u_0, λ_0), ϕ_i, ϕ_i^*, $v_0, i = 1, 2$ with one of various extended systems;*
Step 2) *Setting up the reduced bifurcation equations (2.7) and calculating its solutions α^0;*
Step 3) *Solving $(v^0, 0, 0)$ from the nonsingular system*

$$\begin{pmatrix} D_u G_0^\Sigma & \phi_1 & \phi_2 \\ (\phi_1, \cdot) & 0 & 0 \\ (\phi_2, \cdot) & 0 & 0 \end{pmatrix} \begin{pmatrix} v \\ c_1 \\ c_2 \end{pmatrix} = - \begin{pmatrix} \frac{1}{2} D^2 G_0^\Sigma (\alpha_1^0 \phi_1 + \alpha_2^0 \phi_2 + v_0, 1)^2 \\ 0 \\ 0 \end{pmatrix},$$

one gets solutions $(v^0, \alpha^0, 0)$ of (2.6);
Step 4) *Starting at $(v^0, \alpha^0, 0)$, we do the path following of (2.3) with continuation methods;*
Step 5) *Reconstructing the solution curves of (1.2) by (2.2).*

Example 2.4: *If $D^2 G_0(v_0, 1)^2 \in R(D_u G_0)$, one of the equations in (2.7) is an ellipse and the solution $(0, 0)$ of (2.7) is not degenerate, then $G^\Sigma(u, \lambda) = 0$ has at least two, at most four different solution curves passing through (u_0, λ_0).*

Other situations of solving (2.7) can be studied in a similar manner (cf. Mei [6]).

3. Bifurcation Problems with the Additional Symmetry $Z_2 := \{1, -1\}$

For the product group $Z_2 \times \Gamma$ we define its action on X by

$$(\pm 1, \gamma)u = \pm(\gamma u) \quad \text{for all } \gamma \in \Gamma,\ u \in X.$$

If the mapping G is odd in u, i.e. $G(-u, \lambda) = -G(u, \lambda)$ for all $(u, \lambda) \in X \times \mathbf{R}$, then $\{(0, \lambda),\ \lambda \in \mathbf{R}\}$ is the trivial solution curve of (1.1). For $(u_0, \lambda_0) = (0, \lambda_0)$, if $D_{u\lambda}^2 G_0^\Sigma \phi_i \notin R(D_u G_0^\Sigma)$, $i = 1, 2$ and (1.3) holds, a Σ-branch $(u(t), \lambda(t))$ of (1.1) with $(u(0), \lambda(0)) = (u_0, \lambda_0)$ is necessarily of the form

$$\begin{cases} u(t) = t\alpha_1(t)\phi_1 + t\alpha_2(t)\phi_2 + t^3 v(t), \\ \lambda(t) = \lambda_0 + t^2 \beta(t), \qquad v(t) \in R((D_u G_0^\Sigma)^*). \end{cases} \tag{3.1}$$

Let $\beta(0) \neq 0$. We normalize $\beta(t)$ to 1 and determine (3.1) by the enlarged system

$$F(v, \alpha, t) := \begin{pmatrix} G^\Sigma(t\alpha_1\phi_1 + t\alpha_2\phi_2 + t^3 v,\ \lambda_0 + t^2)/t^3 \\ (\phi_1, v) \\ (\phi_2, v) \end{pmatrix} = 0. \tag{3.2}$$

The reduced bifurcation equations become a system of two cubic polynomials of α_1, α_2:

$$g(\alpha) := \begin{pmatrix} (\phi_i^*, D_{u\lambda}^2 G_0^\Sigma(\alpha_1\phi_1 + \alpha_2\phi_2) + \tfrac{1}{6} D_{uuu}^3 G_0^\Sigma(\alpha_1\phi_1 + \alpha_2\phi_2)^3) \\ i = 1, 2 \end{pmatrix} = 0. \tag{3.3}$$

Theorem 3.1: *The mappings F, g in (3.2), (3.3) are $Z_2 \times \Gamma$-equivariant.*

Compared with Γ, the larger group $Z_2 \times \Gamma$ often makes the solutions of (3.3) easier available than those of (2.7). For example, if there is a $\gamma \in \Gamma$ such that $\gamma\phi_1 = \phi_2$ and $\gamma\phi_2 = \phi_1$, the eight solutions of (3.3) are in the form

$$\pm(a, 0), \quad \pm(0, a), \quad \pm(b, b), \quad \pm(c, -c) \quad \text{with } a \cdot b \cdot c \neq 0 \tag{3.4}$$

and can be solved directly. By Algorithm 2.3 these eight solutions lead to eight solution branches of (3.2). Due to the Z_2-symmetry of G with respect to u, the solution $\pm(v^0, \alpha^0, 0)$ of (3.2) yields the same solution curve for $t \neq 0$. We obtain four different Σ-symmetric solution curves of (1.1) bifurcating at $(0, \lambda_0)$ from the trivial solution curve.

Example 3.2: Consider the simplified buckling problem

$$\begin{cases} \Delta u + \lambda \sin u = 0 \quad & \text{in} \quad \Omega := [0, 1] \times [0, 1], \\ u = 0 \quad & \text{on} \quad \partial\Omega. \end{cases} \tag{3.5}$$

A weak form of (3.5) in $X := H_0^1(\Omega)$ is

$$G(u, \lambda) := u + \lambda T \sin u = 0, \tag{3.6}$$

where $T : g \in L^2(\Omega) \mapsto Tg \in X$ is defined implicitly by

$$\int_\Omega \nabla Tg \nabla v \, dx dy = \int_\Omega gv \, dx dy, \qquad \forall v \in X.$$

T is self-adjoint, compact and $D_u G(0, \lambda) = I + \lambda T$ is a Fredholm operator with index 0 (cf. Vanderbauwhede [8]). Let D_4 be the isometry group of Ω. One sees easily that G is $Z_2 \times D_4$-equivariant and $(0, 50\pi^2)$ is a corank-3 bifurcation of (3.6). D_4-symmetric solution curves of (3.6) are described by the reduced problem

$$G^{D_4}(u, \lambda) = 0 \quad \text{for } (u, \lambda) \in X^{D_4} \times \mathbf{R}. \tag{3.7}$$

Now, $(0, 50\pi^2)$ is a corank-2 bifurcation point of (3.7) and

$$N(D_u G_0^{D_4}) = \operatorname{span}[\phi_1, \phi_2]$$

with $\phi_1 = 2 \sin 5\pi x \sin 5\pi y$, $\phi_2 = \sqrt{2}(\sin \pi x \sin 7\pi y + \sin 7\pi x \sin \pi y)$. The reduced bifurcation equations are very simple

$$\begin{cases} \alpha_1 [1 - 25\pi^2(\frac{3}{4}\alpha_1^2 + \alpha_2^2)] = 0, \\ \alpha_2 [1 - 25\pi^2(\alpha_1^2 + \frac{7}{8}\alpha_2^2)] = 0. \end{cases}$$

It has eight different isolated solutions which lead to four different D_4-symmetric solution branches $(u_i(t), \lambda_i(t))$, $i = 1, \ldots, 4$ bifurcating at $(0, 50\pi^2)$ from $\{(0, \lambda), \lambda \in \mathbf{R}\}$. The figures below show the nodal lines of $u^i(0)$, $i = 1, \ldots, 4$.

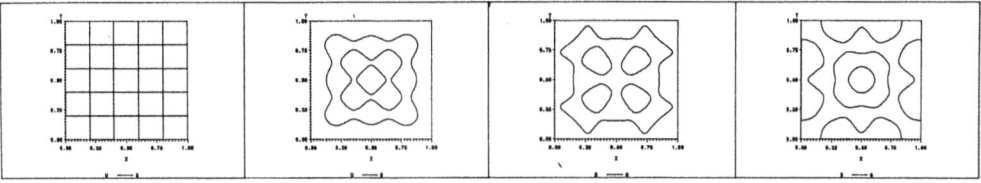

References

[1] Allgower, E. L.; Böhmer, K.; Mei, Z.: A generalized equibranching lemma with application to $D_4 \times Z_2$ symmetric elliptic problems, *Bericht des Fachbereichs Mathematik, University of Marburg* **9**, (1990)

[2] Allgower, E. L.; Georg, K.: **Introduction to numerical continuation methods**, Springer-Verlag, Berlin, Heidelberg, New York 1990

[3] Decker, D. W.; Keller, H. B.: Multiple limit point bifurcation, *J. Math. Anal. Appl.* **75**, 417-430 (1980)

[4] Dellnitz, M.; Werner, B.: Computational methods for bifurcation problems with symmetries-with special attention to steady state and Hopf bifurcation points, *J. Comp. Appl. Math.* **26**, 97-123 (1989)

[5] Golubitsky, M.; Stewart, I.; Schaeffer, D. G.: **Singularities and Groups in Bifurcation Theory**, Vol. II, Springer-Verlag, Heidelberg Berlin New York 1988

[6] Mei, Z.: Numerical approximations of corank-2 bifurcation problems, Dissertation, Department of Mathematics, University of Marburg 1989

[7] Seydel, R.: **From equilibrium to chaos-practical bifurcation and stability analysis**, Elsevier Publishing, New York, Amsterdam, London 1988

[8] Vanderbauwhede, A.: **Local bifurcation theory and symmetry**, Pitman 1982

The work was supported by Deutsche Forschungsgemeinschaft.

References

International Series of Numerical Mathematics, Vol. 97, © 1991 Birkhäuser Verlag Basel

Two-dimensional maps modelling periodically driven strictly dissipative oscillators

U. Parlitz, C. Scheffczyk, T. Kurz and W. Lauterborn

Institut für Angewandte Physik

TH Darmstadt, Schloßgartenstr. 7, D-6100 Darmstadt, FRG

Sinusoidally driven nonlinear oscillators like Duffing's equation reveal a typical pattern of bifurcation curves in parameter space that recurs with the resonances of the systems (see contribution [1] in these Proceedings and References therein). In order to improve our understanding of the underlying mechanisms and to facilitate further numerical investigations we consider in the following simple iterated mappings that show qualitatively the same behaviour. The derivation of such maps from a local (Taylor expansion) and a global (geometry of the flow) point of view and its bifurcation structure will be discussed in the following. Other work related to this topic may be found in Ref.7-11.

Local modelling

Within the period–doubling process an infinite number of periodic orbits are created by period-doubling- (pd) and saddle-node- (sn) bifurcations. It has been shown that in general the knowledge of low-period cycles and their stability can suffice to accurately compute dynamical invariants of the emerging strange attractors [3]. The numerical results concerning the bifurcation structure also suggest that the unstable low-periodic orbits that are generated by the first stages of period–doubling are responsible for the occurence of typical bifurcation patterns. Specifically, the m-fold Poincaré map restricted to a neighborhood of a period–m fixed-point $x_0^{(m)}$ can be expanded into a Taylor series. The third–order truncation contains 18 coefficients depending on various derivatives of the flow map at $x_0^{(m)}$. As in the case of the Hénon–map [4], results of Engel [5] can be applied to show that this polynomial approximation can be transformed into a cubic Hénon–type map

$$x_{n+1} = sx_n^3 + ax_n + b + y_n,$$
$$y_{n+1} = -cx_n. \tag{1}$$

It depends on three parameters a, b, c and a sign $s = \pm 1$, c being the area contraction factor that is assumed to be a constant. The two remaining free parameters are necessary to realize codimension-two bifurcation scenarios also observed in the oscillators investigated in Ref.1 and 2. Fig.1 gives the phase diagrams of the map (1) for both choices of the sign ($s = +1$ and $s = -1$) and $c = 0.1$. Only local bifurcations of orbits with periods up to four are displayed. The figures should be compared with the phase diagrams given in Ref.1

and 2. A detailed discussion and comparison with phase diagrams of nonlinear oscillators
will be given elsewhere [6].

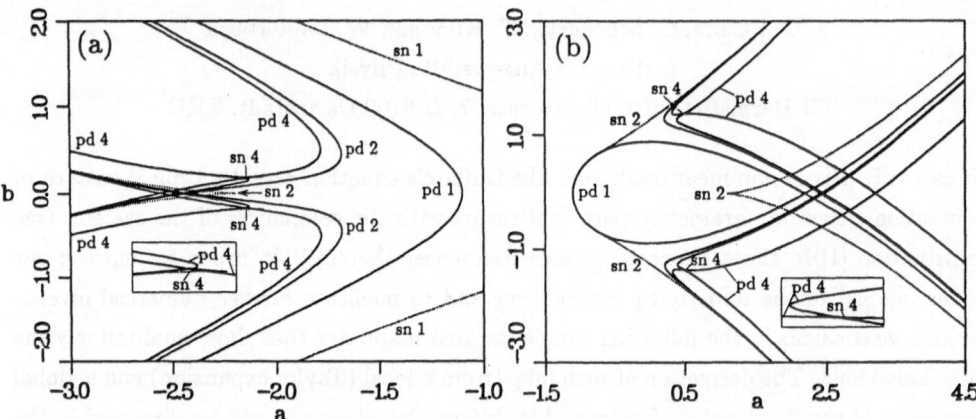

Fig.1 *Phase diagrams of the map (1) for c=0.1 and (a) s=+1 (b) s=−1.*

Global modelling

In the case of free and undamped oscillations the type of oscillators we intend to describe
possesses only a single point of equilibrium at the origin (minimum of the potential) and
closed orbits whose period is a strictly monotonous function of the energy (= integral of
motion). By means of a suitable choice of the coordinate system (action-angle-variables)
the free and undamped oscillations can be described as motions on concentric circles

$$
\begin{aligned}
x(t) &= r\cos(\alpha(t)), \\
y(t) &= r\sin(\alpha(t)).
\end{aligned}
\tag{2}
$$

The angular velocity $\dot{\alpha}$ and thus the period $T = 2\pi/\dot{\alpha}$ is given by a nonlinear function $\dot{\alpha} = f(r)$ of the radius r which is constant ($\dot{r} = 0$). A damped free oscillation is characterized
by a decreasing radius ($\dot{r} = g(r, \alpha) < 0$) and a trajectory that spirals into the only point
of equilibrium. The corresponding vector field v is given by

$$
\begin{pmatrix} \dot{x} \\ \dot{y} \end{pmatrix} = v\begin{pmatrix} x \\ y \end{pmatrix} = \begin{pmatrix} g(r,\alpha)\cos\alpha - yf(r) \\ g(r,\alpha)\sin\alpha + xf(r) \end{pmatrix} \quad \text{with} \quad \begin{array}{l} \alpha(x,y) = \arctan(y/x) \\ r(x,y) = \sqrt{x^2 + y^2} \end{array}
\tag{3}
$$

Motivated by the oscillators to be described we require the divergence $div(v)$ of v to be a
constant $-d$. This implies a damping function $\dot{r} = g(r) = -\frac{d}{2}r$ resulting in an exponential
damping $r(t) = r_0\exp(-\frac{d}{2}t)$. To drive the system it is subjected to T-periodic kicks in
x-direction with amplitude a. A (global) Poincaré map P (4) can be defined referring to
the coordinates (x_0, y_0) immediately after the kicks.

$$P(x_0, y_0) = \sqrt{c} \begin{pmatrix} \cos\beta & -\sin\beta \\ \sin\beta & \cos\beta \end{pmatrix} \begin{pmatrix} x_0 \\ y_0 \end{pmatrix} + \begin{pmatrix} a \\ 0 \end{pmatrix} \qquad (4)$$

with

$$\beta(r_0) = \int_0^T f(r(t))dt \qquad (r_0 = \sqrt{x_0^2 + y_0^2}),$$

$$c = e^{-\frac{d}{2}T} = det(D_x P) \quad \text{contraction rate of } P.$$

The function f contains the only nonlinearity of the map P. As an example we chose $f(r) = 1 + r^\gamma$ which corresponds to a hard spring that becomes asymptotically linear for small amplitudes. Solving the integral for β results in

$$\beta(r_0) = T + T\frac{1 - e^{-b}}{b}r_0^\gamma \quad \text{with} \quad b = \gamma\frac{d}{2}T = -\gamma\ln c. \qquad (5)$$

Scaling the coordinates x, y and the driving amplitude a with the factor ϵ

$$\epsilon = (T\frac{1 - e^{-b}}{b})^{-\frac{1}{\gamma}} > 0 \qquad (6)$$

and omitting the index 0 yields the following family of maps

$$P(x, y) = \sqrt{c} \begin{pmatrix} \cos\beta & -\sin\beta \\ \sin\beta & \cos\beta \end{pmatrix} \begin{pmatrix} x \\ y \end{pmatrix} + \begin{pmatrix} a \\ 0 \end{pmatrix} \quad \text{with} \quad \beta = T + r^\gamma = 2\pi p + (x^2 + y^2)^{\frac{\gamma}{2}} \quad (7)$$

This discrete model possesses all parameters that are typical for periodically driven oscillators, i.e. the amplitude of the driving a, the period of the driving $T = 2\pi p$, the damping constant $d = -2T\ln c$ and the nonlinearity parameter γ (here $\gamma = 2$). For small amplitudes ($r \to 0$) the map becomes linear and describes a harmonic oscillator with eigenfrequency 1. The map (7) shows the same periodic recurrence of bifurcation patterns as observed for driven oscillators upon variation of the driving frequency (see Ref. 1 and 2). This follows from the scaling (6) since in equation (7) the periods p and $p+1$ yield the same dynamical system. For this reason we restrict our numerical investigation of the bifurcation curves in parameter space to a single "unit" (resonance horn) of the bifurcation set. As decribed in Ref.1 and 2 the bifurcation curves are labeled by the torsion number and the period of the bifurcating orbits. The torsion number is well defined for the map (7) due to the knowledge of the underlying flow. Fig.2 shows two patterns of bifurcation curves of the map (7) for diffent values of the contraction rate c (damping). For small damping (i.e. large c) the sn(1,1) and pd(1,2) curves posses a common segment (indicated by the arrow in Fig. 2a) that shrinks when c is decreased until both curves separate as can be seen in Fig.2b. This behaviour has also been observed for the nonlinear oscillators that have been investigated in Ref.1 and 2.

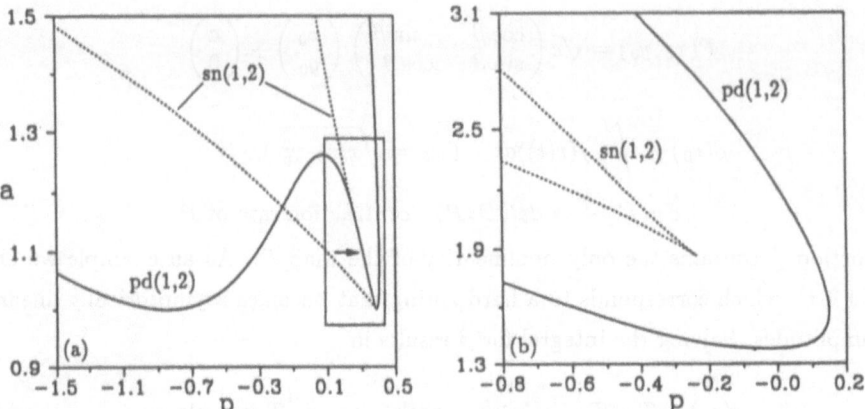

Fig.2: *Phase diagrams of the map (7) for (a) c = 0.5 and (b) c = 0.1.*

Fig.3 shows a sketch of the underlying bifurcation surfaces in the three-dimensional parameter space. The bifurcation curves in Fig.2b and in the frame of Fig. 2a correspond to
plane cross sections through the bifurcation surfaces in Fig.3 for different c-values. (With
respect to p Fig.2 and Fig.3 have opposite orientation !) The bifurcation surfaces separate
at the codimension-three point P.

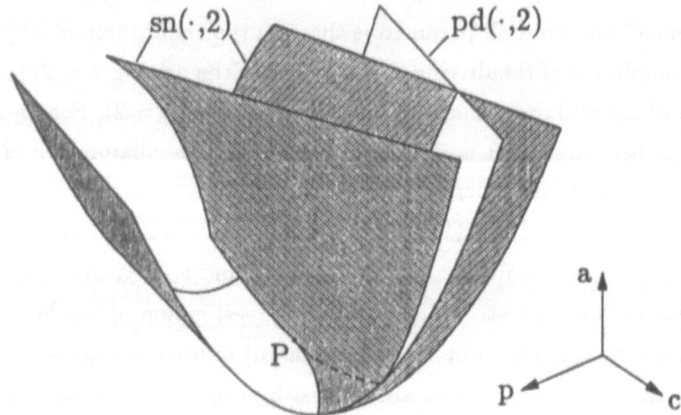

Fig.3: *Sketch of the pd(n,2) and sn(n,2) bifurcation surfaces in parameter space.*

The sn and pd bifurcation curves belonging to the first three steps of the period doubling
hierarchy are given in Fig.4. They agree completely with those computed for the driven
oscillators of Ref.1 and 2 where details and more information about the meaning of the
different curves can be found.

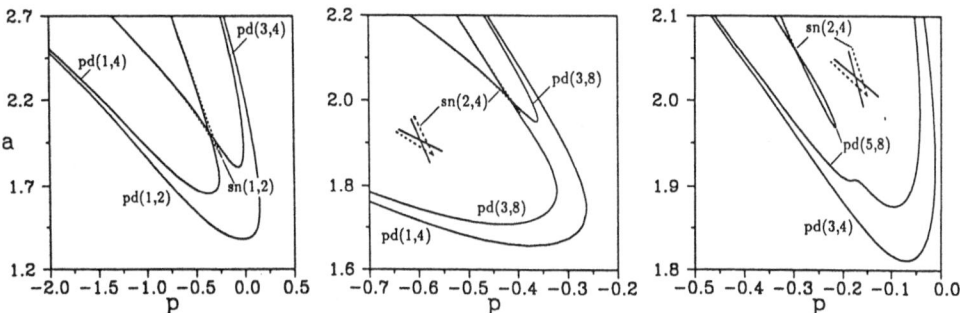

Fig.4: *Phase diagrams of the map (7) for c=0.1.*

It is conjectured that driven strictly dissipative oscillators and the global map models introduced above possess further common properties. This would offer new possibilities to explore the dynamics of nonlinear oscillators.

Acknowledgement

This work was supported by the Deutsche Forschungsgemeinschaft (SFB 185).

References

[1] C.Scheffczyk, U.Parlitz , T.Kurz and W.Lauterborn, this volume

[2] C.Scheffczyk, U.Parlitz, T.Kurz, W.Knop and W.Lauterborn, submitted for publication

[3] P.Cvitanovic, Phys.Rev.Lett. 61(24) (1988) 2729-2732

[4] M.Hénon, Commun. Math. Phys. 50 (1976) 69-74

[5] W.Engel, Math. Annalen 136 (1958) 319-325

[6] U.Parlitz, T.Kurz, C.Scheffczyk and W.Lauterborn, in preparation

[7] I. Procaccia, S. Thomae and C. Tresser, Phys. Rev. A 35(4) (1987) 1884-1900

[8] R. S. MacKay and C. Tresser, Physica D 27 (1987) 412-422

[9] J. Ringland, N. Issa and M. Schell, Phys.Rev.A 41(8) (1990) 4223-4235

[10] H. El Hamouly and C. Mira, Comptes rendus 294 série I (1982) 387-390

[11] D. Fournier, H. Kawakami and C. Mira, Comptes rendus 301 série I (1985) 223-228

Fig. 4. Phase diagram of the map (7) for $a = 4$.

It is conjectured that driven simple dissipative oscillators that have gain, may also fall into traps, or possess intense random properties. This would offer new possibilities to explore the dynamics of nonlinear oscillation.

Acknowledgement

This work was supported by the Deutsche Forschungsgemeinschaft (SFB 185).

References

[1] O.S. Iselings, C. Grebiat, E. Kutz and M. Hantschmann, this volume.

[2] C. Sitton, N.C. Grebiat, T. Kutz, W. Kaup and W. Bauknat, to be submitted for publication.

[3] E.G. Steinmar's problem: 240 questions and answers with solutions.

[4] W. Jaeger, Astron. Nachr. 100 (1932) 219-229.

[5] E.J. Rutherford, and G. Elliott, Population in ga plasma, in preparation.

[6] Louis Poincaré, Comptes rend. re., leaves et l'appartement, A. Kluer (1897) 100-104.

[7] E.J.C. Meiss, and G. Grebiat, Optics 21 (1971) 31-35.

[8] Louis Poincaré and Fritz Bech, Phys. Rev. A 4 (1972) 3-4235.

[9] R.M. Devanant, and F.J. Mira, Complex analysis and appl. (1985) 227-230.

[10] E.M. Comptes., B. Ramanuan and G. Meiss, Complex analysis with applications, 105-190.

ON COMPUTING COUPLED TURNING POINTS OF PARAMETER DEPENDENT NONLINEAR EQUATIONS

Gerd Pönisch

1. THE PROBLEM

Let us given an n-dimensional nonlinear system

$$F(x,p) = 0 \qquad (1.1a)$$

that expresses implicitly the relation between the state variables $x \in \mathbb{R}^{n+1}$ and a selected design parameter $p \in \mathbb{R}^1$. In general, the system (1.1a) describes a real process that depends on some auxiliary parameters. In this paper we consider only one design parameter p. Moreover, one variable of the vector $x \in \mathbb{R}^{n+1}$, say the (n+1)-th component x_{n+1}, is easily controllable and plays the role of the unfolding parameter.

In many applications, besides the numerical approximation of the solution manifold of (1.1a) the computation of turning points w.r.t. x_{n+1} is required for given values of $p \in \mathbb{R}^1$. For determining such turning points by direct methods the underdetermined system (1.1a) is extended by an appropriate turning point condition

$$f(x,p) = 0 \, , \qquad (1.1b)$$

see [2] for example.

Now, the parameter p is not assumed to be given in advance. Then, a well-posed inverse problem consists in determining p such that two turning points $x^{+,*}$ and $x^{-,*}$ satisfy one given additional equation. This problem can be characterized by the coupled system

$$F(x^i,p) = 0 \, , \quad i=+,- \, , \qquad (1.2a)$$

$$f(x^i,p) = 0 \, , \quad i=+,- \, , \qquad (1.2b)$$

$$g(x^+,x^-,p) = 0 \, . \qquad (1.2c)$$

Such a problem arises in the analysis of nonlinear networks which are modeled by (1.1a). If we consider a trigger circuit the transfer characteristic curve is dominated by two turning points w.r.t. x_{n+1} (input voltage) for a fixed value of p. The engineers want to adjust the two turning points x^+, x^- such that, for a given $\sigma > 0$, the so-called gap condition

$$g(x^+,x^-,p) := (e^{n+1})^T(x^+ - x^-) - \sigma = x^+_{n+1} - x^-_{n+1} = 0 \, , \qquad (1.3)$$

is satisfied, where e^{n+1} denotes the (n+1)-th unit vector of the natural basis in $\mathbb{R}^{n+1}$. Based on (1.2) several techniques can be applied to compute the coupled turning points.

In section 3 we shortly describe a direct method for solving (1.2) by a nonstandard block elimination method of Newton-type which considers the special internal structure of (1.2). An indirect method based on the local parametrization of the turning point branch by the parameter p is sketched in the next section.

2. AN INDIRECT METHOD

Let us assume that F be sufficiently smooth. Furthermore we assume that, for a fixed parameter $p = p^*$, there are two turning points x^{+*} and x^{-*} w.r.t. x_{n+1}, i.e.

(A1) $F(x^{i*}, p^*) = 0$, $i=+,-$,

(A2) $\mathfrak{N}(\partial_1 F(x^{i*}, p^*)) = span\{v^{i*}\}$, $(e^{n+1})^T v^{i*} = 0$, $i=+,-$,

(A3) $(\psi^{i*})^T \partial_{11} F(x^{i*}, p^*) v^{i*} v^{i*} \neq 0$, $i=+,-$,.

where $\psi^{i*} \in \mathbb{R}^n$ is the unique solution of

$$\partial_1 F(x^{i*}, p^*)^T \psi^{i*} = e^{n+1} \quad , \quad i=+,- .$$

Here $\mathfrak{N}$ denotes the nullspace of a linear operator.

By varying p we can study the curve of turning points

$$\mathfrak{L} := \{(x,p) \in \mathbb{R}^{n+1} \times \mathbb{R}^1 : F(x,p)=0, \ f(x,p)=0\}. \tag{2.1}$$

The generic situation is sketched in Fig. 1. In the generic case the turning point curve $\mathfrak{L}$ has again a turning point $(\hat{x}, \hat{p})$ w.r.t. p. This critical point $(\hat{x}, \hat{p})$ is called hysteresis point at which the simple turning points x^+ and x^- w.r.t. x_{n+1} coincide in a double turning point w.r.t. x_{n+1}. Hence, $\mathfrak{L}$ can not be smoothly parametrized by p in the neighborhood I of $(\hat{x}, \hat{p})$.

In the following we consider the turning point curve $\mathfrak{L}$ excluding the neighborhood I of $(\hat{x}, \hat{p})$. Under standard assumptions the remaining set $\mathfrak{L} \backslash I$ consists of two branches $\mathfrak{L}^+$ and $\mathfrak{L}^-$ each of those can be parametrized by p, see Figure 2. Let $(x^+(p), p)$ and $(x^-(p), p)$ be the local parametrization of $\mathfrak{L}^+$ and $\mathfrak{L}^-$, resp., i.e.,

$$F(x^+(p),p)=0 \ , \ f(x^+(p),p)=0 \text{ and } F(x^-(p),p)=0 \ , \ f(x^-(p),p)=0 \tag{2.2}$$

in a neighborhood of p^*, where $p^* \in \mathbb{R}^1$ is a root of the equation

$$g(x^+(p), x^-(p), p) = 0 . \qquad (2.3)$$

This description motivates the following procedure for computing $(x^{+*}, x^{-*}, p^*) \in \mathbb{R}^{n+1} \times \mathbb{R}^{n+1} \times \mathbb{R}^1$.

Procedure 2.1:

Compute initial guesses p_0, $x^{+,0}$, $x^{-,0}$ by a continuation process. Choose $\delta \geq 0$ and a sequence $\{\varepsilon_k\}$ with $\varepsilon_k \geq 0$. Set $k := 0$.

<u>repeat</u>   (i) Compute a new parameter estimation p_{k+1} from a linearization of (2.3) w.r.t. p at p_k.

   (ii) <u>for</u> $i := +, -$ <u>do</u>

   Compute $x^{i,k+1}$ as an approximation to $x^i(p_{k+1}) \in \mathbb{R}^i$ from

$$F(x^i, p_{k+1}) = 0 , \quad f(x^i, p_{k+1}) = 0 \qquad \text{according to}$$

$$\|F(x^{i,k+1}, p_{k+1})\| + |f(x^{i,k+1}, p_{k+1})| \leq \varepsilon_k .$$

   (iii) $k := k+1$.

<u>until</u> $|g(x^{+,k+1}, x^{-,k+1}, p_{k+1})| \leq \delta$.   ∎

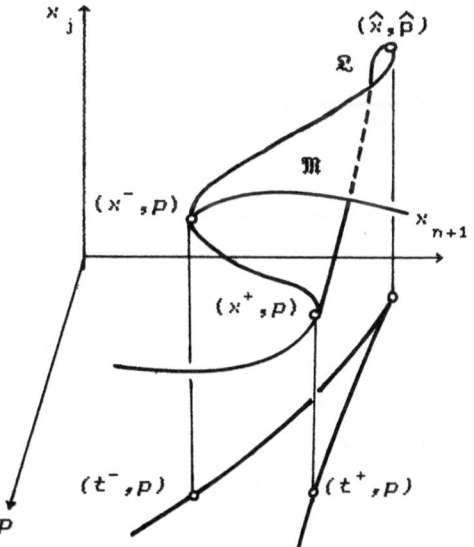

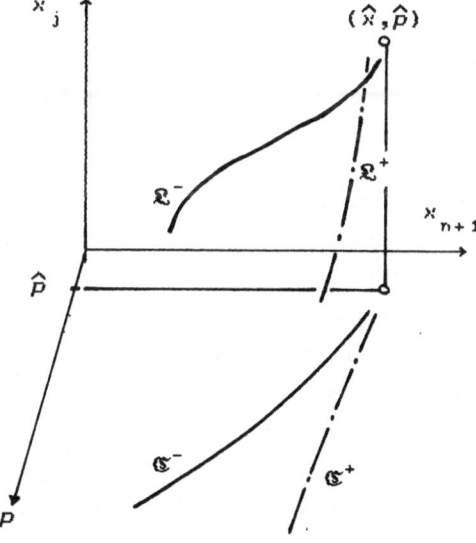

Figure 1:   The Solution set $\mathfrak{M} = \{(x,p): F(x,p) = 0\}$ containing a turning point curve $\mathcal{L}$

Figure 2:   The splitting of the turning point curve $\mathcal{L}$ and their projection on the x_{n+1}, p-plane

We suggest the application of Newton's method at step (i), i.e.

$$P_{k+1} := P_k - g(x^{+,k}, x^{-,k}, P_k) \big/ \frac{d}{dp} g(x^{+,k}, x^{-,k}, P_k) \; . \tag{2.4}$$

The Newton step (2.4) is locally well-defined if we assume that

(A4) $g(x^{+*}, x^{-*}, p^*) = 0$,

and

(A5)
$$0 \neq \frac{d}{dp} g(x^+(p^*), x^-(p^*), p^*) = [\partial_3 g(z^*) + \partial_1 g(z^*)\dot{x}^+(p^*) +$$
$$\partial_2 g(z^*)\dot{x}^-(p^*)]_{z^* = (x^{+*}, x^{-*}, p^*)}$$

where $\dot{x}^+$ and $\dot{x}^-$ denote the derivatives of x^+ and x^- w.r.t. p, resp. Assumption (A5) is natural and means for g given by (1.3) that the projections of the tangent directions to $\mathfrak{L}^+$ and $\mathfrak{L}^-$ into the (x_{n+1}, p)-plane are not parallel, see [1].

Since (2.4) requires the evaluation of $\dot{x}^+(p_k)$ and $\dot{x}^-(p_k)$ and $\dot{x}^i$ can be obtained by implicit differentiation w.r.t. p from (2.2), the choice of the function f influences the computational costs essentially. For this reason and in view of step (ii) we choose the function $f: \mathbb{R}^{n+1} \times \mathbb{R}^1 \to \mathbb{R}^1$ as

$$f(x^i, p) = f_i(x^i, p) := (e^{n+1})^T B_i^{-1} e^{n+1} \qquad , \; i = +, -, \tag{2.5}$$

with

$$B_i := B(x^i, p, r^i) = \begin{bmatrix} \partial_1 F(x^i, p) \\ \cdots\cdots\cdots \\ (r^i)^T \end{bmatrix} \in L(\mathbb{R}^{n+1}) \tag{2.6}$$

where the auxiliary vectors $r^i \in \mathbb{R}^{n+1}$ are chosen such that the $(n+1) \times (n+1)$ matrix B_i is nonsingular in the neighborhood of the considered turning points $x^+(p^*)$ and $x^-(p^*)$, resp.

At step (ii) we apply the direct turning point method [2] that bases on the same two-stage idea employed here for the calculation of $\dot{x}^i$.

In [1] it is shown that, under the assumptions (A1-A5), Procedure 2.1 is well-defined in a neighborhood of $z^* := (x^{+*}, x^{-*}, p^*)$ and the sequence $\{(x^{+,k}, x^{-,k}, p_k)\}$ converges to z^* if, additionally,

$$\lim_{k \to \infty} \varepsilon_k = 0$$

holds. A special choice of $\{\varepsilon_k\}$ even ensures a R-convergence order ≥ 2 .

3. A DIRECT METHOD

In order to solve the system (1.2) directly, we propose a special implementation of the Newton's method. Let $z = (x^+, x^-, p)$ be the current iterate that is supposed to be sufficiently close to the solution z^*. Then the next iterate $\tilde{z} = (\tilde{x}^+, \tilde{x}^-, \tilde{p})$ is defined by

$$\tilde{x}^i := x^i - \Delta x^i \quad , \quad i=+,- \quad , \quad \tilde{p} := p - \Delta p$$

where Δx^i and Δp are the solution of the linear system

$$\partial_1 F(x^i, p) \Delta x^i + \partial_2 F(x^i, p) \Delta p = F(x^i, p) \quad , \quad i=+,- \quad , \quad (3.1a)$$

$$\partial_1 f_i(x^i, p) \Delta x^i + \partial_2 f_i(x^i, p) \Delta p = f_i(x^i, p) \quad , \quad i=+,- \quad , \quad (3.1b)$$

$$\partial_1 g(x^+, x^-, p) \Delta x^+ + \partial_2 g(x^+, x^-, p) \Delta x^- + \partial_3 g(x^+, x^-, p) \Delta p = g(x^+, x^-, p). \quad (3.1c)$$

Due to the regularity of $B(x^i, p, \partial_1 f_i(x^i, p)^T)$ in the neighborhood of (x^{i*}, p^*) we can preeliminate Δx^i from (3.1a,b) in dependence on Δp, where we consider the special structure of (3.1). At first, we represent the general solution $(\Delta x^i, \Delta p)$ of (3.1a) as

$$\Delta x^i := s^i + \lambda_i v^i - \Delta p q^i \quad , \quad \lambda_i \in \mathbb{R}^1 \text{ and } \Delta p \in \mathbb{R}^1 \text{ arbitrary}, \quad (3.2)$$

where s^i, v^i and q^i are defined by

$$B(x^i, p, r^i) v^i = e^{n+1} \quad ,$$

$$B(x^i, p, r^i) \, s^i := \left[\begin{array}{c} F(x^i, p) \\ \hline 0 \end{array} \right] \quad , \quad B(x^i, p, r^i) q^i := \left[\begin{array}{c} \partial_2 F(x^i, p) \\ \hline 0 \end{array} \right] .$$

In order to determine λ_i we insert (3.2) into (3.1b) and obtain

$$\lambda_i \partial_1 f_i(x^i, p) v^i = f_i(x^i, p) - \partial_1 f_i(x^i, p) s^i + [\partial_1 f_i(x^i, p) q^i - \partial_2 f_i(x^i, p)] \Delta p. \quad (3.3)$$

Considering the continuity of $\partial_1 f_i$ assumption (A3) implies

$$\nu_i := \partial_1 f_i(x^i, p) v^i \neq 0 \quad (3.4)$$

in the neighborhood of (x^{i*}, p^*). Therefore we obtain

$$\lambda_i := \gamma_i + \kappa_i \Delta p \quad , \quad (3.5)$$

where

$$\gamma_i := (f_i(x^i, p) - \partial_1 f_i(x^i, p) s^i)/\nu_i \quad , \quad \kappa_i := (\partial_1 f_i(x^i, p) q^i - \partial_2 f_i(x^i, p))/\nu_i.$$

Replacing λ_i in (3.2) by (3.5) we achieve at

$$\Delta x^i := s^i + \gamma_i v^i + \Delta p [\kappa_i v^i - q^i] . \quad (3.6)$$

Now we insert (3.6) into (3.1c) and obtain

$$[\partial_3 g(z) + \partial_1 g(z)(\kappa_+ v^+ - q^+) + \partial_2 g(z)(\kappa_- v^- - q^-)] \Delta p = g(z) + \quad (3.7)$$

$$\partial_1 g(z)(\gamma_+ v^+ + s^+) + \partial_2 g(z)(\gamma_- v^- + s^-).$$

Note, that, for $z=z^*$, the term in brackets is assumed to be non-zero according to (A5). Hence, Δp and Δx^i, can be computed by (3.7) and (3.6) provided that z is close to z^*.

Since the procedure described by (3.2), (3.5), (3.6) and (3.7) is an implementation of one step of Newton's method, the assumptions (A1-A5) imply the method to be locally Q-quadratically convergent when starting from a sufficiently good initial guess z^0.

The computational costs can drastically be reduced if the directional derivatives of f in (3.4) and (3.5) are approximated by efficient divided difference formulas requiring only a few additional values of F, see [3]. In this way one reduces the overall number of algebraic operations essentially and one avoids the preparation of the second order partial derivatives of F.

For sufficiently small discretization stepsizes and under the assumptions (A1-A5) it can be proved that the resulting method has the typical properties of Newton-type methods. For further considerations we refer to Method I in [3], which is a general method for computing m turning points satisfying l coupling equations.

The indirect method as well as the direct method are tested on some examples. In view of the computational expense both mentioned methods are comparable. The indirect method can be recommended when a continuation method for turning point branches is available. Then, only the step (i) has to be implemented additionally. The direct method should be preferred in the case of an interactive simulation program.

REFERENCES

[1]   G. PÖNISCH AND V. JANOVSKY, *Coupled turning points of two-parameter dependent nonlinear equations*, Preprints 07-15-89, 07-16-89, Technical Univ. of Dresden, Dresden, 1989.

[2]   G. PÖNISCH AND H. SCHWETLICK, *Computing turning points of curves implicitly defined by nonlinear equations depending on a parameter*, Computing 26 (1981), pp. 107-121.

[3]   G. PÖNISCH AND H. SCHWETLICK, *Some types of inverse problems for nonlinear equations having coupled turning points,* Preprint 07-01-90,Technical Univ. of Dresden, Dresden,1990.

Author's adress: Gerd Pönisch
                     Technical Univ. of Dresden, Dep. of Mathematics
                     Mommsenstr. 13 , O - 8027 Dresden , Germany

International Series of Numerical Mathematics, Vol. 97, © 1991 Birkhäuser Verlag Basel

Generating Hopf Bifurcation Formulae with MAPLE

E. J. PONCE-NÚÑEZ, E. GAMERO

Dept. Applied Mathematics, University of Sevilla, E.T.S.I.I.

Avda. Reina Mercedes s/n, 41012–SEVILLA, Spain

ABSTRACT. Computing certain relevant coefficients for Hopf bifurcations is of interest to characterize the bifurcation (ie, to determine direction, kind, stability and number of eventual emanating branches). The calculations involved are tedious and very prone to error, thereby being nearly impossible to do them by hand.

Using an algorithm previously developed by Freire *et al*, which is well suited for symbolic computation, a MAPLE procedure has been developed and tested.

In particular, the explicit general expressions of the coefficients of Hopf bifurcation up to fifth degree have been obtained. For specific cases, higher degree results are also reachable.

Introduction

As it is well-known, the Hopf bifurcation theory ([8], [11]) explains the appearance of periodic oscillations in nonlinear dynamical systems, as the bifurcation parameter varies from a critical value. However, to take advantage of this theory long computations are needed to obtain the coefficients which determine the direction, stability, number and local amplitude of periodic orbits in a neighbourhood of the bifurcation parameter value.

As the computations are very error inclined, it is of relevance to develop computational procedures to fulfill this task in an accurate and convenient way. To fix ideas, suppose we are given the system:

$$\begin{pmatrix} \dot{x} \\ \dot{y} \end{pmatrix} = \begin{pmatrix} \alpha(\lambda) & -\omega(\lambda) \\ \omega(\lambda) & \alpha(\lambda) \end{pmatrix} \begin{pmatrix} x \\ y \end{pmatrix} + \begin{pmatrix} f(x,y) \\ g(x,y) \end{pmatrix} \tag{1}$$

where λ is the bifurcation parameter, the linear part is indicated, f, g are smooth, $\omega(0) = \omega_0 > 0$, $\alpha(0) = 0$ and $\alpha'(0) \neq 0$ (transversality condition). Then for $\lambda = 0$, we are dealing with a Hopf bifurcation and –to characterize it– some changes of variables have to be done in order to put the system in the corresponding Poincare-Birkhoff normal form:

$$\begin{aligned} \dot{x} &= -\omega_0 y + \sum_{j\geq 1} \left[a_j(x^2+y^2)^j x - b_j(x^2+y^2)^j y \right] \\ \dot{y} &= \omega_0 x + \sum_{j\geq 1} \left[a_j(x^2+y^2)^j y + b_j(x^2+y^2)^j x \right] \end{aligned} \tag{2}$$

It is easy to realize that the last expression is directly related with the following polar form:

$$\dot{r} = r \sum_{j \geq 1} a_j r^{2j}$$

$$\dot{\theta} = \omega_0 + \sum_{j \geq 1} b_j r^{2j}$$

and as $\omega(0) = \omega_0 > 0$, that the local behaviour will be determined by the equation for r. So, the coefficients a_j give us the information needed to characterize the present bifurcation.

Because of transversality condition, it suffices to study the system (1) for $\lambda = 0$, which is now rewritten in the form:

$$\dot{x} = -\omega_0 y + \sum_{k \geq 2} f_k(x, y)$$

$$\dot{y} = \omega_0 x + \sum_{k \geq 2} g_k(x, y)$$

(3)

where $f_k, g_k \in V(k)$, the space of homogeneous polynomials in x, y of degree k. So, our problem is to pass from (3) to (2).

Problem approach

In applications, one normally deals with n-dimensional systems, $n > 2$. Thus, a first duty is to formulate a 2-dimensional system which summarizes the bifurcation behaviour of the original one. The center manifold theory can help us in a convenient way ([1]) and we can use an efficient algorithm to get the dimensional reduction required ([5], [7]). Here, we suppose that we have a system as in (3) and we restrict ourselves to obtain the corresponding (2) system.

With such a purpose, in [6] a symbolic algorithm was presented making use of the Lie transformation theory ([3], [4]) to avoid unnecessary calculations. A REDUCE ([10]) program was available and some results had been obtained.

Following the same ideas, we have developed AMPHOB, a MAPLE ([2]) program which seems superior to the previous code. The program will appear elsewhere and it is available on request to the authors.

Some results of the MAPLE program

We ran AMPHOB on an Amdhal at UMRCC selecting a virtual machine of 12 MB. In the sequel, first results of the code are presented.

For generic f and g, using the program, it is a matter of seconds to obtain the first coefficient:

$$a_1 = 1/16(f_{yy}g_{yy} - f_{xx}g_{xx} + f_{xx}f_{xy} - g_{xy}g_{xx} - g_{xy}g_{yy}$$
$$+ f_{yy}f_{xy} + g_{yyy}\omega_0 + f_{xyy}\omega_0 + g_{xxy}\omega_0 + f_{xxx}\omega_0)/\omega_0$$

which was wellknown ([8], [9]), but we can also present the next one –of interest in more degenerate cases–, never explicitly computed:

$$
\begin{aligned}
a_2 = \; & 1/1152(10g_{yyy}g_{yy}^2\omega_0 + 3f_{yyyy}g_{yy}\omega_0^2 + 10g_{yyyy}f_{yy}\omega_0^2 - 8g_{xxxy}f_{xy}\omega_0^2 \\
& -10g_{xxxy}g_{xx}\omega_0^2 - 11f_{xxxx}g_{yy}\omega_0^2 - 10f_{xxxx}g_{xx}\omega_0^2 + 5g_{yyyy}g_{xy}\omega_0^2 \\
& +11g_{yyyy}f_{xx}\omega_0^2 + 10f_{xyyy}f_{yy}\omega_0^2 + 8f_{xyyy}g_{xy}\omega_0^2 - 14g_{xxxy}g_{yy}\omega_0^2 \\
& +15f_{xy}f_{xxy}f_{yy}\omega_0 + 21f_{xy}f_{xxy}f_{xx}\omega_0 + 3f_{xy}g_{xxx}f_{yy}\omega_0 + 8f_{xy}g_{xxx}g_{xy}\omega_0 \\
& +5f_{xy}g_{xxx}f_{xx}\omega_0 + 3g_{xx}f_{yyy}g_{xy}\omega_0 + 3g_{xx}f_{yyy}f_{xx}\omega_0 + 15g_{xx}g_{xyy}g_{xy}\omega_0 \\
& -15g_{xx}g_{xyy}f_{xx}\omega_0 - 6g_{xx}f_{xxy}f_{yy}\omega_0 - 21g_{xx}f_{xxy}f_{xx}\omega_0 + 13g_{xx}g_{xxx}g_{xy}\omega_0 \\
& +13g_{xx}g_{xxx}f_{xx}\omega_0 + 13g_{yy}f_{yyy}f_{yy}\omega_0 + 5g_{yy}f_{yyy}g_{xy}\omega_0 - 21g_{yy}g_{xyy}f_{yy}\omega_0 \\
& +21g_{yy}g_{xyy}g_{xy}\omega_0 + 10g_{xxy}g_{xy}^2\omega_0 - 2f_{xxx}g_{xy}^2\omega_0 + 2g_{yyy}g_{xy}^2\omega_0 \\
& -2f_{xyy}g_{xy}^2\omega_0 + 14f_{xyyy}f_{xx}\omega_0^2 - 3g_{xxxx}f_{xx}\omega_0^2 + 6f_{xxxy}f_{yy}\omega_0^2 \\
& +18f_{xxxy}f_{xx}\omega_0^2 - 3g_{xxxx}g_{xy}\omega_0^2 - 6g_{xx}g_{xyy}f_{yy}\omega_0 + 10g_{yyy}f_{yy}^2\omega_0 \\
& +5g_{yyy}f_{xx}^2\omega_0 - 8f_{xyy}g_{yy}^2\omega_0 - 5f_{xyy}g_{xx}^2\omega_0 + 10f_{xyy}f_{yy}^2\omega_0 \\
& +19f_{xyy}f_{xx}^2\omega_0 + 19g_{xxy}g_{yy}^2\omega_0 + 10g_{xxy}g_{xx}^2\omega_0 + f_{xxx}f_{yy}f_{xx}\omega_0 \\
& -5g_{xxy}f_{yy}^2\omega_0 + 5f_{xxx}g_{yy}^2\omega_0 + 10f_{xxx}g_{xx}^2\omega_0 - 5g_{xxy}f_{yy}f_{xx}\omega_0 \\
& +25g_{xxy}g_{yy}g_{xx}\omega_0 + 15f_{xxx}g_{yy}g_{xx}\omega_0 - 5f_{xxx}f_{yy}^2\omega_0 + 10f_{xxx}f_{xx}^2\omega_0 \\
& +3f_{yyyy}f_{xy}\omega_0^2 + g_{yyy}g_{yy}g_{xx}\omega_0 + 15g_{yyy}f_{yy}f_{xx}\omega_0 - 5f_{xyy}g_{yy}g_{xx}\omega_0 \\
& +25f_{xyy}f_{yy}f_{xx}\omega_0 - 5g_{yyy}g_{xx}^2\omega_0 + 6f_{xxxyy}\omega_0^3 + 6g_{xxyyy}\omega_0^3 \\
& +3g_{xxxxy}\omega_0^3 + 3f_{xxxxx}\omega_0^3 + 3g_{yyyyy}\omega_0^3 + 3f_{xyyyy}\omega_0^3 - 10g_{xy}^3g_{yy} \\
& -10g_{xy}^3g_{xx} - 9f_{xx}^3g_{yy} + 10f_{xy}^3f_{yy} + 10f_{xy}^3f_{xx} + 9g_{yy}^3f_{xx} \\
& +10f_{xx}^3f_{xy} - 10f_{xx}g_{xx}^3 - 10f_{xx}^3g_{xx} + 10f_{yy}g_{yy}^3 + 10f_{yy}^3g_{yy} \\
& -10g_{xy}g_{yy}^3 - 10g_{xy}g_{xx}^3 + 10f_{yy}^3f_{xy} - 9f_{xy}^2g_{yy}f_{yy} + 2f_{xy}^2g_{yy}g_{xy} \\
& -13f_{xy}^2g_{yy}f_{xx} - 5f_{xy}^2g_{xx}f_{yy} + 2f_{xy}^2g_{xx}g_{xy} - 9f_{xy}^2g_{xx}f_{xx} \\
& -18g_{yy}f_{xy}g_{xx}g_{xy} + g_{yy}^2f_{xy}g_{xy} - 19f_{xy}g_{xx}^2g_{xy} - 2g_{xy}^2f_{xx}f_{xy} \\
& -g_{xy}f_{xx}^2f_{xy} + 19f_{yy}^2g_{xy}f_{xy} - 2f_{yy}g_{xy}^2f_{xy} + 9g_{xy}^2g_{yy}f_{yy} \\
& +13g_{xy}^2g_{yy}f_{xx} + 5g_{xy}^2g_{xx}f_{yy} + 9g_{xy}^2g_{xx}f_{xx} + 20f_{xx}^2f_{xy}f_{yy} \\
& +5f_{xx}g_{xx}g_{yy}^2 - 10f_{xx}g_{xx}^2g_{yy} + 20f_{xx}f_{xy}f_{yy}^2 + 5f_{xx}g_{xx}f_{yy}^2 \\
& -f_{xx}^2g_{xx}f_{yy} - 5f_{xx}f_{xy}g_{yy}^2 - 4f_{xx}f_{xy}g_{yy}g_{xx} - 9f_{xx}f_{xy}g_{xx}^2 \\
& +f_{yy}g_{yy}^2g_{xx} - 4f_{yy}f_{xy}g_{yy}g_{xx} - 5f_{yy}f_{xy}g_{xx}^2 + 4g_{xy}g_{xx}f_{yy}f_{xx} \\
& +9g_{xy}g_{xx}f_{xx}^2 - 9f_{yy}f_{xy}g_{yy}^2 + 9g_{xy}g_{yy}f_{yy}^2 + 5g_{xy}g_{yy}f_{xx}^2 \\
& -5f_{yy}g_{yy}f_{xx}^2 - 20g_{xy}g_{yy}^2g_{xx} - 20g_{xy}g_{yy}g_{xx}^2 + 10f_{yy}^2g_{yy}f_{xx} \\
& -5f_{yy}g_{yy}g_{xx}^2 + 6g_{xxyy}f_{yy}\omega_0^2 - 6g_{xxyy}g_{xy}\omega_0^2 + 12g_{xxyy}f_{xx}\omega_0^2 \\
& +6f_{xxyy}f_{xy}\omega_0^2 + 10f_{xyy}f_{xy}^2\omega_0 - 2g_{xxy}f_{xy}^2\omega_0 + 2f_{xxx}f_{xy}^2\omega_0 \\
& -2g_{yyy}f_{xy}^2\omega_0 - 5g_{xx}f_{xyy}f_{xy}\omega_0 + 19g_{xx}g_{xxy}f_{xy}\omega_0 + 9g_{xx}f_{xxx}f_{xy}\omega_0
\end{aligned}
$$

$$-7g_{yy}f_{xyy}f_{xy}\omega_0 + 17g_{yy}g_{xxy}f_{xy}\omega_0 + 7g_{yy}f_{xxx}f_{xy}\omega_0 + g_{xx}g_{yyy}f_{xy}\omega_0$$
$$-12g_{yy}f_{xxyy}\omega_0^2 - 6g_{xx}f_{xxyy}\omega_0^2 + 3g_{xyy}f_{xxx}\omega_0^2 - 3f_{xxy}g_{yyy}\omega_0^2$$
$$+9f_{xxy}f_{xyy}\omega_0^2 - 9f_{xxy}g_{xxy}\omega_0^2 + 3f_{xxy}f_{xxx}\omega_0^2 + 3g_{xxx}g_{yyy}\omega_0^2$$
$$+3g_{xxx}f_{xyy}\omega_0^2 - 3g_{xxx}f_{xxx}\omega_0^2 + 3f_{yyy}g_{yyy}\omega_0^2 + 3f_{yyy}f_{xyy}\omega_0^2$$
$$-3f_{yyy}g_{xxy}\omega_0^2 - 3f_{yyy}f_{xxx}\omega_0^2 - 3g_{xyy}g_{yyy}\omega_0^2 + 9g_{xyy}f_{xyy}\omega_0^2$$
$$-9g_{xyy}g_{xxy}\omega_0^2 - g_{yy}g_{yyy}f_{xy}\omega_0 + 17f_{xyy}f_{xx}g_{xy}\omega_0 - 5g_{xxy}g_{xy}f_{yy}\omega_0$$
$$+f_{xxx}f_{yy}g_{xy}\omega_0 - f_{xxx}g_{xy}f_{xx}\omega_0 + 9g_{yyy}g_{xy}f_{yy}\omega_0 + 7g_{yyy}g_{xy}f_{xx}\omega_0$$
$$+19f_{xyy}f_{yy}g_{xy}\omega_0 - 15g_{yy}f_{xxy}f_{yy}\omega_0 + 3g_{yy}f_{xxy}g_{xy}\omega_0 - 30g_{yy}f_{xxy}f_{xx}\omega_0$$
$$+3g_{yy}g_{xxx}f_{yy}\omega_0 + 11g_{yy}g_{xxx}g_{xy}\omega_0 + 8g_{yy}g_{xxx}f_{xx}\omega_0 + 13f_{xy}f_{yyy}f_{yy}\omega_0$$
$$+8f_{xy}f_{yyy}g_{xy}\omega_0 + 11f_{xy}f_{yyy}f_{xx}\omega_0 + 9f_{xy}g_{xyy}f_{yy}\omega_0 + 3f_{xy}g_{xyy}f_{xx}\omega_0$$
$$+4g_{xy}g_{yy}f_{yy}f_{xx} - 18g_{xyyy}g_{yy}\omega_0^2 - 7g_{xxy}g_{xy}f_{xx}\omega_0 + 18g_{xy}f_{xx}f_{xy}f_{yy}$$
$$-8g_{xxy}f_{xx}^2\omega_0 + 5g_{xy}g_{xx}f_{yy}^2 - 30g_{yy}g_{xyy}f_{xx}\omega_0 - 5f_{xxxx}f_{xy}\omega_0^2$$
$$-6g_{xyyy}g_{xx}\omega_0^2 + 9g_{xx}f_{xxy}g_{xy}\omega_0 + 8g_{yy}f_{yyy}f_{xx}\omega_0 - 3g_{xxx}g_{xxy}\omega_0^2)/\omega_0^3$$

No need to say that we are less interested in this kind of formulae than in being able to compute them for specific cases. We have also tried to compute a_3 but MAPLE failed in last steps (in fact, trying to simplify an object too large: more than 400 pages long!).

In presence of a symmetry the expressions involved are shorter. For $f(-x, -y) = -f(x, y)$, $g(-x, -y) = -g(x, y)$, we have obtained up to a_3 whithout any problem, but for the sake of brevity it will not be shown here.

Acknowledgements

This work was possible thanks to an invitation of Prof. Paul Stewart (Manchester University) who also collaborated with helpful discussions. First author also wants to acknowledge the hospitality of Prof. John Brindley (CNLS, University of Leeds) and the financial support of the *Consejería de Educación de la Junta de Andalucía*.

References

[1] Carr, J., *Applications of Centre Manifold Theory*, Appl. Math. Sci. Series, vol. 35, Springer–Verlag, (1981).

[2] Char, B. W. et al, *MAPLE Reference Manual*, 5th edition, WATCOM Publications, (1988).

[3] Chow, S.; Hale, J. K., *Methods of Bifurcation Theory*, Springer–Verlag, (1982).

[4] Deprit, A., *Canonical Transformations Depending on a Small Parameter*, Celest. Mechan., 1 pp. 12–32, (1969).

[5] Freire, E.; Gamero, E.; Ponce, E.; G.–Franquelo, L., *An Algorithm for Symbolic Computation of Center Manifolds*, Symbolic And Algebraic Computation, Lecture Notes in Computer Science, 358, Ed. P. Gianni, pp. 218–230, (1988).

[6] Freire, E.; Gamero, E.; Ponce, E., *An Algorithm for Symbolic Computation of Hopf Bifurcation*, Computers and Mathematics, eds. E. Kaltofen & S. M. Watt, Springer–Verlag, pp. 109–118, (1989).

[7] Freire, E.; Gamero, E.; Ponce, E., *Symbolic Computation and Bifurcations Methods*, Continuation and Bifurcations: Numerical Techniques and Applications, eds. D. Roose, B. de Dier & A. Spence, NATO ASI Series, Kluwer, pp. 105–122, (1990).

[8] Hassard, B. D.; Kazarinoff, N. D.; Wan, Y–H., *Theory and Applications of Hopf Bifurcation*, Cambridge University Press, (1981).

[9] Hassard, B.; Wan, Y. H., *Bifurcation Formulae Derived From Center Manifold Theory*, Journal of Mathematical Analysis and Applications, 63, pp. 297–312, (1978).

[10] Hearn, A. C., *Reduce User's Manual*, The Rand Corporation, (1985).

[11] Marsden, J. E.; McCracken, M., *The Hopf Bifurcation and Its Applications*, Appl. Math. Sci. Series, vol. 19, Springer–Velag, (1976).

ON A CODIMENSION 3 BIFURCATION ARISING IN AN AUTONOMOUS ELECTRONIC CIRCUIT

A.J. RODRÍGUEZ-LUIS, E. FREIRE and E. PONCE

Dept. Appl. Math., E.T.S.I.I., Avda. Reina Mercedes s/n, 41012–SEVILLA, Spain

ABSTRACT. Some aspects of the bifurcations of a modified van der Pol oscillator are considered. We focus our attention on the bifurcations related to a double–zero degeneracy in the linear part of the equilibrium point in the origin. The analysis of the corresponding normal form shows the possibility of additional degeneracy in the nonlinear part, which leads us to study a 3–parameter family of planar vector fields whose bifurcation set is described.

The governing equations of this modified van der Pol oscillator [3], after a suitable rescaling, are:

$$r\dot{x} = -(\nu + \beta)x + \beta y - a_3 x^3 - a_5 x^5 + b_3(y-x)^3 + b_5(y-x)^5$$
$$\dot{y} = \beta x - \beta y - z - b_3(y-x)^3 - b_5(y-x)^5 \qquad (1)$$
$$\dot{z} = y$$

where the parameters ν, β and r take account of the linear physical devices that appear in the circuit and a_i, b_i correspond to the coefficients of the nonlinear terms in the voltage–current characteristics.

Between the great variety of dynamical behaviour that this circuit exhibits [3], let us focus our attention on the bifurcations related to a double–zero degeneracy in the linear part of the equilibrium point at the origin.

A study of the matrix of the linearized system shows that for $\nu_c = -\sqrt{r}$, $\beta_c = \sqrt{r}$, a Takens–Bogdanov bifurcation appears. In order to compute its normal form we translate the critical parameter values to the origin, put the linear part of the vector field (by means of a linear coordinate change) into canonical form, achieve the center manifold reduction and obtain the reduced system. The normal form computed, up to the fifth order, is [5]:

$$\dot{x} = y \qquad \dot{y} = \epsilon_1 x + \epsilon_2 y + \gamma_3 x^3 + \delta_3 x^2 y + \gamma_5 x^5 + \delta_5 x^4 y \qquad (2)$$

where $\gamma_3 = -(a_3 + b_3)r^{-5/2}$, $\delta_3 = 3[a_3(1-r) + b_3(1+r)]r^{-3}$ (we omit the medium–sized expressions of γ_5 and δ_5). Because of the physical constraints r, a_3, b_3 are positive. Then γ_3 will be always negative but δ_3 can be either positive or negative (if $\delta_3 \neq 0$ a classical Takens–Bogdanov bifurcation will appear). But what happens if δ_3 vanishes? We will try to answer this question studying the dynamical and bifurcation behaviour of the three–parameter system

$$\dot{x} = y \qquad \dot{y} = \epsilon_1 x + \epsilon_2 y - x^3 + \epsilon_3 x^2 y + k x^4 y \qquad (3)$$

with $k = \pm 1$, obtained from (2) rescaling in a suitable way and neglecting the term of x^5 (reasoning analogously as Dumortier et al. [2]). It is clear that the bifurcation set corresponding to $k = 1$ is also obtained when $k = -1$ letting $(t, y, \epsilon_2, \epsilon_3) \longrightarrow (-t, -y, -\epsilon_2, -\epsilon_3)$.

We start the study of this three–parameter system with its local analysis. The origin is always an equilibrium point and other two equilibria, $(\pm\sqrt{\epsilon_1}, 0)$, appear after the existence of a pitchfork bifurcation. First, we analyze the Hopf bifurcation the origin exhibits.

We compute its normal form, in polar coordinates, up to the fifth–order coefficient [4]: $\dot{r} = (\epsilon_2/2)r + (\epsilon_3/8)r^3 + (k/16)r^5$. We conclude that this bifurcation appears when $\epsilon_2 = 0$ ($\epsilon_1 < 0$), that the character of its stability depends on the sign of ϵ_3 and that a saddle–node bifurcation of periodic orbits appears on the surface $\epsilon_3^2 = 8k\epsilon_2$ ($\epsilon_3/k < 0$).

Analogously, we study the Hopf bifurcation of nontrivial equilibria computing its normal form:

$$\dot{r} = \frac{\epsilon_2 + \epsilon_1\epsilon_3 + k\epsilon_1^2}{2}r - \frac{\epsilon_3}{4}r^3 - \frac{k(4\epsilon_1^2 k^2 + 17)}{32}r^5 \tag{4}$$

That is, this Hopf bifurcation appears on $\epsilon_2 + \epsilon_1\epsilon_3 + k\epsilon_1^2 = 0$ ($\epsilon_1 > 0$), is supercritical (subcritical) if $\epsilon_3 > 0$ ($\epsilon_3 < 0$) and the saddle–node bifurcation surface is (locally) $\epsilon_2 = -\epsilon_3^2/[k(4\epsilon_1^3 k^2 + 17)] - \epsilon_1\epsilon_3 - k\epsilon_1^2$ ($\epsilon_3/k < 0$).

Up to here we have detected two saddle–node bifurcations related with degenerated Hopf bifurcations, but we will see that the bifurcation set is more complex by means of a semiglobal analysis. Using this rescaling transformation $x = \epsilon^{1/3}X$, $y = \epsilon^{2/3}Y$, $\epsilon_1 = \epsilon^{2/3}\mu_1$, $\epsilon_2 = \epsilon^{4/3}\mu_2$, $\epsilon_3 = \epsilon^{2/3}\mu_3$, $\epsilon \geq 0$, and rescaling time $t \rightarrow \epsilon^{1/3}t$ the system (3) becomes:

$$\dot{X} = Y \quad \dot{Y} = \mu_1 X - X^3 + \epsilon(\mu_2 Y + \mu_3 X^2 Y + kX^4 Y) \tag{5}$$

i.e., it is composed by a Hamiltonian vector field and a small perturbation.

First let us look for saddle connections. Melnikov's method (setting $\mu_1 = 1$) leads to the curve $\mu_2 + (4/5)\mu_3 + (32/35)k = 0$ in the $\mu_2 - \mu_3$ plane and to the surface, in the ϵ–parameter space, $\epsilon_2 + (4/5)\epsilon_1\epsilon_3 + (32/35)k\epsilon_1^2 = 0$ ($\epsilon_1 > 0$). We remark that the homoclinic orbit is attractive (repulsive) when $\epsilon_2 < 0$ ($\epsilon_2 > 0$). As a consequence of this change of stability two saddle–node bifurcation curves of periodic orbits will appear.

We now investigate, by means of the semiglobal analysis, the saddle–node bifurcations of periodic orbits present in the bifurcation set of the system (3). Firstly we do that in the $\mu_3 - \mu_2$ plane, later in the $\mu_1 - \mu_2$ plane. Setting $\mu_1 = 1$ we obtain the Melnikov function

$$M(e, \mu_2, \mu_3) = \left(\mu_2 + \frac{4e}{7}\right) I_0(e) + \left(\mu_3 + \frac{8}{7}\right) I_2(e) \tag{6}$$

where $I_i(e) = \int_{\gamma(e)} (X^i Y, 0)(dX, dY)$, $i = 0, 2$, are calculated along the closed orbit γ and e corresponds to the energy level of the unperturbed Hamiltonian system. Introducing the function $R(e) = I_2(e)/I_0(e)$ we obtain (dropping an $I_0(e)$ factor, if $I_0(e) \neq 0$):

$$G(e, \mu_2, \mu_3) = \mu_2 + \mu_3 R(e) + \frac{4}{7}(e + 2R(e)) \tag{7}$$

Then, the saddle–node bifurcation curve is determined by

$$G(e, \mu_2, \mu_3) = 0, \quad \frac{\partial G}{\partial e}(e, \mu_2, \mu_3) = 0, \quad \frac{\partial^2 G}{\partial e^2}(e, \mu_2, \mu_3) = \frac{-4R''(e)}{7R'(e)} \neq 0 \tag{8}$$

that geometrically give the envelope to the straigth line family $G(e, \mu_2, \mu_3) = 0$. By means of these expressions we get the following parametrization for the saddle–node bifurcation curve:

$$\mu_3(e) = \frac{-4}{7}\left(\frac{1}{R'(e)} + 2\right), \quad \mu_2(e) = \frac{4}{7}\left(\frac{R(e)}{R'(e)} - e\right), \quad \left(\frac{R''(e)}{R'(e)} \neq 0\right) \tag{9}$$

At this moment a deep knowledge of the function $R(e)$ is needed. This involves lengthy calculations with elliptic functions due to the fact that

$$I_i(e) = 4 \int_0^c x^i y(x,e)\, dx \quad (e > 0); \qquad I_i(e) = 2 \int_b^c x^i y(x,e)\, dx \quad \left(\frac{-1}{4} < e < 0\right) \quad (10)$$

$i = 0, 2$, where $y(x,e) = \sqrt{2e + x^2 - x^4/2}$ and c, b are the positive real roots of $y(x,e) = 0$. We compute the limits of $R(e)$ and $R'(e)$ at the endpoints of the two intervals of e where are defined and, furthermore, we find that R verifies this Riccati's differential equation:

$$4e(1 + 4e)R'(e) = 8eR(e) - 4e - 4R(e) + 5R^2(e) \quad (11)$$

Using elementary properties of the calculus of one real variable we proved the following result, partially reported by Carr [1]:
Proposition: There exist $e_2 > e_1 > 0$ such that $R'(e) < 0$ if $-1/4 < e < e_1$ and $R'(e) > 0$ if $e > e_1$; $R''(e) < 0$ if $-1/4 < e < 0$ or $e > e_2$ and $R''(e) > 0$ if $0 < e < e_2$.

There are three important values of e ($e = 0$ (unbounded first derivative), $e = e_1 \approx 0.08899$ (zero first derivative) and $e = e_2 \approx 0.4238$ (zero second derivative)) that split the domain of definition of R into four subintervals (see fig.1).

Using (9) four different curves appear as we move along these four subintervals (see fig.2). We observe that the saddle–node bifurcation curve (SNH2) related with the change of stability ($\epsilon_3 = 0$) of the Hopf bifurcation of the nontrivial equilibria coalesces with one (SN1) of the two saddle–node bifurcations (SN1, SN2) that appear when the homoclinic orbit changes its stability ($\epsilon_2 = 0$). We see also other two saddle–node bifurcation (SN3, SN4) curves that coalesce in a cusp parametrized by $\epsilon_1(s) = s$, $\epsilon_2(s) = a_2 s$, $\epsilon_1(s) = a_3 s^2$, $s \geq 0$, where $a_2 = \mu_2(e_2) \approx 1.5713$, $a_3 = \mu_3(e_2) \approx -3.3484$. This cusp appears more clearly in the ϵ_1–ϵ_2 plane whose study we start now. Again Melnikov function (setting $\mu_3 = -1$) depends on two integrals:

$$M(e, \mu_1, \mu_2) = \left(\mu_2 + \frac{4e}{7}\right) I_0^{\mu_1}(e, \mu_1) + \left(\frac{8\mu_1}{7} - 1\right) I_2^{\mu_1}(e, \mu_1) \quad (12)$$

where $I_i^{\mu_1}(e, \mu_1) = \mu_1^m I_i(t)$ if $\mu_1 > 0$ and $I_i^{\mu_1}(e, \mu_1) = (-\mu_1)^m I_i^*(t)$ if $\mu_1 < 0$ with $m = (i + 3)/2$ and $t = e/\mu_1^2$. When μ_1 is positive we obtain, in an analogous way as we did previously, the parametrization of the saddle–node bifurcation curve:

$$\mu_1(t) = \frac{7R'(t)}{4(1 + 2R'(t))}, \quad \mu_2(t) = \frac{7R'(t)[R(t) - tR'(t)]}{4(1 + 2R'(t))^2} \quad \left(\frac{R'(t)R''(t)}{(1 + 2R'(t))^2} \neq 0\right) \quad (13)$$

We obtain once more four saddle–node bifurcation curves (if $\mu_1 > 0$ and then if $\epsilon_1 > 0$, see fig. 3). When μ_1 is negative the Melnikov function (except a factor $(-\mu_1)^{3/2} I_0^*(t) \neq 0$) can be written as:

$$M^*(t, \mu_1, \mu_2) = \left(\mu_2 + \frac{4}{7}t\mu_1^2\right) - \left(\frac{8}{7}\mu_1 - 1\right)\mu_1 R^*(t) \quad \text{where} \quad R^*(t) = \frac{I_2^*(t)}{I_0^*(t)} \quad (14)$$

$(t \in (0, +\infty))$ is defined by means of $I_i^*(t) = 4 \int_0^c x^i y(x,t)\, dx$ $(t > 0)$, $i = 0, 2$, where $y(x,t) = \sqrt{2t - x^2 - x^4/2}$ and c is the positive real root of $y(x,t) = 0$. The parametrization of the saddle–node bifurcation curve is given in this case ($\mu_1 < 0$) by:

$$\mu_1(t) = \frac{-7R^{*'}(t)}{4\left(1 - 2R^{*'}(t)\right)}, \quad \mu_2(t) = \frac{7R^{*'}(t)[R^*(t) - tR^{*'}(t)]}{4\left(1 - 2R^{*'}(t)\right)^2} \quad \left(\frac{R^{*'}(t)R^{*''}(t)}{\left(1 - 2R^{*'}(t)\right)^2} \neq 0\right) \quad (15)$$

Computing the limits of $R^*(t)$ and $R^{*'}(t)$ at the endpoints of its domain of definition and finding that verifies this Riccati's differential equation:

$$4t(1 + 4t)R^{*'}(t) = 8tR^*(t) + 4t - 4R^*(t) - 5R^{*2}(t) \quad (16)$$

we prove this proposition (partially reported by Carr)[1]:
Proposition: $R^{*'}(t) > 0$ and $R^{*''}(e) < 0$ for all $t > 0$. (see fig.1).

Using the parametrization (15) we obtain a curve of saddle–node (for $\mu_1 < 0$) that coalesces with other saddle–node curve that appeared in the analysis of the $\mu_1 > 0$ case at the point $(\mu_1, \mu_2) = (0, 7\lambda^2/16)$ where $\lambda = \lim_{e \to \infty} e^{-1/2}R(e) = \lim_{t \to \infty} t^{-1/2}R^*(t) = 48\pi^2/5\Gamma^4(1/4)$ where the gamma function appears. In figure 3 we show the complete bifurcation set in the $\epsilon_1 - \epsilon_2$ plane for a negative value of ϵ_3. We remark several bifurcations: the Hopf subcritical of nontrivial equilibria (H2); the saddle–connection (HOM) that changes its stability ($\epsilon_2 = 0$) and then two saddle–node bifurcation curves (SN1, SN2) appear, one at each side of the homoclinic curve; a saddle–node bifurcation (SN3) related with the Hopf bifurcation of the origin (H1) (supercritical in this case) that coalesces in a cusp (parametrized before) with the saddle–node bifurcation (SN4) that appears from the origin as in the classical Takens–Bogdanov bifurcation.

Conclusions. We have studied the bifurcation set in the ϵ–parameter space of the system (3) finding five codimension 2 curves: two degenerated Hopf bifurcations, a degenerated saddle–connection, a cusp of saddle–node bifurcations of periodic orbits and a straigth line (the ϵ_3 axis) of Takens–Bogdanov bifurcation.

REFERENCES

1. Carr, J. (1981) Applications of Centre Manifold Theory, Springer–Verlag, Heidelberg Berlin New–York.
2. Dumortier, F., Roussarie, R. and Sotomayor, J. (1987) 'Generic 3-parameter families of vector fields on the plane, unfolding a singularity with nilpotent linear part. The cusp case of codimension 3', Ergod. Th. & Dynam. Sys., **7**, 375–413.
3. Freire, E., G.Franquelo, L., Aracil, J. (1984) 'Periodicity and Chaos in an Autonomous Electronic System', IEEE Transactions on Circuits and Systems, Vol. CAS–31, No. 3, pp. 237–247.
4. Freire, E., Gamero, E., Ponce, E. (1989) 'An Algorithm for Symbolic Computation of Hopf Bifurcation', Computers and Mathematics, Eds. E. Kaltofen & S.M. Watt, Springer–Verlag, pp. 109–118.
5. Gamero, E., Freire, E., Ponce, E. (1991) 'On the Normal Forms for Planar Systems with Nilpotent Linear Part' (see this volume).

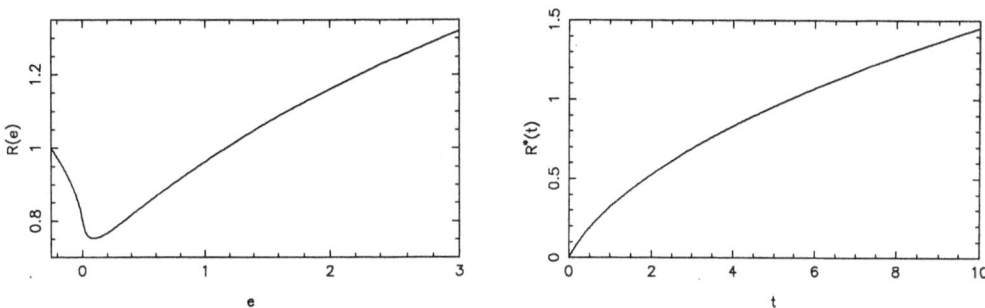

Figure 1. The graphs of $R(e)$ and $R^*(t)$.

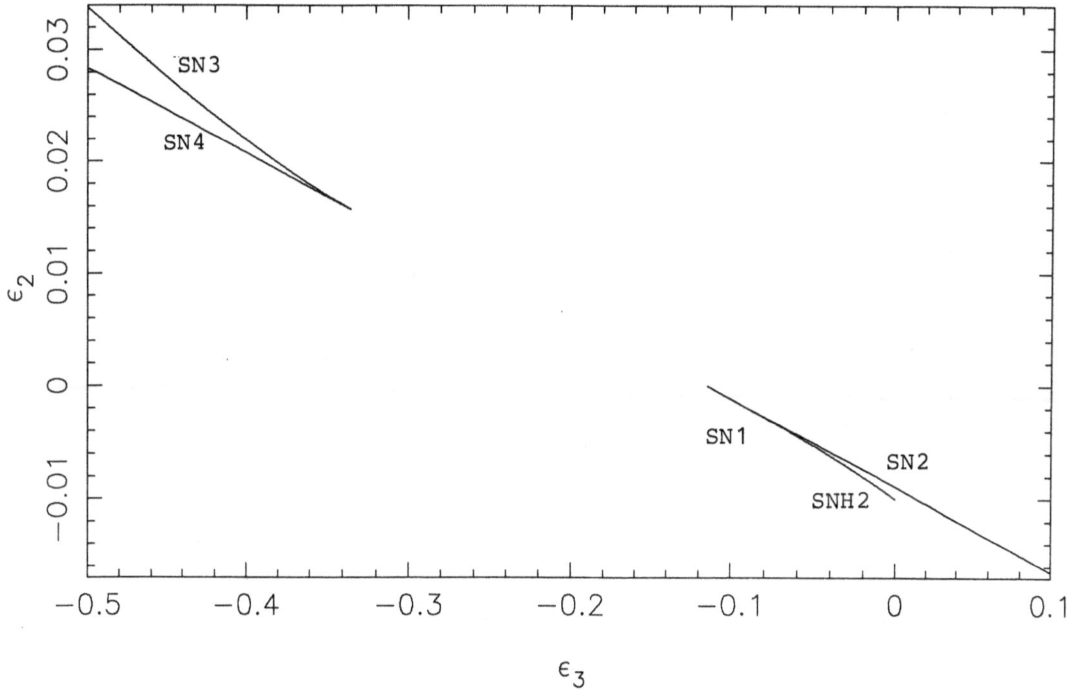

Figure 2. Partial bifurcation set (saddle–node bifurcations) in the $\epsilon_3 - \epsilon_2$ plane for system (3), $\epsilon_1 = 0.1$, $k = +1$. We haven't drawn the straight lines corresponding to Hopf and to homoclinic bifurcations for the sake of clearness.

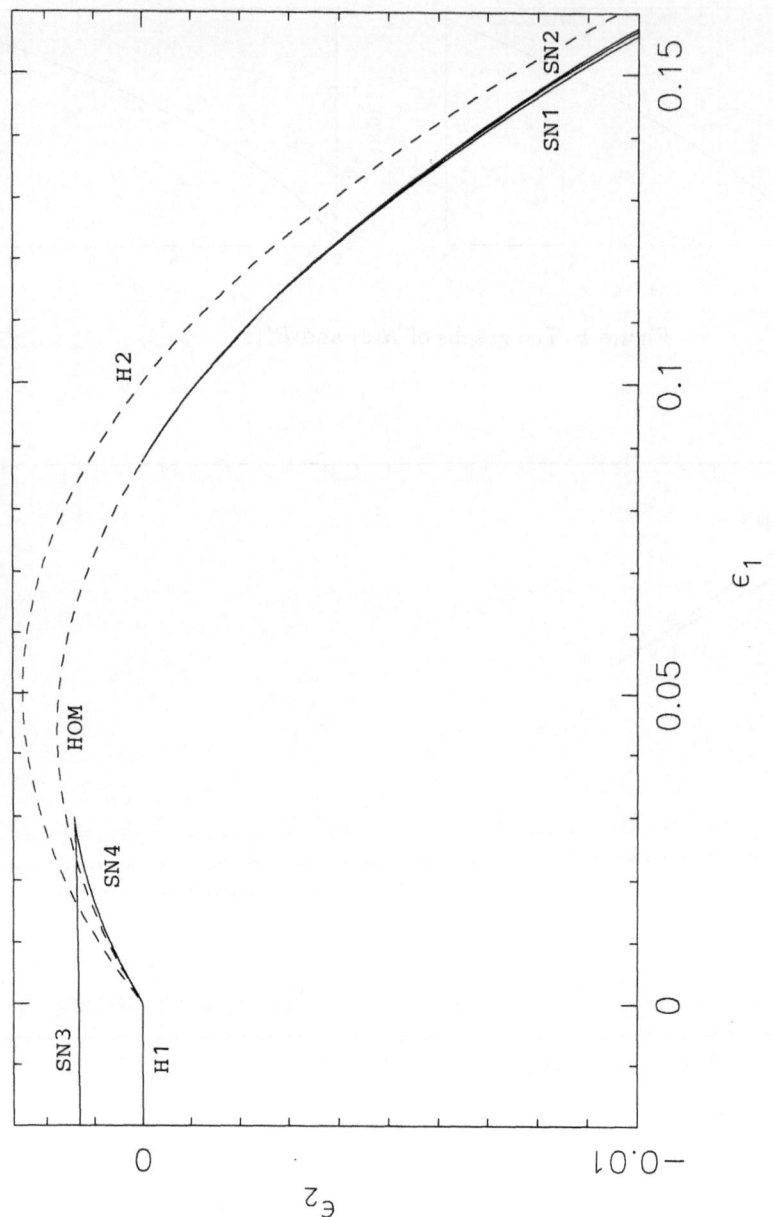

Figure 3. Bifurcation set in the $\epsilon_1 - \epsilon_2$ plane for system (3), $\epsilon_3 = -0.1$, $k = +1$.

International Series of Numerical Mathematics, Vol. 97, © 1991 Birkhäuser Verlag Basel

EFFICIENT PARALLEL COMPUTATION OF PERIODIC SOLUTIONS
OF PARABOLIC PARTIAL DIFFERENTIAL EQUATIONS

Dirk Roose and Stefan Vandewalle

Dept. of Computer Science, K. U. Leuven
Celestijnenlaan 200A, B-3001 Heverlee - Leuven, Belgium

Abstract. Periodic solutions of autonomous parabolic partial differential systems can be computed by using a shooting method, applied to the (large) system of ordinary differential equations that results after spatial discretization. Many time integrations of this system may be required in each shooting iteration step. Hence these calculations can be extremely expensive, especially in the case of fine spatial discretizations and/or higher dimensional problems.
This paper deals with techniques to make the shooting method more feasible by reducing the arithmetic complexity of the standard approach and by exploiting parallelism and vectorization. The former is achieved by providing a 'coarse grid approximation' to the Jacobian matrix, that arises in the shooting algorithm. This leads to a substantial reduction of the number of required time integrations. The latter is based on the application of the Multigrid Waveform Relaxation algorithm, a recent technique for time-integration of parabolic partial differential equations that is well suited for vector and parallel computers. Results are given for the time-integration and the calculation of periodic solutions of the two-dimensional Brusselator-model on a distributed memory parallel computer.

1. The computation of periodic solutions

The computation of periodic solutions of nonlinear parameter-dependent evolution equations is a basic operation during bifurcation analysis. In this paper we are interested in periodic solutions of autonomous parabolic partial differential systems of the form

$$\frac{\partial u(x,t)}{\partial t} = L(u (x,t)), \quad (x,t) \in \Omega \times \mathbb{R} \quad \text{with} \quad \Omega \subset \mathbb{R}^2 \tag{1}$$

with suitable boundary conditions. In Eq. (1), L is a general non-linear elliptic operator, e.g. consisting of a linear diffusion and a non-linear reaction term, and $u(x,t)$ is a q-vector of functions defined on $\Omega \times \mathbb{R}$. For notational convenience we neglect any dependence on a parameter. Note however that the methods that we introduce can easily be incorporated into a continuation procedure.

After spatial discretization and incorporation of the boundary conditions, the parabolic problem is transformed into a system of n ordinary differential equations with q equations defined at each grid point

$$\frac{dU}{dt} = L(U), \quad U(t) \in \mathbb{R}^n \tag{2}$$

where U denotes the vector of unknown functions defined at the grid points.

To determine a stable periodic solution, one can perform a 'brute-force' time integration [13,14] starting from an arbitrary initial condition. The periodic solution is then obtained when the transient part of the solution has decayed. If the periodic solution is not strongly attracting, integration over a long time interval is required, which leads to high computing times. Furthermore it is difficult to detect in an automatic way that the periodic behaviour is attained and finally, only stable solutions can be found.

Other numerical techniques are based on the reformulation of the problem as a two-point boundary value problem in time, with the period being an additional variable. Indeed, if (2) has a periodic solution with period T, then using the transformation $t = T\tau$, the periodic solution satisfies

$$\frac{dU}{d\tau} = T L(U); \quad \frac{dT}{d\tau} = 0 \quad \text{with boundary conditions} \quad U(1) - U(0) = 0; \quad P(U,T) = 0 \tag{3}$$

where the latter denotes the phase condition, e.g. $P(U,T) = U_k(0) - c = 0$ (c being a constant) [8,10,14]. We refer to [4] for a description of other phase conditions that are appropriate in the context of continuation of periodic solutions.

The two-point boundary-value problem (3) can be solved by using a finite difference or collocation discretization in time [4,10] or by using a shooting technique [8,9,13,14]. As we will be using the latter approach, we present a simple shooting algorithm below. We denote by $\phi(U(0), T)$ the solution at $\tau = 1$ of $\frac{dU}{d\tau} = T L(U)$ with initial condition $U(0)$ at $\tau = 0$. The shooting algorithm is based on the use of a *Newton (-like) iteration* for solving the residual equation

$$r(U(0), T) = \begin{bmatrix} \phi(U(0), T) - U(0) \\ P(U, T) \end{bmatrix} = 0 \tag{4}$$

---

choose starting values $U^0(0)$, T^0

repeat for $k = 0, 1, \cdots$

    compute the residual $r(U^k(0), T^k)$

    solve $\mathcal{J} \begin{bmatrix} \Delta U \\ \Delta T \end{bmatrix} = -r$

    $U^{k+1}(0) := U^k(0) + \lambda \Delta U$; $T^{k+1} := T^k + \lambda \Delta T$ (λ: damping factor)

until convergence

The computation of the residual in each iteration step requires a time-integration of $\frac{dU}{d\tau} = T L(U)$ from $\tau = 0$ to $\tau = 1$ with $T = T^k$ and initial condition $U^k(0)$. The matrix $\mathcal{J} \in \mathbb{R}^{(n+1)\times(n+1)}$ represents (an approximation of) the Jacobian matrix J of first-order partial derivatives of r. This matrix can be computed by integration of variational equations [9,13] or can be approximated by numerical differencing. In both cases, the computational effort is equivalent to $n+1$ time-integrations of an n-dimensional system. To reduce the computational cost, $\mathcal{J}$ can be kept fixed during a number of shooting iterations, which leads to a Chord-Newton iteration, or $\mathcal{J}$ can be updated in the following steps by using a rank-one update [15]. Classical continuation techniques can easily be incorporated into this approach [8,9,14].

The standard shooting approach can be prohibitively expensive, e.g. in the case of two-dimensional parabolic problems or when many periodic solutions have to be computed during a continuation procedure. In the next section we will indicate how the Jacobian matrix J can be approximated with less than $n + 1$ time-integrations. In section 3 we will describe Multigrid Waveform Relaxation, a recent technique for time-integration of parabolic partial differential equations. We show that this method can be more efficient than classical time-integration techniques, when a good approximation of the solution over the whole time interval is available. Moreover, the method can be vectorized and parallelized easily and efficiently, which is not the case for classical time-integration techniques.

2. Efficient approximation of the Jacobian matrix within shooting

We denote the k-th column of the Jacobian matrix by $J_{.k}$, with

$$J_{.k} = \frac{\partial r}{\partial U_k} \quad (k = 1,...,n) \quad \text{and} \quad J_{.n+1} = \frac{\partial r}{\partial T} \tag{5}$$

The matrix J can be approximated by numerical differencing, by using the formula

$$J_{.k} \approx \frac{1}{\delta} [r(U(0)+\delta e_k, T) - r(U(0),T)] = \frac{1}{\delta} [\phi(U(0)+\delta e_k, T) - \phi(U(0),T) - \delta e_k] \quad (k=1,...,n) \tag{6}$$

A similar formula holds for $J_{.n+1}$. The calculation of each column of the Jacobian corresponds to one time integration of the parabolic problem. Consequently, the calculations in (6) make out the major computational cost of the shooting method when n is large. As it shows, the 'brute-force integration' technique may be faster than shooting, namely if the former obtains a sufficiently accurate periodic solution after integration over a time interval of length mT, when $m < n$. Thus for large n the shooting approach will only be competitive if a Jacobian approximation can be obtained with less than $n+1$ time-integrations. We now describe a possible strategy for doing so, based on a *coarse grid approximation* of the Jacobian.

For notational convenience, we assume a one-dimensional parabolic problem on the unit interval, discretized in space with mesh size $h = 1/2^p$, i.e. $n = 2^p - 1$. Eq. (6) indicates that $J_{.k}$ $(k = 1,...,n)$ is (approximately) proportional to the perturbation of the state of the system after integration over time T, caused by a perturbation of the initial condition localized in grid point x_k. In particular $\delta J_{i,k}$ gives the resulting perturbation in grid point x_i. If the mesh size h is sufficiently small, a perturbation of the initial condition in point x_k will have nearly the same effect in grid point x_i, as the effect in grid points x_{i+1} and x_{i-1} caused by a perturbation of the initial condition in the grid points x_{k+1} and x_{k-1} respectively. In other words, the values $J_{i,k}$, $J_{i-1,k-1}$ and $J_{i+1,k+1}$ are closely related.

A *coarse grid approximation of the Jacobian matrix* can now be obtained by computing $J_{.k}$ for k odd with (6) and by approximating $J_{.k}$ for k even by using a shifted interpolation between $J_{.k-1}$ and $J_{.k+1}$. When linear interpolation is used, this leads to the formulae :

$$J_{i,k} = \frac{1}{2} [J_{i-1,k-1} + J_{i+1,k+1}] \quad (i = 2,...,n-1) ; \quad J_{1,k} = J_{2,k+1} ; \quad J_{n,k} = J_{n-1,k-1} \tag{7}$$

Column $J_{.n+1} = \partial r/\partial T$ must be computed by numerical differencing.

Alternatively, the even columns of J can be computed by numerical differencing and the uneven columns by (shifted) interpolation. However, in this case the first and the last column require some extrapolation, which may cause rather large approximation errors. One can even further reduce the number of time-integrations required, by using numerical differencing only for the columns that are associated with an even coarser grid, e.g. having $2^{p-l} - 1$ equidistant points.

This technique can be generalized immediately to partial differential equations defined on a two-dimensional space domain. In that case the columns associated with grid point $x_{k,l}$ can be computed by a shifted bilinear interpolation of the columns associated with the neighbouring grid points, i.e. $x_{k-1,l}$, $x_{k+1,l}$, $x_{k,l-1}$ and $x_{k,l+1}$.

Currently, we investigate whether a sufficiently good approximation of J can be obtained by a *probing technique* [2,3]. Compared with a coarse grid approximation of J, the advantage of probing is that the number of time-integrations to be performed does not depend on n, the number of unknowns.

3. Multigrid Waveform Relaxation

3.1 Standard Waveform Relaxation

Numerical methods for time-integration of parabolic problems are usually based on a time stepping approach. Both explicit methods (e.g. Euler, Runge Kutta) and implicit discretization techniques (e.g. Crank-Nicolson) belong to this class. A different approach is used in the family of Waveform Relaxation methods (WR), also called dynamic iteration or Picard-Lindelöf iteration methods, which may be used for solving large systems of ordinary differential equations. We will explain this method by its application to the system (2), supplemented with an initial condition at time t_0, $U_i(t_0) = U_{i0}$, $i=1,...,n$. The Jacobi variant of the WR algorithm can be formulated as follows.

$k := 0$;   choose $U_i^{(0)}(t)$ for $t \in [t_0, \infty)$ and $i=1,...,n$

repeat

    for all i:

        solve $\dfrac{d}{dt} U_i^{(k+1)} = L_i(U_1^{(k)},...,U_{i-1}^{(k)}, U_i^{(k+1)}, U_{i+1}^{(k)},...,U_n^{(k)})$ with $U_i^{(k+1)}(t_0) = U_{i0}$

    $k := k+1$

until convergence.

The general idea of the WR algorithm is to start from an initial guess for the solution of the system and then to solve each equation in sequence as part of a relaxation loop until convergence is achieved. In the iteration step each differential equation is solved as an equation in one variable. These simple first order differential equations can be solved with any standard, stiff ODE integrator, e.g. the trapezoidal rule. As such the method is very similar to the iterative techniques for solving algebraic systems except that each variable is a function of time rather than a scalar unknown. The adaptation of the algorithm to obtain a Gauss-Seidel type iteration is straightforward. The theoretical foundations of Waveform Relaxation (WR) for linear and nonlinear systems of ordinary differential equations have been discussed in e.g. [12, 20]. Attempts to use WR in the way described above, to solve parabolic problems have not been very successful. This is due to the slow convergence of the method. Indeed, as was shown in [12], the convergence rates for the Jacobi and Gauss-Seidel scheme are of order $1 - O(h^2)$.

3.2 Multigrid acceleration

Standard relaxation techniques can be accelerated if they are combined with the multigrid idea [6]. The multigrid method uses a set of nested grids, with the finest one corresponding to the one on which the solution is desired. Its superior convergence characteristics are based on the interplay of fine grid *smoothing*, which annihilates high frequency errors, and *coarse grid correction*, which is applied to reduce the low frequency errors.

In the *Multigrid Waveform Relaxation* algorithm (MWR) the multigrid method is extended to time dependent problems. Each of the multigrid operations is adapted to operate on the *functions* $U_i(t)$ instead of on *single scalar values*, see [11,16,17].

- The *smoothing* is performed by applying one or more Gauss-Seidel or damped Jacobi waveform relaxations. Smoothing rates for these relaxations have been given in [11].
- The *defect* or residual of an approximation $\overline{U}$ is defined as $D = \dfrac{d}{dt}\overline{U} - L(\overline{U})$.

- The *restriction* and *prolongation* are calculated using identical formulae as in the elliptic case (e.g. full-weighting restriction). However these formulae now define linear combinations of functions.

We can now state the Multigrid Waveform Relaxation algorithm for solving nonlinear parabolic problems, see [16]. The algorithm is easily derived from the well-known multigrid full approximation scheme. We consider autonomous initial value problems, described by Eq. (2), as well as the more general non-autonomous problem,

$$\frac{dU}{dt} = L(U) + F(t) \quad \text{with} \quad U(t_0) = U_0 \tag{8}$$

Let G_i, $i = 0, 1, ..., k$ be the hierarchy of grids with G_k the finest grid and G_0 the coarsest grid. Eq. (8), or equivalently,

$$\frac{dU_k}{dt} = L_k(U_k) + F_k \quad \text{with} \quad U_k(t_0) = U_{k0} \tag{9}$$

where the subscript k now denotes the grid level, is solved by iteratively applying the following algorithm to an initial approximation of U_k.

---

procedure *fas* (k, F_k, U_k)

if $k = 0$ *solve* $\dfrac{d}{dt} U_0 = L_0(U_0) + F_0$ *exactly*

else

   - *perform* ν_1 *smoothing operations by Gauss-Seidel WR*

   - *project* U_k *onto* G_{k-1}: $\overline{U}_{k-1} := I_k^{k-1} U_k$

   - *calculate the coarse problem right hand side:* $F_{k-1}' := \dfrac{d\overline{U}_{k-1}}{dt} - L_{k-1}(\overline{U}_{k-1}) - I_k^{k-1}(\dfrac{dU_k}{dt} - L_k(U_k) - F_k)$

   - *solve on* G_{k-1}: $\dfrac{d}{dt} U_{k-1} = L_{k-1}(U_{k-1}) + F_{k-1}$

      repeat γ_k **times** *fas* $(k-1, F_{k-1}, U_{k-1})$, *starting with* $U_{k-1} := \overline{U}_{k-1}$.

   - *interpolate and correct* U_k : $U_k := U_k + I_{k-1}^k (U_{k-1} - \overline{U}_{k-1})$.

   - *perform* ν_2 *smoothing operations by Gauss-Seidel WR*

endif

The algorithm is completely defined by specifying the grid sequence G_i, $i = 0, ..., k$, the discretized operators L_i, the inter-grid transfer operations I_{k-1}^k and I_k^{k-1}, the nature of the smoothing relaxations, and by assigning a value to the constants ν_1, ν_2 and γ_k. So-called V- and W-multigrid-cycles are obtained with the values 1 and 2 for γ_k. Note that the calculation of the derivative in the definition of the defect and the determination of the coarse grid right hand side can be avoided [19]. The algorithm can be combined with the idea of nested iteration. This leads to the waveform equivalent of the *full multigrid method*. In this case, the algorithm proceeds from the coarsest grid to the finest grid. At each intermediate gridlevel a few multigrid cycles are performed before going to the next finer level. The initial approximate solution to the problem on G_{i+1} is derived by a suitable interpolation of the solution obtained on G_i. On the finest level, further V- or W-cycles are performed to fulfil the convergence requirements.

In general, comparable computing times are required on a sequential computer to achieve a certain accuracy with MWR and with standard time stepping techniques, e.g. the Crank-Nicolson method [17].

However, in the context that we consider in this paper, MWR has a number of advantages.

a) MWR versus classical methods on parallel and vector computers

As the computation of periodic solutions of parabolic PDEs is very time consuming it is worthwhile to take advantage of the computational power offered by parallel and vector computers. In our experiments we used the Intel iPSC/2 distributed memory multiprocessor. The iPSC/2 consists of a number of nodes, each containing a processor and memory. Communication between nodes is achieved by message passing. For problems defined on a geometric domain, the domain – and the associated data – can be decomposed into subdomains, that are treated by different nodes, see e.g. [5]. Each node executes the algorithm for its part of the data. This usually requires communication between nodes at some stages of the algorithm. For the problem under consideration, data corresponding to grid points lying at the boundaries of the subdomains must be exchanged with nodes that treat neighbouring subdomains. This introduces a 'parallel overhead', which decreases the 'speed-up' and the 'parallel efficiency' of a parallel computation [5]. The communication time consists of a 'start-up time' and a part that increases linearly with the message length. On the iPSC/2 – as on most other parallel computers – this start-up time is relatively high [1]. Hence the speed-up and the parallel efficiency depends not only on the total amount of communication, but also on the number of messages.

In [17] it is shown that a much higher speed-up can be obtained on a parallel computer with MWR than with classical time-stepping integration methods (e.g. Crank-Nicolson). This is due to the different 'grain-size' of the parallelism. Within MWR communication only occurs after treating $n_s \times n_t$ grid points, where n_s denotes the number of grid points in each subdomain and n_t denotes the number of discretization points in the time direction. Within time-stepping methods communication occurs after treating n_s grid points, i.e. at each time level. Especially when a multigrid method is used at each time level, this may lead to a very low speed-up and parallel efficiency. For a more detailed analysis we refer to [17,18].

The MWR algorithm operates on functions, which are represented by vectors of function values. As such, MWR can be vectorized easily and efficiently *in the time direction*. Standard time-stepping algorithms can only be vectorized in the spatial direction, which does not lead to a performance improvement unless the number of grid points is very large.

Note that the different columns of the Jacobian matrix J can be calculated independently, once the time-integration for determining the residual has been performed. Hence this phase of the shooting algorithm can be parallelized with almost perfect speed-up, regardless of the integration method being used. The use of MWR is advantageous also here as it allows to exploit vectorization.

b) MWR versus classical methods within a shooting procedure

Within the shooting approach a sequence of time-integrations are performed for the same differential equations with slightly perturbed initial conditions. In this case, MWR can be started with the solution profile over the whole time interval, obtained in the previous step. The availability of a good starting profile will significantly reduce the number of MWR-iterations (see section 4). Classical time-stepping methods do not allow to exploit the availability of a good approximation of the solution over the whole time interval.

Also for the time-integrations needed to approximate the Jacobian matrix good starting values are available. When a branch of periodic solutions is calculated by continuation, good starting values for MWR are available even for the first step of the shooting iteration.

c) MWR for computing periodic solutions of non-autonomous problems

In [19] it was shown that a slight modification of the MWR algorithm leads to a very efficient technique for calculating the periodic solution of a non-autonomous parabolic problem. The arithmetic complexity of this *Periodic Multigrid Waveform Relaxation* method is similar to the complexity of solving *one* initial-boundary value problem. In addition it has the same favourable parallelization and vectorization characteristics. We refer to [19] for further details and for a comparison of the method with the 'best' standard approach (multigrid of the second kind [6]).

4. Numerical results for the two-dimensional Brusselator model

The two-dimensional Brusselator model is described by the following system of nonlinear partial differential equations defined over the unit square $\Omega = [0,1] \times [0,1]$:

$$\frac{\partial X}{\partial t} = \frac{D_X}{L^2} \left[\frac{\partial^2 X}{\partial r^2} + \frac{\partial^2 X}{\partial s^2} \right] - (B+1)X + X^2 Y + A$$

$$\frac{\partial Y}{\partial t} = \frac{D_Y}{L^2} \left[\frac{\partial^2 Y}{\partial r^2} + \frac{\partial^2 Y}{\partial s^2} \right] - X^2 Y + BX$$

(10)

The functions $X(r,s,t)$ and $Y(r,s,t)$ denote chemical concentrations. The homogeneous concentrations A and B and the diffusion coefficients D_X and D_Y are considered to be fixed control parameters and L, the reactorlength, can be used as bifurcation parameter. For the control parameters the following values are used : $D_X = 0.004$, $D_Y = 0.008$, $A = 2.0$ and $B = 5.45$. We consider Dirichlet boundary conditions $X(r,s) = A$, $Y(r,s) = B/A$, $(r,s) \in \partial\Omega$. For all values of L, a homogeneous steady state solution exists, equal to the boundary conditions. The first bifurcation, occurring at $L \approx 0.72$ [7], is a supercritical Hopf bifurcation and stable periodic spatially symmetric solutions exist for $L > 0.72$. Close to the Hopf point, the periodic solution has a period $T \approx 3.44$ and can be approximated by a perturbation of the homogeneous solution with a function of the form $C \cos\omega t \sin\pi r \sin\pi s$, with $\omega = 2\pi/T$. A bifurcation diagram of periodic solutions for the one-dimensional Brusselator with similar values for the control parameters have been presented by M. Holodniok et al. [8].

We used a second order finite difference discretization on an equidistant mesh, with mesh sizes $h = 1/8, 1/16, 1/32$, resulting in a system of ODEs of dimension 98, 450 and 1922 respectively. The symmetry of the solutions is not exploited in the calculations. For $L = 0.9$ we have first computed the time-evolution of the system with a perturbation of the homogeneous solution as the initial condition. Fig. 1a illustrates the convergence behaviour of Red-Black Gauss-Seidel Waveform Relaxation. For all grid points, a constant profile in time was used as starting value. The approximations for the function $Y(0.5,0.5,t)$, obtained for consecutive iteration steps, are shown. The corresponding results obtained with Multigrid Waveform Relaxation, in which Red-Black Gauss-Seidel is used as smoother, are given in Fig. 1b. They illustrate the improved convergence properties. Each V-cycle requires $\approx 8/3$ workunits. (A workunit is equivalent to the computational effort of one WR Gauss-Seidel iteration.) The convergence properties can be further improved by using the Full Multigrid approach, see Fig. 1c. The full multigrid step, needed to obtain a first approximation on the finest grid, requires $\approx 2/3$ workunits. These figures also illustrate the typical WR phenomenon that the largest error occurs at the end of the integration interval. This error can be reduced by a partitioning of the time interval into several smaller 'windows' that are treated in sequence.

In Fig. 2, the performance of MWR and the Crank-Nicolson (C-N) methods are compared for the integration of (10). The accuracy of the solution is plotted versus execution time. With the annotation 'WR V(1,1)' we mean 'MWR using V-cycles with 1 pre- and 1 post-smoothing step'; with 'C-N, 2 V(1,1)' we mean 'Crank-Nicolson with 2 V(1,1) cycles per time step'. For MWR the error decreases linearly when more multigrid cycles are used. The C-N results show up as discrete points, depending on the accuracy of the solution procedure at each time level.

For this particular problem and mesh-size $h = 1/16$, C-N is more efficient than MWR on 1 processor when a high accuracy is required, but MWR clearly outperforms C-N on 16 processors, due to the superior parallel characteristics. The use of vectorization would make the difference even larger in favour of MWR [19]. Further experiments indicate that for $h = 1/32$, MWR is as efficient as C-N on 1 processor. On 16 processors, the relative difference between the execution times for MWR and C-N is somewhat smaller than for $h = 1/16$, due to the increased parallel efficiency of C-N on this larger problem.

For the calculations reported in Fig. 2, constant starting profiles in time are used. When good starting values over the whole time interval are available, the initial MWR error will be reduced, causing a shift downwards of the MWR curves. This leads to a substantial reduction of the execution time needed with MWR to achieve a certain accuracy, due to the linear multigrid convergence rate.

Finally, we have used MWR within a shooting procedure to compute the periodic solution of (10) for $L = 1.1$ with $h = 1/8$ and $1/16$. As starting values for $U^{(0)}(0)$ and $T^{(0)}$ we used the values corresponding to the periodic solution for $L = 0.9$. For both grid sizes, a 'coarse grid approximation' of the Jacobian was obtained by computing the columns corresponding to a 3×3 grid by numerical differencing. These columns may be calculated in parallel. The other columns were computed by interpolation. A total of 19 time-integrations were needed to compute J, which was kept fixed during the remainder of the damped Chord-Newton iteration. The convergence history is shown in Table 1. As the shooting iteration proceeds, better starting values for the MWR procedure become available. This leads to a reduction of the number of MWR V(1,1)-cycles needed to achieve the requested accuracy for the time-integration (10^{-7}). For the time-integrations needed to compute J (using Eq. (6) with $\delta = 10^{-4}$) very good starting values are available, hence only a few MWR cycles are required. The periodic solution for $L = 0.9$ is shown in Fig. 3 as a series of plots of $X(r,s,t_i)$, at equidistant points along the time axis.

$h = 1/8$ (7 × 7 gridpoints)				$h = 1/16$ (15 × 15 gridpoints)			
iter. step	|| residual ||	|| correction ||	number of MWR cycles	iter. step	|| residual ||	|| correction ||	number of MWR cycles
0	0.26 E-01		5	0	0.51 E-01		9
	coarse grid approx. of J		3 †		coarse grid approx. of J		5 †
1	0.15 E-01	0.42 E-01	5	1	0.30 E-01	0.12 E 00	8
2	0.11 E-01	0.31 E-01	5	2*	0.37 E-01	0.80 E-01	8
3	0.78 E-02	0.20 E-01	5	2	0.74 E-02	0.40 E-01	8
4	0.53 E-02	0.14 E-01	5	3	0.72 E-02	0.15 E-01	8
5	0.34 E-02	0.91 E-02	4	4*	0.92 E-02	0.17 E-01	7
				4	0.13 E-02	0.85 E-02	7
...	...	...		5*	0.17 E-02	0.31 E-02	6
10	0.24 E-03	0.69 E-03	4	5	0.30 E-03	0.15 E-02	6
11	0.12 E-03	0.37 E-03	3	6*	0.38 E-03	0.70 E-03	6
12	0.62 E-04	0.19 E-03	3	6	0.67 E-04	0.35 E-03	6
13	0.29 E-04	0.94 E-04	3				

Table 1 : Computation of periodic solutions by shooting (two-dimensional Brusselator with $L = 1.1$): convergence history of the damped Chord-Newton iteration & required number of MWR V(1,1)-cycles.

† : average number of MWR cycles for the 19 time-integrations needed to compute J.

* : increase of \|\| residual \|\| : result rejected ; damping factor decreased.

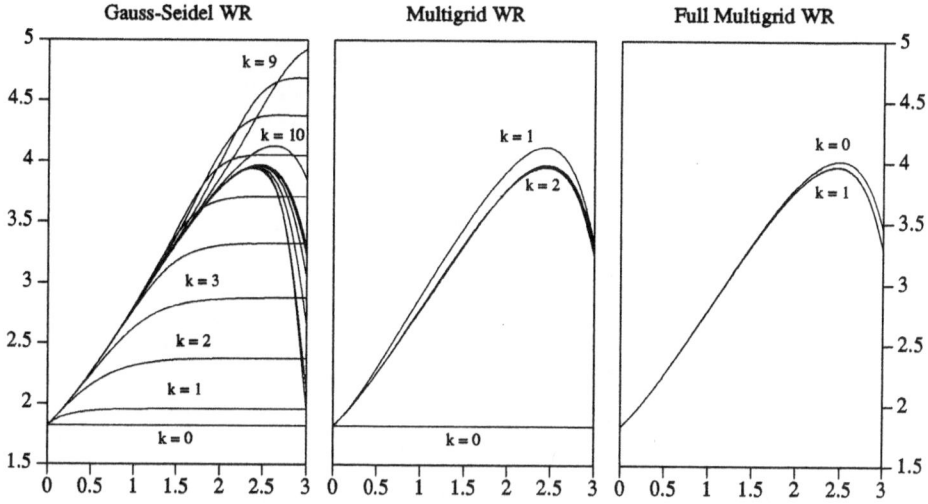

Fig. 1 : Waveform Relaxation for time-integration of (10) with $L = 0.90$ and $h = 1/16$. Approximations $Y^{(k)}(0.5, 0.5, t)$ obtained after k iteration steps. Starting value : $Y^{(0)}(0.5, 0.5, t) =$ constant.

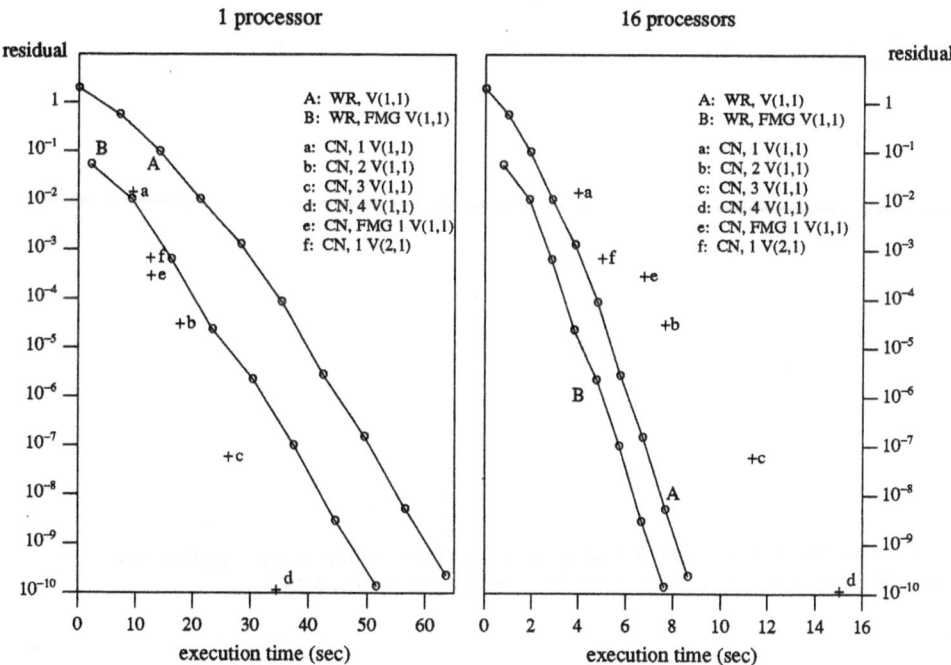

Fig. 2 : Comparison of Multigrid Waveform Relaxation and Crank-Nicolson for time-integration of (10) with $L = 0.75$ and $h = 1/16$: execution times on iPSC/2.

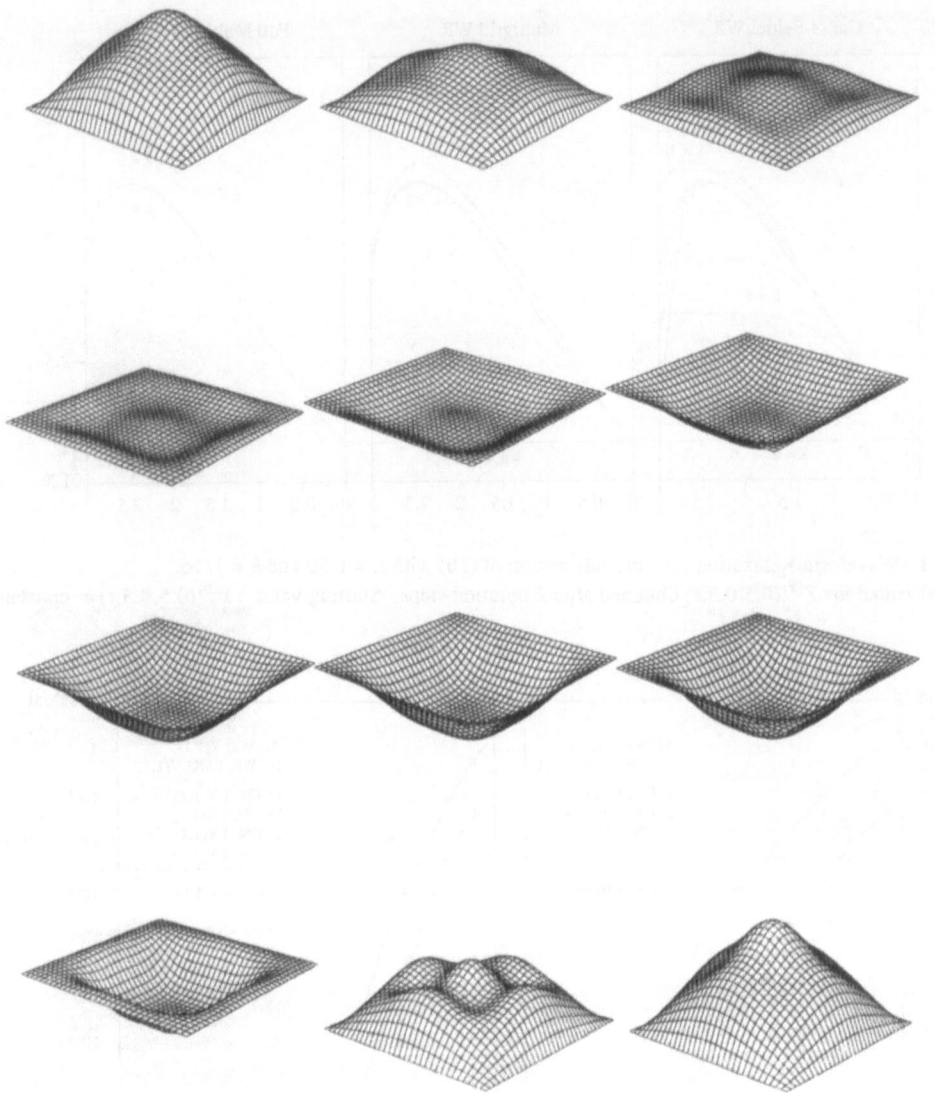

Fig. 3 : Periodic solution of the two-dimensional Brusselator model with $L = 0.90$ and $h = 1/32$, represented by $X(r,s,t_i)$ ($i=0,...,11$), with $\Delta t_i = 0.29$ (period $T \approx 3.7$).

References

[1]   Bomans, L., and Roose, D., *Benchmarking the iPSC/2 hypercube multiprocessor*, Concurrency: Practice and Experience, 1 (1), 1989, pp. 3-18.

[2]   Curtis, A.R., Powell, M.J.D., and Reid, J.K., *On the estimation of sparse Jacobian matrices*, J. Inst. Math. Appl. 13, 1974, pp. 117-119.

[3]   Chan, T.F., and Resasco, D., *A survey of preconditioners for domain decomposition*, In: Proc. IV Coloquio de Matematicas, Taller de Analisis Numerico y sus Applicationes, Taxco, Guerrero, Mexico, Springer-Verlag, Berlin, 1985.

[4]   Doedel, E., *AUTO: Software for continuation and bifurcation problems in ordinary differential equations*, Report Applied Mathematics, CalTech, Pasadena, 1986.

[5]   Fox, G.C., Johnson, M.A., Lyzenga, G.A., Otto, S.W., Salmon, J.K. and Walker, D.W., *Solving Problems on Concurrent Processors*, Prentice Hall, 1988.

[6]   Hackbusch, W., *Multi-grid methods and Applications*, Springer Series in Comp. Math. 4, Springer-Verlag, Berlin, 1985.

[7]   Hassard, B.D., Kazarinoff, N., and Wan, Y.H., *Theory and Applications of Hopf Bifurcation*, Cambridge University Press, Cambridge, 1981.

[8]   Holodniok, M. and Kubicek, M., *DERPER - an algorithm for continuation of periodic solutions in ordinary differential equations*, J. Comput. Phys. 55, 1984, pp. 254-267.

[9]   Holodniok, M., Knedlik, P., and Kubicek, M., *Continuation of periodic solutions in parabolic differential equations*, In : Bifurcation : Analysis, Algorithms, Applications (T. Küpper, R. Seydel and H. Troger, eds.), ISNM 79, Birkhäuser, 1987, pp. 122-130.

[10] Keller, H.B., and Jepson, A.D., *Steady state and periodic solution paths: their bifurcations and computations*, In: Numerical Methods for Bifurcation Problems (T. Küpper, H.D.Mittelmann, H. Weber, eds.), ISNM 70, Birkhäuser, 1984, pp. 219-246.

[11] Lubich, Ch. and Ostermann, A., *Multi-Grid Dynamic Iteration for Parabolic Equations*, BIT, 27 (1987), pp. 216-234.

[12] Miekkala, U. and Nevanlinna, O., *Convergence of Dynamic Iteration Methods for Initial Value Problems*, SIAM J. Sci. Stat. Comput., 8 (4), 1987, pp. 459-482.

[13] Parker, T.S, and Chua, L., *Practical Numerical Algorithms for Chaotic Systems*, Springer-Verlag, New-York, 1989.

[14] Seydel, R., *From Equilibrium to Chaos. Practical Bifurcation and Stability Analysis*, Elsevier, New York, 1988.

[15] Stoer, J., and Bulirsch, R., *Introduction to Numerical Analysis*, Springer-Verlag, New York, 1980.

[16] Vandewalle, S. and Roose, D., *The Parallel Waveform Relaxation Multigrid Method*, In: Parallel Processing for Scientific Computing (G. Rodrigue, ed.), SIAM, Philadelphia, 1989, pp. 152-156.

[17] Vandewalle, S. and Piessens, R., *A Comparison of the Crank-Nicolson and Waveform Relaxation Multigrid Methods on the Intel Hypercube*, in: Proceedings of the Fourth Copper Mountain Conference on Multigrid Methods (J. Mandel, S. McCormick, J. Dendy, C. Farhat, G. Lonsdale, S. Parter, J. Ruge and K. Stüben, eds), SIAM, Philadelphia, 1990, pp. 417-434.

[18] Vandewalle, S., Van Driessche, R. and Piessens, R., *The Parallel Implementation of Standard Parabolic Marching Schemes*, submitted to Int. J. of High Speed Computing, 1990.

[19] Vandewalle, S. and Piessens, R., *Efficient parallel algorithms for solving initial-boundary value and time periodic parabolic partial differential equations*, submitted to SIAM J. Sci. Stat. Comput., 1990.

[20] White, J., Odeh, F., Sangiovanni-Vincentelli, A.S. and Ruehli, A., *Waveform Relaxation: Theory and Practice*, Memorandum No. UCB/ERL M85/65, Electronics Research Laboratory, College of Engineering, University of California, Berkeley, 1985.

International Series of Numerical Mathematics, Vol. 97, © 1991 Birkhäuser Verlag Basel

Comparison of bifurcation sets
of driven strictly dissipative oscillators

C. Scheffczyk, U. Parlitz, T. Kurz and W. Lauterborn

Institut für Angewandte Physik

TII Darmstadt, Schloßgartenstraße 7, D-6100 Darmstadt, FRG

The bifurcation sets of different periodically driven strictly dissipative nonlinear oscillators are compared in terms of phase diagrams and fixed-point diagrams. The oscillators studied are given in Table 1.

System	Properties	Equation	Potential
cavitation bubble model	soft spring	$\left(1 - \dfrac{\dot{R}}{c}\right)R\ddot{R} + \dfrac{3}{2}\left(1 - \dfrac{\dot{R}}{c}\right)\dot{R}^2 = \left(1 + \dfrac{\dot{R}}{c}\right)\dfrac{P}{\rho} + \dfrac{R}{\rho c}\dfrac{dP}{dt}$ with : $P(R, \dot{R}, t) = \left(P_{\text{stat}} - P_{\text{v}} + \dfrac{2\sigma}{R_{\text{n}}}\right)\left(\dfrac{R_{\text{n}}}{R}\right)^{3\kappa} - \dfrac{2\sigma}{R} - 4\mu\dfrac{\dot{R}}{R} - P_{\text{stat}} + P_{\text{v}} - P_{\text{s}}\sin(2\pi\nu t)$	
Toda	soft spring	$\ddot{x} + d\dot{x} - e^{-x} + 1 = f\cos(\omega t)$	
Morse	soft spring	$\ddot{x} + d\dot{x} + 8\left[e^{-x} - e^{-2x}\right] = f\cos(\omega t)$	
modified Toda	hard spring	$\ddot{x} + d\dot{x} - e^{-x} + x + 1 = f\cos(\omega t)$	
soft symmetric oscillator	symmetric soft spring	$\ddot{x} + d\dot{x} + \dfrac{x}{\sqrt[3]{(1 + x^2)}} = f\cos(\omega t)$	
Duffing	symmetric hard spring	$\ddot{x} + d\dot{x} + x + x^3 = f\cos(\omega t)$	

Table 1: *Systems investigated*

All oscillators can be written as three dimensional autonomous systems of ODE's and are assumed to be driven by a sinusoidal force. They define a vector field on $R^2 \times S^1$ that induces a flow map ϕ^t. A global Poincaré map is given as a time-T map of ϕ^t with period $T = 2\pi/\omega$ (ω driving frequency) [1]. All models are chosen to be strictly dissipative, i.e. they possess an everywhere contracting Poincaré map. Therefore Hopf bifurcations and the existence of invariant tori are excluded.

From previous investigations of different nonlinear oscillators [2-8] it is well known that the bifurcation curves form a recurrent structure in parameter space. A phase diagram showing the typical alternating sequence of saddle-node (sn) and period-doubling (pd) bifurcation curves is given in Fig. 1.

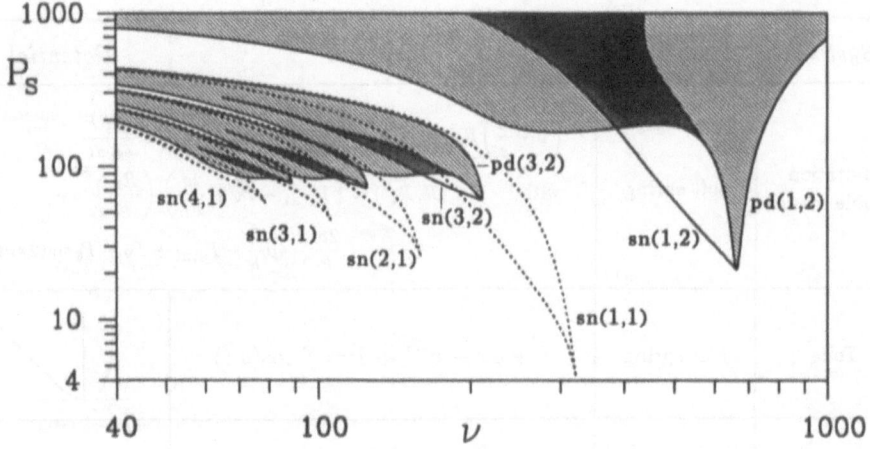

Fig. 1: *Phase diagram of the cavitation bubble model ($R_n = 10\mu m$, P_s in kPa, ν in kHz, the values of the other parameters are the same as in Ref. [4]).*

In Fig. 1 sn-bifurcation curves (dotted lines) and pd-bifurcation curves (enclosing shaded areas) of the basic period-one solution as well as sn-bifurcation curves of period-two solutions can be seen (the regions of coexistence of three period-two solutions are shaded darkly). The curves are labeled by the bifurcation type and a pair of two integers, the torsion number n and the period number m, where n counts the number of rotations of a neighbouring orbit around a given periodic trajectory and is constant along bifurcation curves [4,9]. For the computation of bifurcation curves see [4,6,10,11].

In the following we turn to the discussion of the inner structure of period-doubling resonance horns. The first and second column of Fig. 2 show three phase diagrams of the cavitation bubble model and the Duffing oscillator located in the pd(3,2) and the sb(8,1) resonance horn, respectively. Equivalent patterns of bifurcation curves have also been found for the other systems of Table 1 [11]. A scheme of these patterns is given in the third column of Fig. 2.

For the cavitation bubble model the outermost line is a pd-bifurcation curve with an odd torsion number n. For the symmetric Duffing oscillator the pd-cascade has to be preceeded by a symmetry breaking (sb), i.e. the outermost curve then denotes a sb-curve with even torsion number (first row). The pd-bifurcation from period 2 to period 4 (period 1 to period 2 in the case of the Duffing oscillator) can take place on two different bifurcation curves (second and third row).

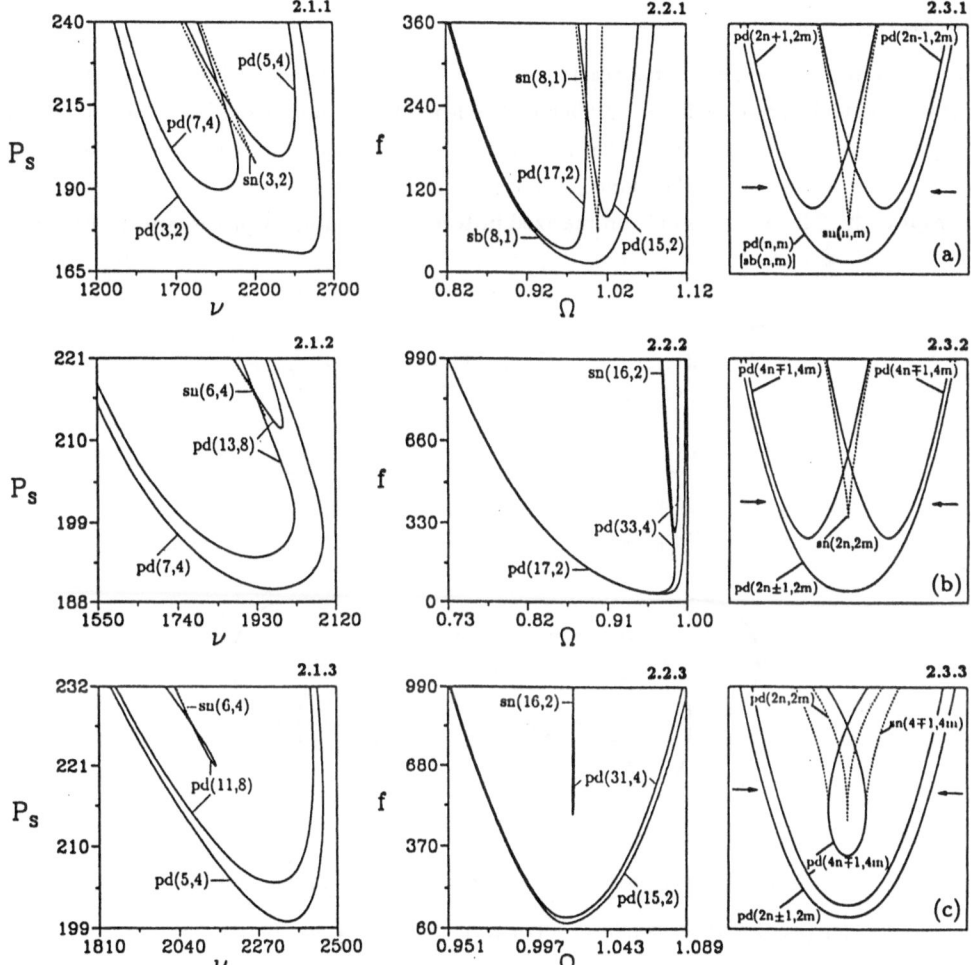

Fig. 2: *Phase diagrams of pd-resonance horns. Column one: cavitation bubble model, $R_n = 1\mu m$; column two: Duffing oscillator, $d = 0.2$; column three: scheme of the bifurcation patterns. For the Duffing oscillator a nonlinear transformation $\omega \mapsto \Omega$ with $\Omega = \omega + 1 - \sqrt[3]{f/220}$ has been applied to yield better discernible curves.*

These curves are the first stages of two pd-cascades with different torsion number sequences. Within these two pd-bifurcation curves further period-doublings take place. Two subpatterns of pd-bifurcation curves can be identified which can be seen in the second and third row of Fig. 2. They both consist of one sn-bifurcation curve and two pd-bifurcation curves. In contrast to the first row of Fig. 2 these two pd-curves possess the same torsion number. The general rules of construction for the torsion numbers are given in the third column of Fig. 2. The patterns recur with the resonances of the system and are essentially the same for all systems investigated.

To understand how different bifurcating solutions are connected by families of stable or unstable periodic orbits a large number of fixed-point diagrams have been computed for all systems given in Table 1. It turned out that the topological structure of the fixed-point curves is essentially the same for all systems if the parameter sections are chosen appropriately. These results are summarized in terms of fixed-point pictograms in Fig. 3.

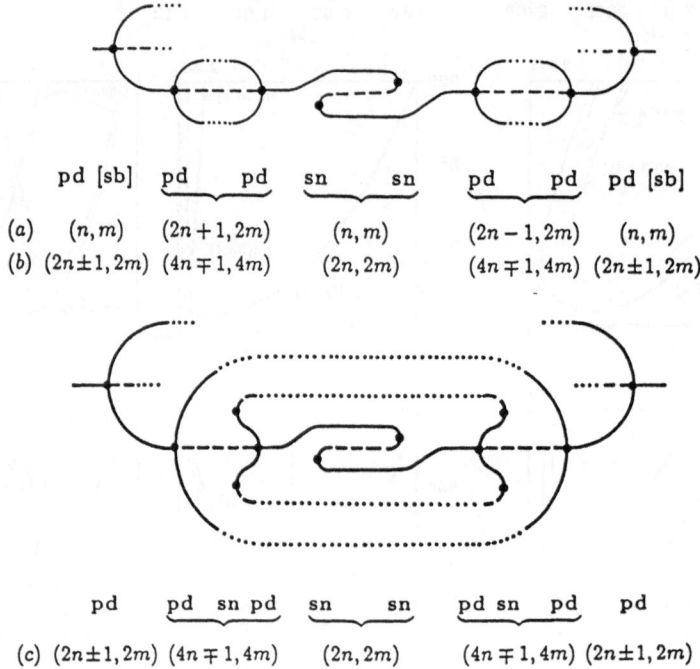

$$
\begin{array}{ccccccc}
\text{pd [sb]} & \text{pd} & \text{pd} & \text{sn} & \text{sn} & \text{pd} & \text{pd} & \text{pd [sb]} \\
\end{array}
$$

(a)	(n,m)	$(2n+1,2m)$	(n,m)	$(2n-1,2m)$	(n,m)
(b)	$(2n\pm1,2m)$	$(4n\mp1,4m)$	$(2n,2m)$	$(4n\mp1,4m)$	$(2n\pm1,2m)$

$$
\begin{array}{ccccccc}
\text{pd} & \text{pd sn pd} & \text{sn} & \text{sn} & \text{pd sn} & \text{pd} & \text{pd} \\
\end{array}
$$

(c)	$(2n\pm1,2m)$	$(4n\mp1,4m)$	$(2n,2m)$	$(4n\mp1,4m)$	$(2n\pm1,2m)$

Fig. 3: *Fixed-point pictograms for the parameter sections denoted by the arrows in the third column of Fig. 2.*

The solid and dashed lines in Fig. 3 denote stable and unstable solutions. The interconnection of the branches is indicated by dotted lines disregarding the stability or further bifurcations. The upper pictogram shows the succession of bifurcations and the arrangement of periodic solutions corresponding to Fig. 2.3.1 and Fig. 2.3.2. For the identification of the bifurcation curves in both figures one has to chose different period numbers m in the upper fixed-point pictogram. The lower pictogram corresponds to the parameter section of Fig. 2.3.3. Beneath each pictogram the bifurcations are characterized by their bifurcation type and their torsion and period number. For symmetric systems the outer bifurcation points in the upper pictogram belong to symmetry-breaking bifurcations (denoted by sb).

The comparison of oscillators has revealed that there exist two typical patterns of bifurcation curves that describe the succession of pd-bifurcations within a plane parameter section. They are expected to constitute elementary building stones of the bifurcation set at each level of the pd-hierarchy. These two patterns can occur in different combinations, enclosed by the primary pd-bifurcation curve. Their appearence and the recurrent structure can be modelled by maps based on simple assumptions concerning the global structure of the flow (see U. Parlitz *et al.* in these proceedings) [12].

Acknowledgement

This work was supported by the Deutsche Forschungsgemeinschaft (SFB 185).

References

[1] J. Guckenheimer and P. Holmes. *Nonlinear oscillations, dynamical systems and bifurcation of vector fields*, volume 42 of *Appl. Math. Sci.* Springer, New York, third edition, 1990.

[2] U. Parlitz and W. Lauterborn. *Phys. Lett. A*, 107(8):351–355, 1985.

[3] T. Kurz and W. Lauterborn. *Phys. Rev. A*, 37(3):1029–1031, 1988.

[4] U. Parlitz, V. Englisch, C. Scheffczyk, and W. Lauterborn. *J. Acoust. Soc. Am.*, 88(2):1061–1077, 1990.

[5] W. Knop and W. Lauterborn. *J. Chem. Phys.*, 1990. To appear.

[6] H. Kawakami. *IEEE Transaction on Circuits and Systems*, CAS-31(3):248–260, 1984.

[7] S. Sato, M. Sano, and Y. Sawada. *Phys. Rev. A*, 28(3):1654–1658, 1983.

[8] J. M. T. Thompson. *Proc. R. Soc. Lond. A, Math. Phys. Sci.*, 421:195–225, 1989.

[9] U. Parlitz and W. Lauterborn. *Z. Naturforsch.*, 41 a:605–614, 1986.

[10] R. Seydel. *From equilibrium to chaos: practical bifurcation and stability analysis.* Elsevier, New York, 1988.

[11] C. Scheffczyk, U. Parlitz, T. Kurz, W. Knop, and W. Lauterborn. submitted for publication.

[12] U. Parlitz, C. Scheffczyk, and T. Kurz W. Lauterborn. This volume.

International Series of Numerical Mathematics, Vol. 97, © 1991 Birkhäuser Verlag Basel

Echo Waves in Reaction-Diffusion Excitable Systems

Sevcikova H., Marek M.

Department of Chemical Engineering, Prague Institute of Chemical Technology, Technicka 5, 166 28 Prague 6, Czechoslovakia

Introduction

An excitable system is usually characterized by such dynamical behaviour, where a small finite stimulation of a given stationary state starts an excitation cycle - a large trajectory in the phase space returning to the neighbourhood of the perturbed stationary state. The excitability is a general property common to many systems of different physical nature among which the biological and chemical systems have recently become the subject of wide interest [1,2]. The significant part of studies is devoted to investigation of generation and propagation of waves of excitation travelling through spatially distributed excitable media, as the wave propagation forms the basis of the signal transmission in different excitable functional units of living organisms.

One of the interesting questions is how the specific dynamic properties of a single excitable cell (a lumped parameter system) - the basic unit of the spatially distributed excitable medium - affects the wave propagation and the response of the distributed system to a local stimulation. In papers [3,4,5] we have reported on the behaviour of spatially distributed reaction - diffusion media with two different types of excitability represented by two different phase portraits of a single cell. The first phase portrait contained one stable and two unstable stationary states while the second one consisted of two stable and one unstable stationary states. In both cases the stable separatrices of the middle unstable state (saddle point) were responsible for the threshold behaviour of the cell. The responses of both types of excitable media to the locally applied periodic stimulation showed both qualitative similarities (existence of periodic and aperiodic trains of pulse-like waves, entrainment by the external stimulation, threshold behaviour, etc.) and the important differences (e.g., the extinction of the wave propagation in the system with the second type of excitability [4]).

In this paper we report on the behaviour of the excitable medium with another type of excitability represented by the mathematical model of the spatially one-dimensional reaction-diffusion system.

Mathematical model

The studied model consists of two dimensionless partial differential equations where the reaction kinetics is described by the SH model [5] :

$$\frac{\partial X}{\partial t} - \frac{D_X}{L^2} \cdot \frac{\partial^2 X}{\partial z^2} + \alpha \cdot \frac{v_0 + X^\gamma}{1 + X^\gamma} - X \cdot (1 + Y)$$

$$\frac{\partial Y}{\partial t} - \frac{D_Y}{L^2} \cdot \frac{\partial^2 Y}{\partial z^2} + X \cdot (\beta + Y) - \delta \cdot Y \tag{1}$$

The local applied periodic stimulation is modelled by the time dependent boundary conditions at z=0 [5] :

$$
\begin{aligned}
X(0,t) &- A_f & &\text{for } t \in <nT_f; nT_f + \Delta\tau) \\
\partial X(0,t)/\partial z &- 0 & &\text{for } t \in <nT_f + \Delta\tau; (n+1)T_f) \\
n &- 0,1,2... & & \\
\partial Y(0,t)/\partial z &- 0 & &\text{for } t \geq 0
\end{aligned}
\tag{2}
$$

Here X and Y represent the concentrations of two reaction species, t denotes time, $z \in <0,1>$ spatial coordinate, L the characteristic length of the system and D_X and D_Y the diffusion coefficients of the corresponding reaction species. α ,β ,γ ,δ and v_0 are dimensionless positive kinetic constants, $\Delta\tau$ denotes the time interval of the stimulus application and A_f and T_f stand for the stimulus amplitude and the period, respectively. In this study, the following fixed values of parameters were used: $L = 1.0$, $D_X = 0.008$, $D_Y = 0.004$, $\Delta\tau = 0.6$, $\beta = 1.5$, $\gamma = 3.0$ and $v_0 = 0.01$.

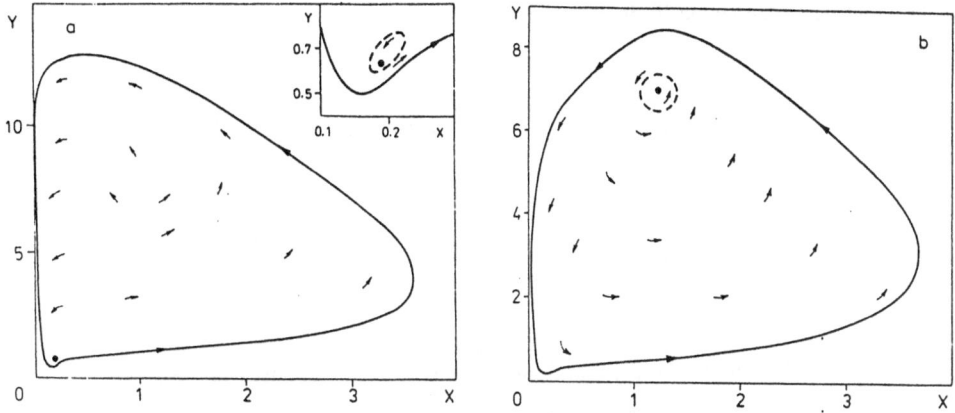

Figure 1:   Phase portraits of lumped parameter systems. Full line - stable
limit cycle, broken line - unstable limit cycle, • - stable
stationary state. a) $\alpha = 18.5$, $\delta = 0.65$, $X_s = 0.19391$,
$Y_s = 0.63773$, the period of the stable limit cycle $T_{LC} = 14.1$;
b) $\alpha = 15.0$, $\delta = 1.5$, $X_s = 1.23545$, $Y_s = 6.97596$, $T_{LC} = 10.2$.

Results

Two pairs of the remaining parameters α, δ have been varied to obtain required dynamics
of the excitable cell (i.e. the lumped parameter system arising from Eq. (1) when $D_X=D_Y=0$)
of the type shown on the phase portraits in Fig. 1a and b. Both portraits are qualitatively
similar consisting of one stable stationary state of a focus type surrounded by an unstable limit
cycle and by a stable limit cycle. The unstable limit cycle determines the basin of attraction
of the stable steady state. The response of both excitable systems to the single stimulus
$(T_f \rightarrow \infty)$ applied to the system in the stable steady state is also qualitatively similar.
Stimulations with the subthreshold amplitude (located inside the unstable limit cycle) decay
in an oscillatory way to the stable steady state, cf. Fig. 2a, while the superthreshold
stimulations (located outside the unstable limit cycle) result finally in sustained oscillations

corresponding to the stable limit cycle, cf. Fig. 2b.

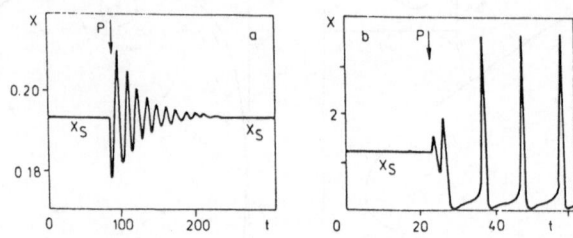

Figure 2:   The subthreshold (a) and superthershold (b) responses to the single
               stimulus (P). a) $\alpha = 18.5$, $\delta = 0.65$, $P = (X_P = 0.18, Y_P = 0.65)$;
               b) $\alpha = 15.0$, $\delta = 1.5$, $P = (X_P = 1.0, Y_P = 6.0)$.

The situation changes when the spatially distributed medium with the non-zero diffusion
is considered. The response to a single stimulus of the medium with the excitability
corresponding to the phase portrait in Fig. 1a is illustrated in Fig. 3. The single stimulus
evokes a wave which while travelling along the system initiates the local limit cycle type
behaviour at every spatial position. The established spatial phase shift appears as several phase
waves. However, the spatial phase shift decreases in time and, finally, spatially homogeneous
limit cycle with the same period as in the lumped parameter system settles in.

Different type of the response has been observed when the excitable medium corresponding
to the phase portrait in Fig. 1b was subjected to a single stimulus. The stimulus evokes a
spatially damped oscillatory response in the neighbourhood of the stimulated boundary as it
is depicted in Fig. 4. However, the spatial concentration gradient established for a certain time
interval between the part of the system with stagnant stationary concentration profiles and the
part of the system with oscillatory variations of concentrations (profile No. 7 in Fig. 4) serves
as a source of front waves propagating from the right to the left (profiles No.7 to 11 in Fig.4).
The front waves are annihilated by local oscillations at the left boundary of the system. The
entire process repeats periodically once the local limit cycle behaviour has been evoked by
a superthreshold stimulation. The process of an interplay between the oscillatory and stationary
concentration profiles creating a source of waves travelling backwards can be called "echo
mechanism of wave generation" and the generated waves we then call "echo waves".

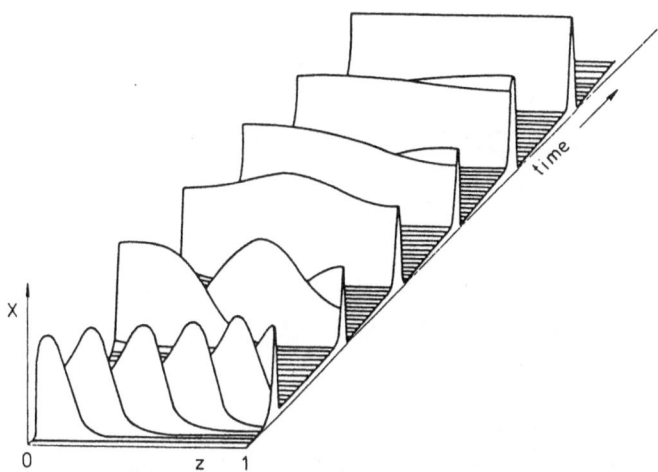

Figure 3: The evolution of the spatially homogeneous limit cycle after a single superthreshold stimulus has been applied. The solution of Eqs. (1) and (2) with $\alpha = 18.5$, $\delta = 0.65$, $A_f = 2.0$, $T_f \rightarrow \infty$.

The periodic superthreshold stimulation of the distributed medium with the excitability given by the phase portrait shown in Fig. 1a gives two basically different types of wave responses. For some values of stimulation parameters (A_f, T_f) the wave trains with different firing numbers [6] propagating down from the point of stimulation have been observed, similarly as in the excitable media with the types of excitability studied previously [4,5]. The period T of wave trains fully synchronized with the period of stimulation (firing number $v = 1{:}1$) is much shorter ($T = T_f = 5.0$) then the period of underlying stable limit cycle ($T_{LC}=14.1$). Thus, the system exhibits the excitability on the stable limit cycle, the phenomenon observed in chemical excitable medium of the Belousov-Zhabotinski type [7].

The second type of the response which differentiates this excitable medium from the previously studied ones [4,5] is shown in Fig. 5. The first stimulus evokes a wave propagating from the left to the right, but several following stimuli do not succeed in generation of waves and the system thus in principal preserves the state with the low X concentration. After a certain time interval the local spontaneous excitation occurs at the right end of the system

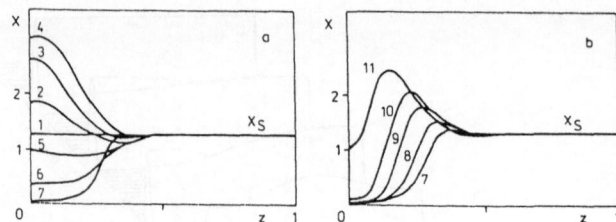

Figure 4:   Time sequence of spatial profiles of X concentration in the system
of Eqs. (1) and (2) with $\alpha = 15.0$, $\delta = 1.5$, $A_f = 2.0$, $T_f \to \infty$.

initiating the wave propagating to the left end where it is annihilated. After an application of
additional stimuli another wave is generated at the left boundary ($z = 0$) and propagates along
the system. In the described case the wave generated by the stimulation at $z = 0$ alternates

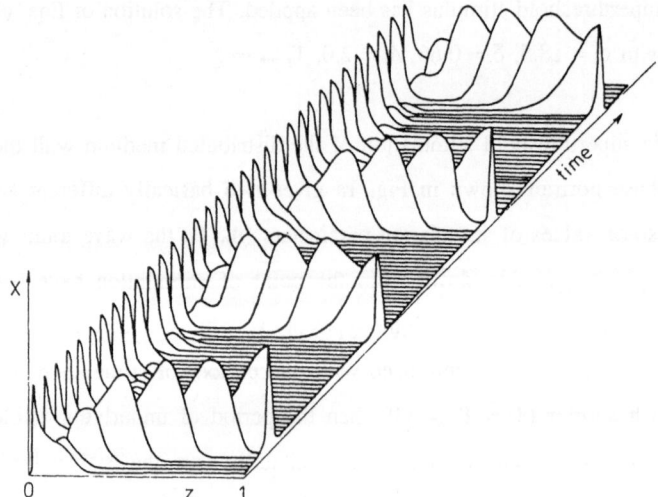

Figure 5:   The evolution of echo waves in the system of Eqs. (1) and (2)
with $\alpha = 18.5$, $\delta = 0.65$, $A_f = 3.0$, $T_f = 1.5$.

regularly with the echo wave generated by the pacemaker formed at $z=1$. The period of the
overall process is $T=28.5$, hence, approximately twice the period of the stable limit cycle.

The periodic stimulation of the medium of the excitability related to the phase portrait in

Fig. 1b gives similar type of the response as the single perturbation, i.e., the complex spatially nonhomogeneous time dependent concentration profiles arise.

Conclusions

The described mechanism of the generation of echo waves is, as far as we know, reported for the first time and can have a profound significance for the studies of irregular functions of biological excitable media, as the echo waves can have different properties (frequency, wave velocity) then the original waves resulting from the primary stimulation.

References:

1.   A.V. Holden, M. Markus, H.G. Othmer, eds., Nonlinear Wave Processes in Excitable Media (Plenum Press, London 1990) (in press).

2.   V. S. Zykov, Modelling of Wave Processes in Excitable Media, (Nauka, Moscow 1984).

3.   L. Ketnerova, H. Sevcikova and M. Marek, "Periodic Forcing of a Spatially One-Dimensional Excitable Reaction-Diffusion System", in ref. [1].

4.   H. Sevcikova and M. Marek, "Wave Patterns in an Excitable Reaction-Diffusion System", to appear in Physica D.

5.   H. Sevcikova and M. Marek, Physica 26D (1986) 61-77.

6.   M. Marek, I. Schreiber and L. Vroblova, in Structure, Coherence and Chaos, eds. R.D. Parmentier and P.L. Christiansen (Manchester Univ. Press 1988), pp. 233-244.

7.   A. Pagola, J. Ross and C. Vidal, J. Phys. Chem. 92 (1982) 163.

International Series of Numerical Mathematics, Vol. 97, © 1991 Birkhäuser Verlag Basel

THE LOCAL STABILITY OF INACTIVE MODES IN CHAOTIC MULTI-DEGREE-OF-FREEDOM SYSTEMS

Steven W. Shaw and Shang-Rou Hsieh
Department of Mechanical Engineering
Michigan State University
East Lansing, MI 48824 U.S.A.

1. INTRODUCTION

In this paper we consider a class of systems for which motions are possible in which a small number of modes undergo chaotic oscillations while the remaining modes, if started with zero energy, remain at a zero energy state for all time. Of interest here is the local stability of these modes, which are parametrically driven by the active modes. We specify the class of systems to be considered, present a general stability result, and then sketch the analysis and results for a specific mechanical device. Details of the results can be found in Reference [1].

2. THE CLASS OF SYSTEMS

The equations of motion of interest here arise from modal truncations of mechanical systems; these are typically nonlinear, coupled second order ordinary differential equations. We assume they have representation in terms of modal amplitudes of the form:

$$\ddot{x}_i + 2\xi_i \dot{x}_i + \omega_i^2 x_i + f_i(\mathbf{x};t) = 0 \qquad i=1,2,\ldots,n \qquad (2.1)$$

where $\mathbf{x} = \left[x_1, x_2, \ldots, x_n\right]$ is the modal displacement vector, ξ_i are the modal damping coefficients, ω_i are the modal undamped natural frequencies and the f_i are the nonlinear coupling and external excitation terms. In order to keep the analysis simple we assume that only the first mode is active; the ideas can be generalized to other cases. If the f_i have the following property

$$f_i\left[x_1, \ldots, x_i=0, \ldots, x_n; t\right] = 0 \qquad i=2,3,\ldots,n \qquad (2.2)$$

then the equations of motion admit a solution of the form

$$x_1(t) = \bar{x}_1(t) \quad , \quad x_i(t) = 0 \qquad i=2,3,\ldots,n \qquad (2.3)$$

where $\bar{x}_1(t)$ satisfies the following equation which governs the first mode when it alone is active:

$$\ddot{x}_1 + 2\xi_1 \dot{x}_1 + \omega_1^2 \bar{x}_1 + f_1(\bar{x}_1, \; 0 \ldots, 0; t) = 0 \tag{2.4}$$

The dynamics of small peturbations of the solution given in equation (2.3)

can be studied by assuming $x_1 = \bar{x}_1 + \mu q_1$, $x_i = \mu q_i$ for $i=2,3,\ldots,n$ (2.5) where $|\mu| \ll 1$. Substitution of this into equation (2.1) and utilizing equation (2.4) yields equations which give the linearized dynamics of the disturbance:

$$\ddot{q}_i + 2\xi_i \dot{q}_i + \omega_i^2 q_i + \sum_{j=1}^{n} \frac{\partial f_i}{\partial x_j} \left(\bar{x}_1, 0, \ldots, 0; t\right) q_j = 0 \qquad i=2,3,\ldots,n \tag{2.6}$$

Here we track only the disturbances of the inactive modes since the behavior

of q_1 is irrelevant for the stability of the first mode. In fact, for $\bar{x}_1$

chaotic, small disturbances in $\bar{x}_1$ will grow exponentially due to sensitive dependence on initial conditions. If, in addition, the f_i have the property that

$$\frac{\partial f_i}{\partial x_j} \left(\bar{x}_1, 0, \ldots, 0; t\right) = \delta_{ij} \phi_j(t) \qquad\qquad i,j=2,3,\ldots,n \tag{2.7}$$

then the linearized equations uncouple to

$$\ddot{q}_i + 2\xi_i \dot{q}_i + \left[\omega_i^2 + \phi_i(t)\right] q_i = 0 \qquad\qquad i=2,3,\ldots,n \tag{2.8}$$

Condition (2.7) occurs quite naturally for polynomial type nonlinearities which satisfy condition (2.2).

The stability of the q_i ($i=2,3,\ldots,n$) for the case when the ϕ_i are periodic can be determined by Floquet theory and has been widely studied. In contrast, we consider the case where the ϕ_i are chaotic; this occurs,

through the f_i, when $\bar{x}_1$ is chaotic.

Examples of systems which fit into this category include beams, plates, surface waves in fluids, and other mechanical systems which are parametrically driven or are excited in such a way that the projection of the excitation onto modal coordinates results in it driving only a single mode.

3. ALMOST-SURE STABILITY FOR LINEAR STOCHASTIC SYSTEMS

In order to estimate the parameter values $\left(\xi_i,\omega_i\right)$ for which the q_i decay to zero, we need to know certain features of the excitations ϕ_i, which depend on $\bar{x}_1$. For the case when the ϕ_i have the following properties: continuous with probability one, weakly stationary, and ergodic, there exist several methods for computing stability bounds for the damping parameters ξ_i. Of these, we employ the result due to Infante [2] which guarantees the stability of $q_i=0$ with probability one if

$$E\left[\phi_i^2\right] - \left(E\left[\phi_i\right]\right)^2 \leq 4\xi_i^2 \left(\omega_i^2+E\left[\phi_i\right]\right) \tag{3.1}$$

where $E[z]$ is the expected value of z. Tighter bounds can be obtained if one has more information about ϕ_i. It should be noted that expression (3.1) a sufficient, but not a necessary condition for stability, and is typically quite conservative.

One approach to these problems is to directly simulate equation (2.4) and numerically compute the required statistical properties of $\phi_1 = \frac{\partial f_i}{\partial x_i}\left(\bar{x}_1,0,\ldots,0;t\right)$. Estimates of the level of damping required to stabilize the remaining modes can then be made. Here we present a special example which allows for a more analytical approach.

4. THE POWER SPECTRUM FOR DUFFING'S EQUATION

Typically one cannot analytically compute the quantities required in the stability condition (3.1). A special oscillator for which this can be done is the nonlinearly damped Duffing oscillator

$$\ddot{u} - u + u^3 = \epsilon\left[-\xi\dot{u}-\delta u^2\dot{u}+\gamma\cos(\omega t)\right] \tag{4.1}$$

studied by Brunsden et al. [3,4]. For $5\xi \simeq 4\delta$ and $|\epsilon|\ll 1$, solutions of equation (3.2) remain bounded in a small neighborhood of the two homoclinic motions which connect the $u=0$ saddle point to itself when $\epsilon=0$:

$$h_+(t) = + \sqrt{2} \text{ sech } (t) = -h_-(t). \tag{4.2}$$

Here the linear and nonlinear damping effects cancel along h_+. In this case, Brunsden et al. show how an approximate solution can be constructed by a simple superposition of h_+ and h_- motions, in random order, each shifted in time by some random amount:

$$u(t) \simeq z(t) = \sum_{j=-\infty}^{\infty} (-1)^{d_j} h_+(t-T_j) .$$ (4.3)

Here d_j and $\left[T_{j+1}-T_j\right]$ are assumed to be discrete, uncorrelated random variables. The variable d_j is equal to 0 or 1 with equal probability depending on whether the jth component follows h_+ or h_-, respectively, and T_j is the peak time of the jth component. It is shown in [3,4] that the power spectrum $S_z(f)$ for a motion given by (4.3) is given by that of a single component, given here by h_+, divided by the mean time between events, $T=E\left[T_{j+1}-T_j\right]$. The power spectrum is thus given by:

$$S_z(f) = \frac{2\pi^2}{T} \operatorname{sech}^2(\pi f) .$$ (4.4)

T can be estimated by noting that it is determined primarily by how close the solution is to the stable manifold of the saddle type periodic solution of equation (4.1) as it passes near u=0. A local analysis near the saddle, which uses the fact that the width of the region in which the motion is trapped is proportional to the supremum of the Melnikov function, allows one to show that

$$T \simeq K - \frac{1}{\lambda_+} \log_e\left[\epsilon\gamma\omega \sqrt{2} \pi \operatorname{sech}\left(\frac{\pi\omega}{2}\right)\right]$$ (4.5)

where K is a constant which can be determined from a single numerical experiment of equation (4.1), and λ_+ is the unstable eigenvalue of the $\epsilon=0$ saddle point. Brunsden et al. [3,4] show convincing evidence of the usefulness of this approximate analysis, even in the case when the nonlinear damping term is absent, $\delta = 0$.

We are now in a position to estimate stability bounds for systems which have a single active mode governed by the special Duffing oscillator. The required ingredients in the stability condition (3.1) are:

$$E[u] = 0 \quad \text{since there is no bias in } d_j$$ (4.6a)

$$E\left[u^2\right] = \int_0^\infty S_u(f) \, df \simeq \int_0^\infty S_z(f) \, df$$ (4.6b)

5.   AN EXAMPLE

A simple mass-spring-damper system with an attached pendulum is depicted in Figure 1.   The translational part of the system is assumed to be governed by equation of motion (4.1).   When coupled to the pendulum, the rescaled equations of motion can be written as

$$\ddot{x}_1 - \rho \sin(x_2)\ddot{x}_2 - \rho \dot{x}_2^2 \cos x_2 - x_1 \tag{5.1a}$$

$$+ x_1^3 + \epsilon \xi \dot{x}_1 - \epsilon \delta x_1^2 \dot{x}_1 = \epsilon \gamma \cos (\omega t)$$

$$\ddot{x}_2 + c \dot{x}_2 + \sin(x_2) \left(\omega_2^2 - \ddot{x}_1\right) = 0 \tag{5.1b}$$

where x_1 is the mass displacement and x_2 is the pendulum angle.   The parameter ρ is a mass ratio, ξ and δ are the linear and nonlinear damping coefficients, respectively, for the mass, γ and ω are the amplitude and frequency of the forcing applied to the mass, c is the linear damping constant for the pendulum pivot and ω_2^2 is the ratio of natural frequencies of the uncoupled systems.   Equation (5.1) admit a solution of the form $x_1 = \bar{x}_1$, $x_2 = 0$ where $\bar{x}_1$ satisfies equation (4.1) with u replaced by $\bar{x}_1$.   Note that in equation (5.1b), the pendulum is parametrically excited by the acceleration of the mass x_1.   The local stability of the $x_2 = 0$ solution is dictated by the behavior of q_2 which solves

$$\ddot{q}_2 + c \dot{q}_2 + q_2 \left(\omega_2^2 - \ddot{\bar{x}}_1\right) = 0 \tag{5.2}$$

Here ϕ_2 is simply $-\ddot{\bar{x}}_1$.   Stability condition (3.1) becomes, for the present case:

$$E\left[\ddot{\bar{x}}_1^2\right] \leq c^2 \omega_2^2 \tag{5.3}$$

since $\ddot{\bar{x}}_1$ has zero mean: $E\left[\ddot{\bar{x}}_1\right] = 0$.   Now

$$E\left[\ddot{\bar{x}}_1^2\right] \approx \int_0^\infty (2\pi f)^4 \, S_z(f) \, df \tag{5.4}$$

by approximating $\ddot{\bar{x}}_1$ by z (z given by equation (4.3)). Completing the required integral results in the final stability bound

$$\frac{14}{15} \frac{1}{T} \leq c^2 \omega_2^2 \tag{5.5}$$

where T is given by equation (4.5). One numerical experiment, used to evaluate K, allows for a bound which holds for all values of ρ, ξ, δ, γ, ω, c, and ω_2 provided $4\delta \simeq 5\xi$ and $0 < \epsilon << 1$.

The dependence of the stability boundaries on some of the system parameters are shown in Figure 2. Comparisons with extensive numerical simulations of the full coupled system (5.1) are shown in Figure 2a for a range of c and ω_2 values with all other parameters fixed. Simulation results from the full coupled system (5.1) which indicate a stable and an unstable case near the actual stability boundary are presented in Figure 3. Figure 4 shows a portion of the chaotic motion given by a solution of equation (4.1); it clearly shows the structure which motivated the approximate solution, z, for $\bar{x}$, (or, equivalently, u).

6. CLOSING REMARKS

The example presented is special in that we were able to estimate the required statistical properties of the chaotic mode. In general, this will not be so, but it will still be possible to obtain upper stability bounds in terms of ξ_i and ω_i (i = 2,3,...,n) for a given set of those parameters which influence $\bar{x}_1$.

Also, the bound obtained is very conservative; the damping level required to stabilize the pendulum in the example system is much lower than the bound obtained by the analysis. This is simply due to the use of only partial information about the driving term $\ddot{\bar{x}}_1$. More information would allow for tighter bounds to be obtained.

In addition, a class of systems where such results would be potentially useful are finite (typically two) mode models of systems with internal resonance, e.g. see [5]. However, in these cases, the chaos is observed for small levels of damping, $\xi_i - 0(\epsilon)$, in which case the stability bounds indicate that the parametric driving terms of the higher modes must be smaller than a quantity of $0(\epsilon^2)$. This seems to present a formidable difficulty.

Finally, it has been rather boldly assumed in this work that the chaotic motion of the active mode is stationary and ergodic, of which we have no proof. The satisfactory results from the example indicate that such assumptions are, if not precisely true, are at least reasonable to use for obtaining the desired stability bounds.

REFERENCES

1. S.R. Hsieh and S.W. Shaw, 1990, Journal of Sound and Vibration 138, 421-431. The stability of modes at rest in a chaotic system.

2. E. F. Infante, 1968, American Society of Mechanical Engineers Journal of Applied Mechanics 35, 7-12. On the stability of some linear nonautonomous random systems.

3. V. Brunsden, J. Cottrel and P. Holmes, 1989, Journal of Sound and Vibration 130, 1-25. Power spectra of chaotic vibrations of a buckled beam.

4. V. Brunsden and P. Holmes, 1987, Physical Review Letters 58, 1699-1702. Power spectra of strange attractors near homoclinic orbits.

5. A.K. Bajaj, 1990, Preprint, Department of Mechanical Engineering, Purdue University, Complex dynamics of whirling strings.

Figure Captions

1. The example system.

2. Stability bounds   a) c vs ω_2^2 with results from simulation

$$\epsilon\gamma = 0.01, \ \omega = 1.00, \ \rho = 0.10,$$
$$\epsilon\xi = 0.4, \ \epsilon\delta = 0.498$$

b) $c\omega_2$ vs ω for values of γ
c) $c\omega_2$ vs γ for values of ω

3. Simulation results. Initial conditions $\left(x_1, \dot{x}_1, x_2, \dot{x}_2\right) = \left(0, 0, 10^{-4}, 0\right)$ at $t = 0$. Parameters as in Figure 2a with $\omega_2^2 = 0.9806$

a) $c = 0.07$, unstable } Note difference in
b) $c = 0.08$, stable   } vertical scales

4. A chaotic solution of equation (4.1). $\epsilon\gamma = 0.01$, $\omega = 1.00$, $\epsilon\xi = 0.4$, $\epsilon\delta = 0.498$.

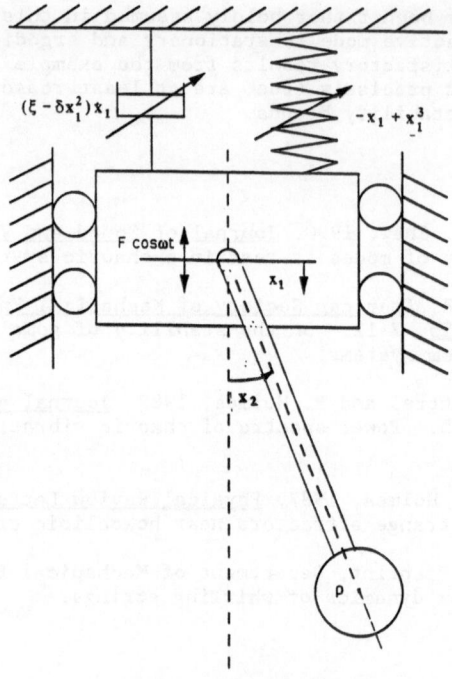

Figure 1. The Example System

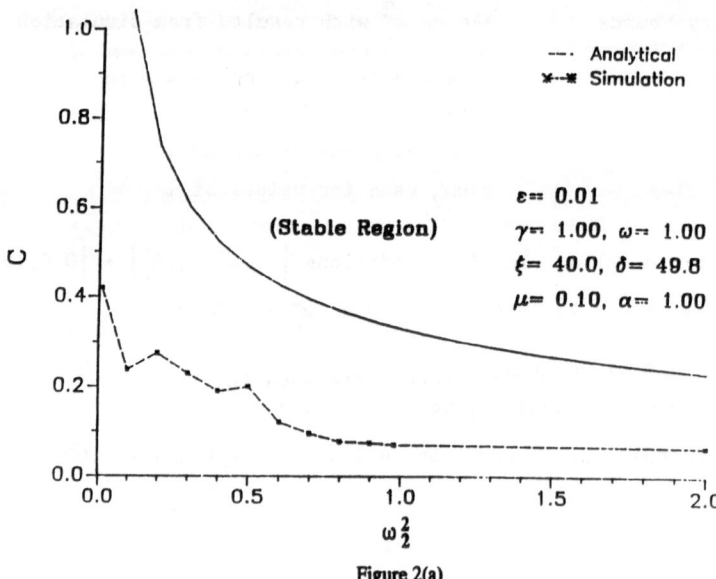

Figure 2(a)

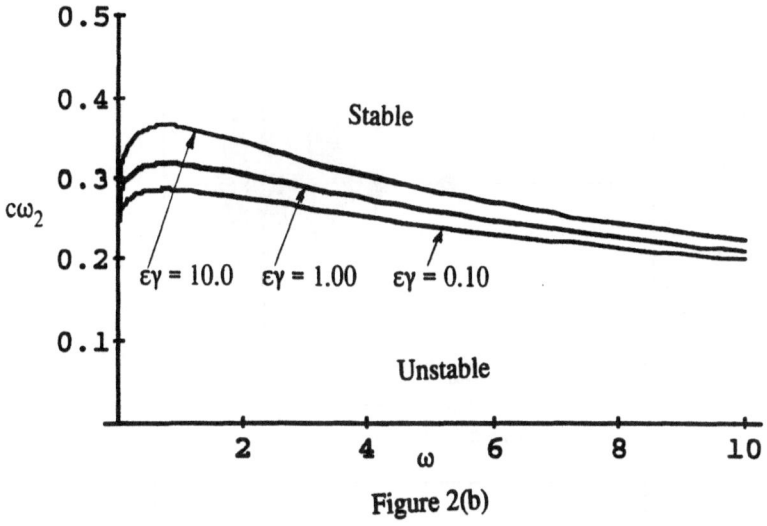

Figure 2(b)

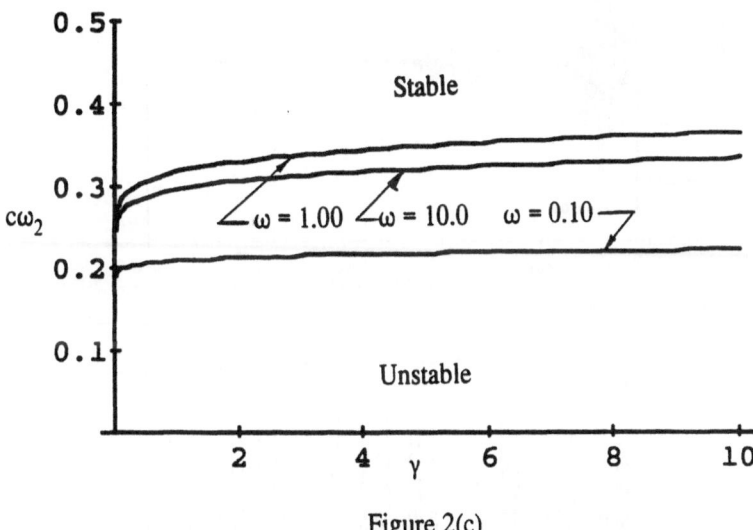

Figure 2(c)

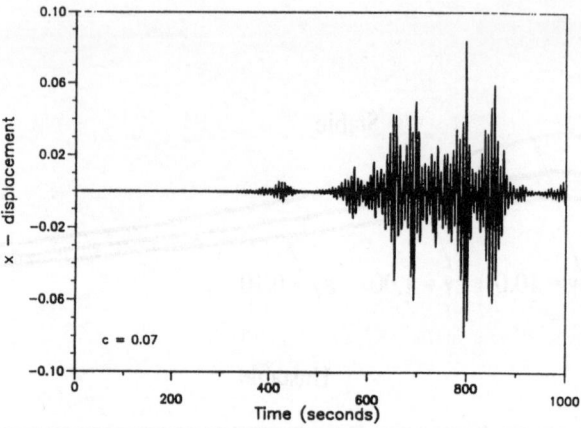

Figure 3(a) Unstable response

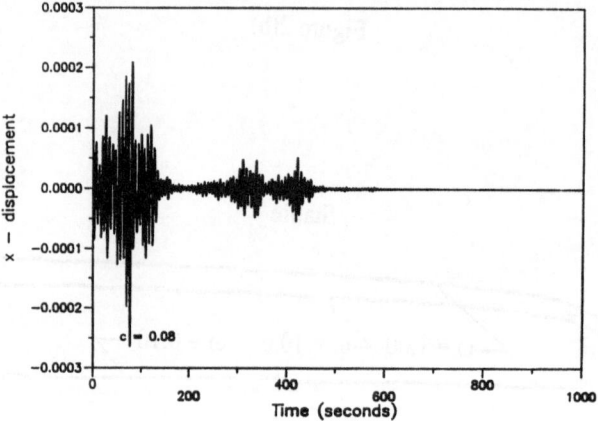

Figure 3(b) Stable response

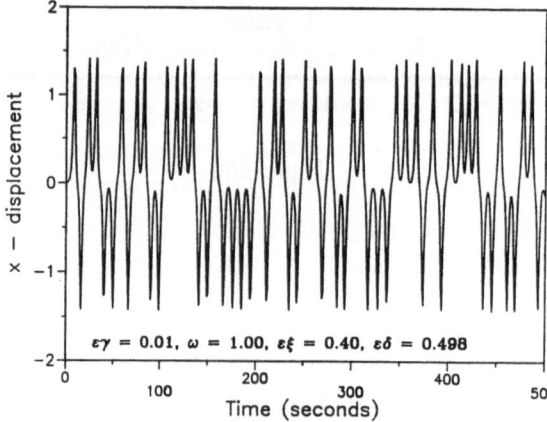

Figure 4. Chaotic response

International Series of Numerical Mathematics, Vol. 97, © 1991 Birkhäuser Verlag Basel

Bifurcations in Dynamic Systems
with Dry Friction

by

Peter Stelter and Walter Sextro

Institute of Mechanics
University of Hannover
FRG

Summary

Dry friction is a main factor of self-sustained oscillations in dynamic systems. The mathematical modelling of dry friction forces result in strong nonlinear equations of motion. The bifurcation behaviour of a deterministic system has been investigated by the bifurcation theory. The stability of stationary solutions has been analyzed by the eigenvalues of the Jacobian. Period doublings and Hopf-bifurcations as well as turning points could be determined with the program package BIFPACK. Phase plane plots of periodic and chaotic motions have been shown for a better understanding of the bifurcation diagrams. Both, unstable branches and stable coexisting solutions have been calculated. Several jumping effects, which are typical for nonlinear systems, have been found.

1 Introduction and Equation of Motion

Systems with dry or Coulomb friction often appear in the field of mechanical engineering. The mathematical modelling of dry friction forces, which are dependent on many parameters such as the relative velocity or the normal force, leads to nonlinear equations of motion. Numerical and experimental investigations have shown, that beside of periodic solutions, more complicated motions are possible. When these motions are generated by deterministic equations, deterministic chaos may occur. The routes to chaos may be via period doublings, torus-bifurcations or intermittency cf. *Kreuzer* [4]. One aim of the investigations was to calculate the bifurcations, where the solution changes without warning. The classification of the bifurcations is possible by the Floquet theory, cf. *Seydel* [9] and *Iooss* [3]. Furthermore, the typical bifurcation scenarios are most important for the understanding of self-sustained oscillations. In order to show the basic phenomena of dynamic systems with dry friction, a simple model has been taken into account.

$$
\left.
\begin{aligned}
x_1' &= x_2 \\
x_2' &= \gamma^{-1}\left\{ -(1+\kappa)x_1 - 2D(1+\delta)x_2 + x_3 + 2Dx_4 + B\rho[\mu(x_2 = 0) - \mu(v_{r1})] \right\} \\
x_3' &= x_4 \\
x_4' &= x_1 + 2Dx_2 - x_3 - 2Dx_4 + B[\mu(x_4 = 0) - \mu(v_{r2})]
\end{aligned}
\right\} \quad (1)
$$

For generalization the following abbreviations have been introduced: the mass ratio $\gamma :=$ m_1/m_2, the damping ratio $\delta := d_1/d_2$, the stiffness ratio $\kappa := c_1/c_2$, the normal force ratio

$\rho := F_{N1}/F_{N2}$, and the load parameter $B := F_{N2}/c_2$. The dimensionless damping is given by $D := d_2/2\sqrt{c_2 m_2}$. The chosen parameter values are: $\omega_2 = 1.0\,\mathrm{s}^{-1}, \gamma = 2.5, \delta = 1.0, \kappa = 2.0$ and $\rho = 1.0$. This system has already been mentioned by *Miyamoto* [7]. The nonlinear structure of equation (1) becomes obvious by the vector notation:

$$x' = Ax + r(x), \qquad x \in \mathbb{R}^4, \tag{2}$$

where A is the linear system matrix and r is the vector of the nonlinear friction forces. Equation (1) represents a two-masses-spring-damper system, which is excited by friction forces exerted by a running band. Self-excitation due to dry friction is only possible, when the friction force has a decreasing characteristic, cf. *Magnus* [6]. For the use of the program package BIFPACK developed by *Seydel*, cf. [10], the function of the friction force has to be continuously differentiable. Thus, for the numerical simulations the following model for the friction characteristic has been used:

$$\left. \begin{array}{rcl} F_{Ri} &=& -F_{Ni}\mu(v_{ri}), \\ \mu(v_{ri}) &=& a_1[1 + a_2\exp(-b_1|v_{ri}|)]\arctan(b_2 v_{ri}), \end{array} \right\}, \, i = 1(1)2 \tag{3}$$

with the constants $a_1 = 0.14, a_2 = 1.14, b_1 = 2.0\,\mathrm{s/m}, b_2 = 100.0\,\mathrm{s/m}$, where μ denotes the friction coefficient, which depends on the relative velocity v_r, and F_N denotes the normal force. The relative velocities v_{ri} are given by $v_{ri} = \omega_2 x'_{2i} - v_0$. To be able to characterize the solution nearby the equilibrium the Jacobian J has to be calculated by differentation with regard to x

$$J(x) = \left[A + \frac{\partial r(x)}{\partial x^T}\right]. \tag{4}$$

Both the equations of motion (eq. 1) and the Jacobian (eq. 4) have been installed in the program package BIFPACK.

2 Bifurcation Behaviour

The important parameters of the system are the load parameter B, the band velocity v_0 and the damping D. The parameter dependencies of the solution of equation (1) have been calculated with the program package BIFPACK. Furthermore, the bifurcation behaviour could be investigated with the use of the Floquet theory, cf. *Seydel* [9], *Hagedorn* [2] and *Iooss* [3]. With the Floquet theory a unique classification of the *global* bifurcations is possible. The amplitude x_3 has been used to show the bifurcation behaviour. Within the bifurcation diagrams, Hopf-bifurcations, turning points and period doublings occur, while stationary bifurcations do not appear. To determine the Hopf-bifurcations, one has to calculate the eigenvalues of the Jacobian. They occur when a complex pair crosses the imaginary axis. The equilibrium is stable, when all eigenvalues are within the left side of the complex plane. A subcritical Hopf-bifurcation (H1) arises at a parameter value of $B = 1.12\,\mathrm{m}$, while a special Hopf-bifurcation (H2), cf. [9], occurs at $B = 7.09\,\mathrm{m}$, which are also shown in the

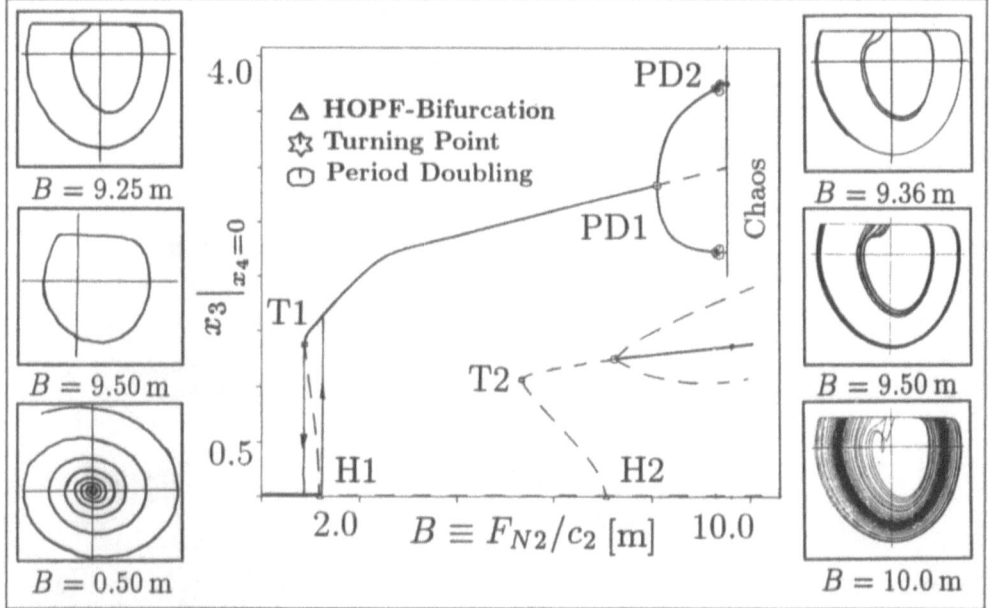

Figure 1: Bifurcation diagram of the load parameter B at $D = 0.1$ and $v_0 = 1.0\,\text{m/s}$ with corresponding phase plane plots

bifurcation diagram of the load parameter, see figure 1. Starting from the equilibrium the amplitude is jumping from the subcritical Hopf-bifurcation (H1) to the stable periodic branch. On the other hand, coming from the periodic branch, the amplitude is jumping from the turning point (T1) to the equilibrium. This jumping phenomenon is typical for systems with dry friction. The unstable branch between the turning point and the subcritical Hopf-bifurcation can be understood as a borderline between the stable attractors. Here, a stable periodic attractor and a stable equilibrium coexist within a parameter range of $0.88\,\text{m} < B < 1.12\,\text{m}$. Following the periodic attractors several period doublings (PD) occur, which end in a chaotic motion. The calculated period doublings are at the load parameters of $8.09\,\text{m}, 9.26\,\text{m}$ and $9.36\,\text{m}$. Beside the bifurcation scenario via period doublings a stable periodic solution coexists, starting with a subcritical period doubling at a load parameter of $B = 7.25\,\text{m}$. Furthermore, the bifurcation diagrams of the band velocity v_0, cf. figure 2, and the damping ratio D, cf. figure 3, have been obtained by means of the program package BIFPACK. They have also shown turning points, Hopf-bifurcations and period doublings. The routes to chaos are also via period doublings. Coexisting solutions, which are limited by subcritical period doublings and turning points could be determined. In figure 2, three stable attractors coexist within the parameter range of $3.57\,\text{m/s} < v_0 < 4.09\,\text{m/s}$.

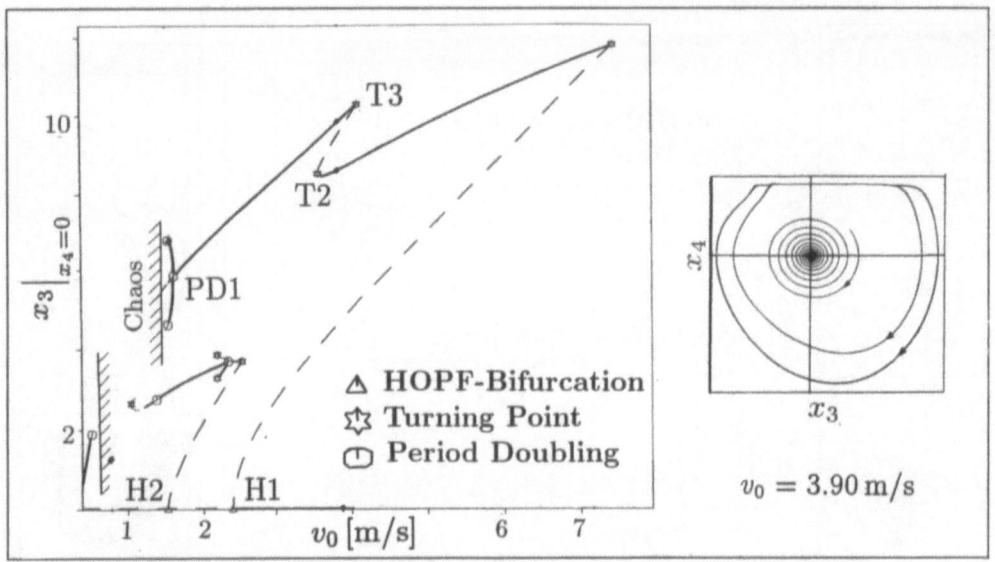

Figure 2: Bifurcation diagram of the band velocity v_0 at $D = 0.1$ and $B = 2.0$ m with a coexisting solution at $v_0 = 3.9$ m/s

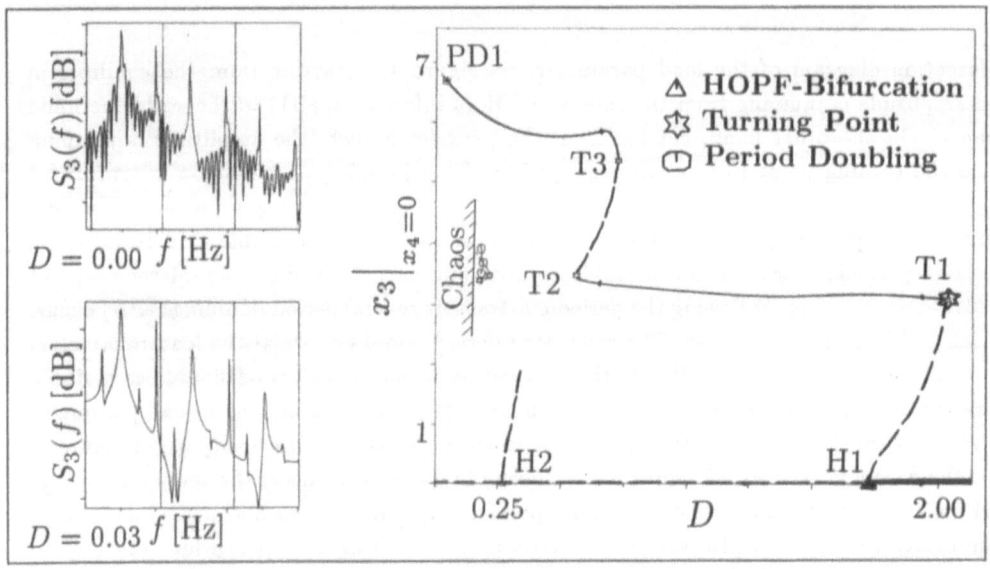

Figure 3: Bifurcation diagram of the damping D at $v_0 = 1.0$ m/s and $B = 20$ m with power spectra at $D = 0.00$ and $D = 0.03$

3 Concluding Remarks

, Bifurcation diagrams have been analyzed by means of the program package BIFPACK. In the case of a two degrees-of-freedom system, the transition to chaotic motions is via period doublings. Quasiperiodic motions do not appear within the investigated parameter range. Several coexisting solutions, which are limited by turning points or subcritical period doublings, occur. The unstable solutions can be seen as the limit between two stable solutions. The phenomenon of the jumping phenonema of the amplitude is typical for self-sustained systems with dry friction. Within all bifurcation diagrams, subcritical Hopf-bifurcations are responsible for the jumping of the amplitude by starting from the equilibrium.

References

[1] Feigenbaum, M.J.: Universal Behavior in Nonlinear Systems. In: *Los Alamos Sci.* **1** (1980), S. 4–27

[2] Hagedorn, P.: *Nichtlineare Schwingungen.* Wiesbaden: Akademische Verlagsgesellschaft, 1984

[3] Iooss, G.; Joseph, P.: *Elementary Stability and Bifurcation Theory.* New York: Springer-Verlag, 1980

[4] Kreuzer, E.: *Numerische Untersuchung nichtlinearer dynamischer Systeme.* Berlin: Springer-Verlag, 1987

[5] Leven, R.W.; Koch, B.-P.; Pompe, B.: *Chaos in dissipativen Systemem.* Berlin: Akademie-Verlag, 1982.

[6] Magnus, K.: *Schwingungen.* 3. Auflage, Stuttgart: Teubner, 1976

[7] Miyamoto, M.: Effect of Dry Friction in Link Suspension on Forced Vibration of Two-Axle Car. In: *Quarterly Reports Vol. 14* No. **2** (1973), S. 99–103

[8] Popp, K.; Stelter, P.: Nonlinear Oscillations of Structures Induced by Dry Friction. In: *Proceedings of IUTAM Symposium on nonlinear dynamics in engineering systems.* Stuttgart, (1989)

[9] Seydel, R.: *From Equilibrium to Chaos; Practical Bifurcation and Stability Analysis.* Amsterdam: Elsevier, 1988

[10] Seydel, R.: *BIFPACK—A Program Package for Continuation, Bifurcation and Stability Analysis.* Mathematische Institute der Julius-Maximilians-Universitaet Wuerzburg, Version 2.3 (1988)

International Series of Numerical Mathematics, Vol. 97, © 1991 Birkhäuser Verlag Basel

THE APPROXIMATE ANALYTICAL METHODS IN THE STUDY OF

BIFURCATIONS AND CHAOS IN NONLINEAR OSCILLATORS

W. Szemplińska-Stupnicka

IPPT, Świętokrzyska 21, 00-049 Warsaw, Poland

We consider the third order dynamical systems which are reduced to the form:

$$\ddot{x} + \Omega_0^2\, x + f(x,\dot{x},\omega t) = 0 \ , \qquad\qquad T = \frac{2\Pi}{\omega} \ , \qquad\qquad (1)$$

$$f(x,\dot{x},\omega t) = h\dot{x} + \alpha_2 x^2 + \alpha_3 x^3 - F \cos \omega t \ ; \quad h > 0 \ ,$$

and we focus on the question of prediction of the bifurcations which are related to, and which precede the escape from a potential well. This class of nonlinear oscillators models a wide spectrum of physical problems and has an extensive literature. Complex bifurcations phenomena in systems characterized in Fig. 1 (a-c) have been reported since 1979, the results being based mostly on computer based or experimental investigations [e.g. 1 - 5].

In this study we derive predictive approximate criteria for the escape from a potential well by making use of the approximate analytical methods and the theory of nonlinear oscillations. The applications appear to be possible due to the following observations and assumptions [6 - 9]:

- the main escape/chaotic region develops from the principal resonance i.e. is close to $\omega \approx \Omega_0 = 1$;
- prior to the first bifurcation the response is T-periodic and is close to the fundamental harmonic with frequency ω;
- the complex bifurcations which separate T-periodic solution from the escape region occur in a very narrow zone of the system parameters, so that the approximate criterion can be based on the first bifurcation ω_{BIF}.

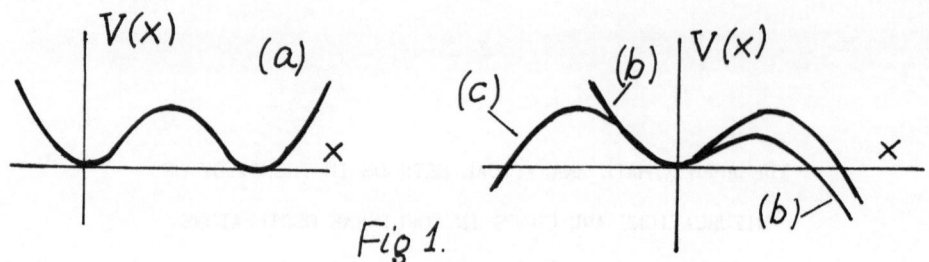

Fig 1.

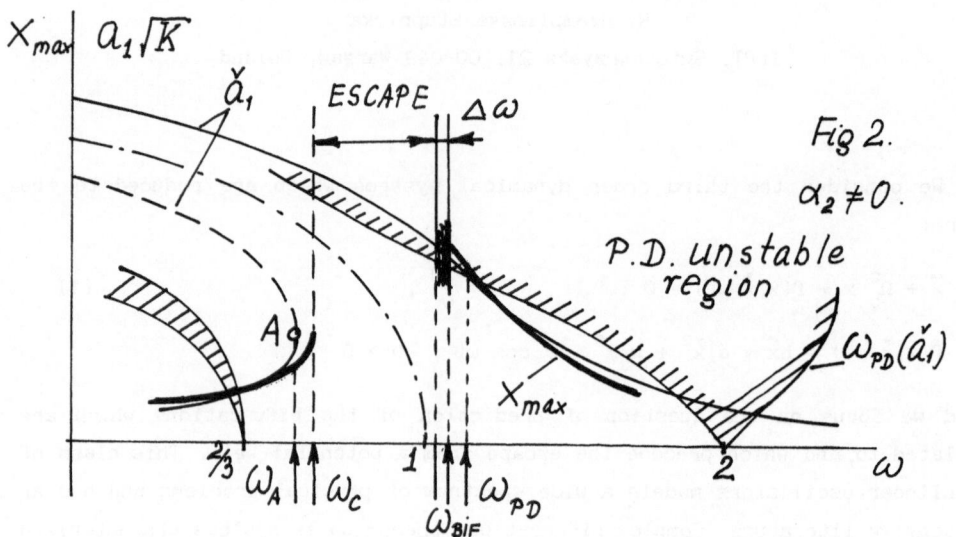

Fig 2.

$\alpha_2 \neq 0$.

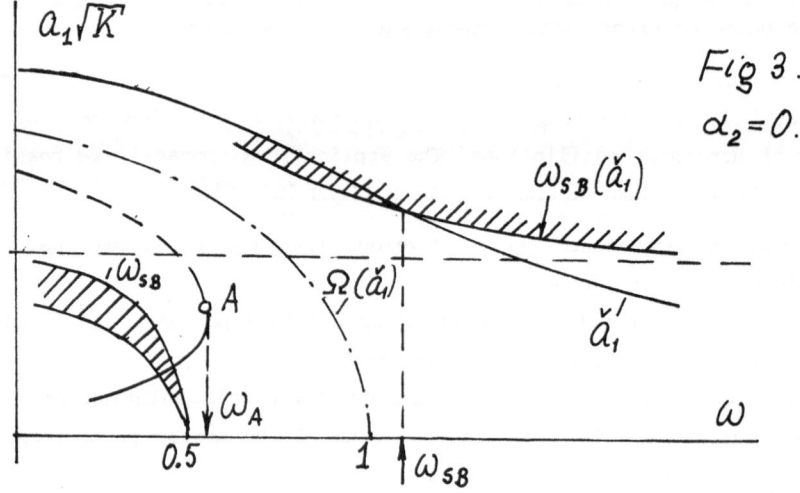

Fig 3.

$\alpha_2 = 0$.

Moreover, from the computer simulation results in the neighborhood of the principal resonance $x_{max}(\omega)$, depicted in Fig.2, we learn that at the left hand-side boundary at ω_c the crisis type transition occurs, while on the right hand side of the escape region a sequence of bifurcations is observed within a narrow band $\Delta\omega$. The first bifurcation, denoted by ω_{BIF} is a period doubling bifurcation in unsymmetric systems (a) and (b), and symmetry breaking bifurcation in the symmetric system (c). In the two-well potential system the escape from the potential well $x = 0$ results in chaotic motion, while in the single well systems (b), (c) it leads to unbounded solution.

The approximate value for the desired ω_c and ω_{BIF} are sought in the following way:

1. The T-periodic solution prior to the first bifurcation is determined by means of a perturbation technique in the second approximation, with the result:

(a) and (b), $\alpha_2 \neq 0$:

$$x(t) = x(t + T) = a_1 \cos(\omega t + \varphi_1) + a_0 + a_2 \cos 2(\omega t + \varphi_1),$$

$$a_0 = -\frac{1}{2}\alpha_2 a_1^2, \qquad a_2 = \frac{1}{6}\alpha_2 a_1^2 ; \tag{2}$$

(c) , $\alpha_2 = 0$;

$$x(t) = x(t + T) = a_1 \cos(\omega t + \varphi_1) + a_3 \cos 3(\omega t + \varphi_1), \tag{3}$$

where for (a), (b), (c) :

$$\overset{\vee}{a}_1^2 [(\Omega^2(\overset{\vee}{a}_1) - \omega^2)^2 + h^2\omega^2] = \overset{\vee}{F}, \qquad \Omega^2 = 1 - \overset{\vee}{a}_1^2 , \tag{4}$$

$$\overset{\vee}{a}_1 \equiv a_1\sqrt{k} =, \quad \overset{\vee}{F} = F\sqrt{k}, \qquad k = \frac{5}{6}\alpha_2^2 - \frac{3}{4}\alpha_3 ;$$

2. Instabilities on the resonant branch of the resonance curve $\overset{\vee}{a}_1(\omega)$ at $\overset{\vee}{F} > \dfrac{h}{\sqrt{4}}$ are sought by considering Hill's type variational equation:

$$\ddot{\delta x} + h\,\dot{\delta x} + \delta x\,[\lambda_0 + G(t)] = 0 , \tag{5}$$

where in (a) and (b)

$$G(t) = G(t + T) = \sum_{n=1,2}^{4} \lambda_n \cos(n\omega t + \varphi_n) ,$$

and in symmetric system (c):

$$G(t) = G(t + \frac{T}{2}) = \sum_{n=2,4,6} \lambda_n \cos(n\omega t + \varphi_n) ;$$

$$\lambda_n \equiv \lambda_n(a_1) ;$$

The approximate solutions for $\delta x(t)$ in the unstable regions are sought as:

$$\delta x(t) = e^{\varepsilon t}\Phi(t) = e^{\varepsilon t}\Phi(t + 2T) \approx e^{\varepsilon t}[b_1\cos(\frac{\omega}{2}t + \gamma_1) +$$

$$+ b_3\cos(\frac{3}{2}\omega t + \gamma_3)] ,\qquad\qquad\qquad\qquad (6)$$

to determine period doubling instability in (a) and (b), and

$$\delta x(t) = e^{\varepsilon t}\Phi(t) = e^{\varepsilon t}\Phi(t + \frac{T}{2}) \approx e^{\varepsilon t}[b_0 + b_2\cos(2\omega t + \gamma_2)],\qquad (7)$$

to determine symmetry breaking instability in the system (c).

The instability limits are found by applying harmonic balance method with the result:

$$\omega^4 - 2B\omega^2 + C = 0 ,\quad B \equiv B(a_1) ,\quad C \equiv C(a_1) ;\qquad\qquad (8)$$

This gives us the desired instability limit $\omega_{PD} \equiv \omega_{PD}(a_1)$ in (a), (b) and $\omega_{SB} = \omega_{SB}(a_1)$ in system (c) (Figs. 2 and 3).

3. Approximate value for the crisis type (or jump) transition value ω_c is identified with the first order instability limit ω_A, which coincides with the point of $a_1(\omega)$ which has vertical tangent

$$\frac{da_1}{d\omega}\Big|_{\omega_A} = \infty ;\qquad\qquad\qquad\qquad (9)$$

The corresponding critical value of the forcing parameter $\check{F}$ is then calculated from eqs. (4). The three systems considered show, qualitatively, similar results (see Fig. 4). In the F - ω plane the theoretical curves $\omega_A(F)$ and $\omega_{PD}(F)$, or $\omega_{SB}(F)$ at $\omega_A < \omega_{PD}$, or $\omega_A < \omega_{SB}$, respectively, form V-shaped region, the region which is proposed as the approximate predictive criterion for the escape or chaotic region. The computer simulation results confirm, that the theoretical criterion gives good estimation of the exact results.

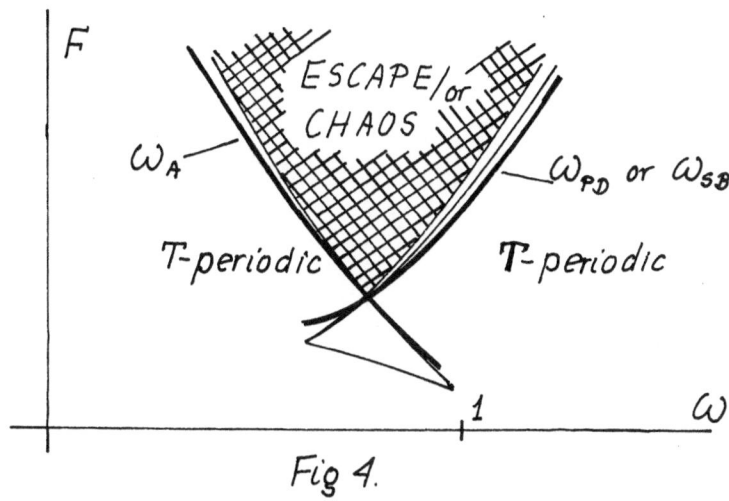

Fig 4.

References:

1. Holmes P. (1979) A nonlinear oscillator with a strange attractor. Phil. Trans. Royal Soc. Ser. A 292, 1394, 419-448.

2. Hubermann B.A. and Crutchfield J.P. (1979) Chaotic states of anharmonic system in periodic field, Physical Rev. Letters 43, 1743-1747.

3. Guckenheimer J. and Holmes P. (1983) Nonlinear Oscillations, Dynamical Systems, and Bifurcations of Vector Fields, Springer Verlag, New York.

4. Moon F.C. (1987) Chaotic Vibrations, J.Willey & Sons, New York

5. Thompson J.M.T. (1989) Chaotic phenomena triggering the escape from a potential well, Proc. Royal Soc. London, A 421, 195-225

6. Szemplińska-Stupnicka W. (1988) The refined approximate criterion for chaos in a two-state mechanical oscillator, Ing. Archiv 58, 354-366

7 Szemplińska-Stupnicka W. (1988) Bifurcations of harmonic solution leading to chaotic motion in the softening type Duffing oscillator, Int. J. Non-Linear Mech. 23, 257-277

8. Szemplińska-Stupnicka W. and Rudowski J. (1990) Local methods in predicting an occurrence of chaos in the two-well potential system: superharmonic frequency region, Journal of Sound and Vibration, (in the press)

9. Szemplińska-Stupnicka W. Period doubling instability and an approximate criterion for the escape from a potential well, (in the press)

International Series of Numerical Mathematics, Vol. 97, © 1991 Birkhäuser Verlag Basel

Periodic and homoclinic orbits in conservative and reversible systems

André Vanderbauwhede

1. INTRODUCTION

One of the main features of the phase portrait of scalar second order equations of the form

$$\ddot{y} + g(y) = 0 \tag{1}$$

is the stable appearance of families $\{\gamma_\tau \mid \tau \geq \tau_0\}$ of τ-periodic orbits γ_τ which are symmetric with respect to the y-axis and which accumulate on a homoclinic orbit as $\tau \to \infty$. Now the equation (1) is both conservative and reversible, and it is the main purpose of this note to show that similar period blow-ups to a homoclinic orbit appear stably in general conservative or reversible systems. For certain subcases the result is not new : Devaney [1] gave a proof for the reversible case, using slightly stronger hypotheses than ours, and Strömgren [4] conjectured the result for Hamiltonian systems. A proof of Strömgren's conjecture for the case of two degrees of freedom was given by Henrard [2]. Here we indicate how a technique introduced recently by X.-B. Lin [3] can be used to give a unified proof for both the conservative and the reversible case. The results stated in this note are based on joint work with Bernold Fiedler, and full details will be given in a forthcoming paper [5].

2. HYPOTHESES AND MAIN RESULTS

Let $x_0 \in \mathbf{R}^n$ be a hyperbolic equilibrium of the system

$$\dot{x} = f(x), \tag{2}$$

where $f : \mathbf{R}^n \to \mathbf{R}^n$ is of class C^k ($k \geq 2$), and let $q : \mathbf{R} \to \mathbf{R}^n$ be a homoclinic solution of (2) asymptotic to x_0, i.e. $\lim_{t \to \pm\infty} q(t) = x_0$. Then the orbit $\gamma = \{q(t) \mid t \in \mathbf{R}\}$ clearly belongs to the intersection of the stable and unstable manifolds (denoted by $W^s(x_0)$ and $W^u(x_0)$) at x_0. Since the sum of the dimensions of $W^s(x_0)$ and $W^u(x_0)$ equals n this is a non-generic situation, i.e. homoclinic orbits do not appear in generic systems (2). However, they can appear generically in certain subclasses of systems, such as conservative or reversible systems (see further). We say that $q(t)$ is a *non-degenerate* homoclinic solution if the space of globally bounded solutions of the variational equation

$$\dot{x} = Df(q(t))x \tag{3}$$

is one-dimensional, and hence spanned by $\dot{q}(t)$.

Our hypotheses imply that (3) has an exponential dichotomy on both $\mathbf{R}_+$ and $\mathbf{R}_-$; combining these dichotomies with the techniques of Lin [3] one then proves the following technical result.

Theorem 1. Let $Y \subset \mathbf{R}^n$ and $Z \subset Y$ be subspaces such that

$$\mathbf{R}^n = Y \oplus (T_{q_0}W^s(x_0) \cap T_{q_0}W^u(x_0)) \tag{4}$$

and

$$\mathbf{R}^n = Z \oplus (T_{q_0}W^s(x_0) + T_{q_0}W^u(x_0)), \tag{5}$$

where $q_0 := q(0)$. Let $\Sigma := \{q_0 + y \mid y \in Y\}$.

Then there exist numbers $\omega_0 > 0$ and $\epsilon_0 > 0$ such that for each $\omega \geq \omega_0$ the equation (2) has a unique solution $x_\omega : [0, 2\omega] \to \mathbf{R}^n$ satisfying the following conditions

(i) $x_\omega(0) \in \Sigma$ and $x_\omega(2\omega) \in \Sigma$;

(ii) $\xi(\omega) := x_\omega(2\omega) - x_\omega(0) \in Z$;

(iii) $\sup_{-\omega \leq t \leq 0} \|x_\omega(t + 2\omega) - q(t)\| \leq \epsilon_0$ and $\sup_{0 \leq t \leq \omega} \|x_\omega(t) - q(t)\| \leq \epsilon_0$.

Moreover, the mapping $\xi : [\omega_0, \infty) \to Z$ is of class C^k, and $\lim_{\omega \to \infty} \xi(\omega) = 0$. Finally we also have

$$\lim_{\omega \to \infty} \sup_{-\omega \leq t \leq 0} \|x_\omega(t + 2\omega) - q(t)\| = \lim_{\omega \to \infty} \sup_{0 \leq t \leq \omega} \|x_\omega(t) - q(t)\| = 0. \tag{6}$$

$\square$

As shown in [3] this and similar results can be used to discuss the existence of various types of solutions (in particular periodic solutions) near the homoclinic orbit γ.

Now consider the conservative case, i.e. one assumes the existence of a C^1-function $H : \mathbf{R}^n \to \mathbf{R}$ such that $DH(x) \cdot f(x) = 0$ for all $x \in \mathbf{R}^n$. Then H is a first integral of (2), and it is not difficult to see that both $W^s(x_0)$ and $W^u(x_0)$ will be contained in the level set $\{x \in \mathbf{R}^n \mid H(x) = H_0 := H(x_0)\}$. If $DH(q_0) \neq 0$ then this level set will be (locally near q_0) a submanifold of codimension 1, and hence $W^s(x_0)$ and $W^u(x_0)$ can, within this submanifold, intersect transversely along the one-dimensional homoclinic orbit γ; they will do so if and only if $q(t)$ is non-degenerate. It follows that non-degenerate homoclinic orbits can appear generically in conservative systems. The following theorem shows that each such non-degenerate homoclinic orbit has an associated family of periodic orbits accumulating on the homoclinic orbit.

Theorem 2. Suppose that (2) is conservative, with first integral $H : \mathbf{R}^n \to \mathbf{R}$. Let $q : \mathbf{R} \to \mathbf{R}^n$ be a non-degenerate homoclinic solution of (2), asymptotic to a hyperbolic equilibrium $x_0 \in \mathbf{R}^n$. Assume also that $DH(q_0) \neq 0$, where $q_0 := q(0)$. Finally let $\Sigma :=$ $\{q_0 + y \mid y \in Y\}$, where Y is a codimension one subspace complementary to span $\{\dot{q}(0)\}$. Then there exist numbers $\bar{\omega} > 0$ and $\bar{\epsilon} > 0$ such that for each $\omega \geq \bar{\omega}$ the equation (2) has a unique 2ω-periodic solution $x_\omega : \mathbf{R} \to \mathbf{R}^n$ satisfying

(i) $x_\omega(0) \in Z$

and

(ii) $\sup_{|t| \leq \omega} \|x_\omega(t) - q(t)\| \leq \bar{\epsilon}$.

Moreover we also have

$$\lim_{\omega \to \infty} \sup_{|t| \leq \omega} \|x_\omega(t) - q(t)\| = 0. \tag{7}$$

$\square$

Next we consider reversible systems. These are systems (2) such that

$$f(Rx) = -Rf(x) \quad , \quad \forall x \in \mathbf{R}^n, \tag{8}$$

where R is a linear involution on $\mathbf{R}^n$, i.e. $R \in \mathcal{L}(\mathbf{R}^n)$ and $R^2 = I$. We say that (2) is R-reversible. An immediate consequence is that the flow $\phi_t(x)$ of (2) satisfies the identity $R\phi_t(x) = \phi_{-t}(Rx)$; or still : if γ is an orbit, then so is $R(\gamma)$. We say that an orbit γ is symmetric if $R(\gamma) = \gamma$. It is easy to see that an orbit γ is symmetric if and only if γ intersects $\text{Fix}(R) = \{x \in \mathbf{R}^n \mid Rx = x\}$; moreover, such symmetric orbit is periodic if and only if it intersects $\text{Fix}(R)$ in two different points. We conclude that a symmetric homoclinic orbit intersects $\text{Fix}(R)$ in exactly one point, and it is also clear that its asymptotic equilibrium must belong to $\text{Fix}(R)$. Now it is easily shown that a symmetric equilibrium x_0 of an R-reversible system (2) can only be hyperbolic if n is even and $\dim \text{Fix}(R) = n/2$. Therefore we assume from now on that the involution R in (8) is such that

$$\dim \text{Fix}(R) = m \quad \text{and} \quad n = 2m. \tag{9}$$

At a symmetric hyperbolic equilibrium x_0 the stable and unstable manifolds then satisfy $W^u(x_0) = R(W^s(x_0))$ and $\dim W^s(x_0) = \dim W^u(x_0) = m$. Moreover, if $q_0 \in W^s(x_0) \cap \text{Fix}(R)$ then $q_0 = Rq_0 \in R(W^s(x_0)) = W^u(x_0)$, and hence q_0 generates a symmetric homoclinic orbit $\gamma = \{\phi_t(q_0) \mid t \in \mathbf{R}\}$. We will say that such symmetric homoclinic orbit γ is *elementary* if $W^s(x_0)$ and $\text{Fix}(R)$ intersect transversely at q_0 (remember that both $W^s(x_0)$ and $\text{Fix}(R)$ are m-dimensional). Clearly such transverse intersection is stable under small R-reversible perturbations of the vectorfield, i.e. elementary symmetric homoclinic orbits can appear generically in R-reversible systems.

The following theorem gives the analogue of theorem 2 for R-reversible systems.

Theorem 3. Let (2) be R-reversible, with R a linear involution satisfying (9). Let γ be an elementary symmetric homoclinic orbit of (2.1), asymptotic to a hyperbolic equilibrium

x_0, and let q_0 be the unique intersection point of γ and Fix (R).
Then there exist numbers $\omega_0 > 0$ and $\epsilon_0 > 0$ such that for each $\omega \geq \omega_0$ the equation (2)
has a unique 2ω-periodic solution $x_\omega : \mathbf{R} \to \mathbf{R}^n$ satisfying

(i) $Rx_\omega(-t) = x_\omega(t)$, $\forall t \in \mathbf{R}$,

and

(ii) $\sup_{|t| \leq \omega} \|x_\omega(t) - \phi_t(q_0)\| \leq \epsilon_0$.

Moreover we also have

$$\lim_{\omega \to \infty} \sup_{|t| \leq \omega} \|x_\omega(t) - \phi_t(q_0)\| = 0. \tag{10}$$

□

So also in the reversible case each elementary (and in particular each non-degenerate) symmetric homoclinic orbit has an associated family of symmetric periodic orbits accumulating
on the homoclinic orbit.

3. Sketch of the proofs.

For details of the proof of theorem 1 we refer to [3] and [5].

To prove theorem 2 we observe that since $q(t)$ is a non-degenerate homoclinic solution we
have $T_{q_0}W^s(x_0) \cap T_{q_0}W^u(x_0) = \operatorname{span}\{\dot{q}(0)\}$, and hence Y satisfies (4); moreover we can
find $\psi_0 \in Y$ such that (5) holds for $Z := \operatorname{span}\{\psi_0\}$. Since both $W^s(x_0)$ and $W^u(x_0)$ are
contained in the level set $\{H = H_0\}$ the condition $DH(q_0) \cdot \psi_0 \neq 0$ implies that

$$DH(q_0) \cdot \psi_0 \neq 0. \tag{11}$$

Let $x_\omega : [0, 2\omega] \to \mathbf{R}^n$ be the solution given by theorem 1. Both $x_\omega(0)$ and $x_\omega(2\omega)$ converge
to q_0 for $\omega \to \infty$ (by (6)), and $\xi(\omega) := x_\omega(2\omega) - x_\omega(0) \in \operatorname{span}\{\psi_0\}$, by Theorem 1(ii).
Then (11) implies that $H(x_\omega(2\omega)) \neq H(x_\omega(0))$ if $\xi(\omega) \neq 0$. However, H is a first integral,
and hence $H(x_\omega(2\omega)) = H(x_\omega(0))$; we conclude that $\xi(\omega) = 0$ and that x_ω extends to a
2ω-periodic solution, which proves theorem 2.

Under the conditions of theorem 3 one can find R-invariant subspaces Y and Z which
satisfy (4) and (5) and such that

$$Z \subset \operatorname{Fix}(R) \subset Y. \tag{12}$$

Then we apply theorem 1 (with $q(t) := \phi_t(q_0)$) to obtain a solution $x_\omega : [0, 2\omega] \to \mathbf{R}^n$
satisfying the requirements of that theorem. Using the R-reversibility of (2) and the
property (12) one then easily shows that also $\tilde{x}_\omega : [0, 2\omega] \to \mathbf{R}^n$, $t \mapsto \tilde{x}_\omega(t) := Rx_\omega(2\omega - t)$
defines a solution satisfying these requirements. Invoking the uniqueness part of theorem 1
it follows that $Rx_\omega(2\omega - t) = x_\omega(t)$ for all $t \in [0, 2\omega]$. In particular we have $R\xi(\omega) = -\xi(\omega)$,
where $\xi(\omega) := x_\omega(2\omega) - x_\omega(0)$. But $\xi(\omega) \in Z \subset \operatorname{Fix}(R)$, by theorem 1 and (12). It

follows that $\xi(\omega) = 0$, and hence x_ω extends to a 2ω-periodic solution of (2), which proves theorem 3.

REFERENCES.

1. R. Devaney, Blue Sky Catastrophes in Reversible and Hamiltonian Systems. Indiana Univ. Math. J. **26** (1977), 247–263.

2. J. Henrard, Proof of a Conjecture of E. Strömgren. Celest. Mech. **7** (1983), 449–457.

3. X.-B. Lin, Using Melnikov's Method to Solve Silnikov's Problems. Preprint North Carolina State University, 1990.

4. E. Strömgren. Connaissance Actuelle des Orbites dans le Problème des Trois Corps. Bull. Astron. **9** (1933), 87–130.

5. A. Vanderbauwhede and B. Fiedler, Period Blow-ups in Reversible and Conservative Systems. In preparation.

Institute for Theoretical Mechanics
State University Gent
Krijgslaan 281
B–9000 Gent (Belgium)

International Series of Numerical Mathematics, Vol. 97, © 1991 Birkhäuser Verlag Basel

On the Dynamics of a Horizontal, Rotating, Curved Shaft

J. Wauer and H. Wei

Institut für Technische Mechanik

Universität Karlsruhe, FRG

Abstract

Flexural vibrations of a slender rotating shaft, which centre line is assumed to be naturally curved, are considered. The shaft simply supported and axially restrained rotates at a constant rate about a horizontal axis connecting the centroids of the end cross-section supported. A modal truncation of the governing nonlinear boundary value problem results in a pair of ordinary, gyroscopic differential equations of the Duffing type. For the cases of a cross-section with extremely different bending stiffnesses and a circular cross-section and a vanishing eccentricity in each case, the vibrational behaviour is analyzed in detail. The steady state response (for a neglected influence of gravity) in form of a synchronous whirling and its stability are studied first. A numerical investigation of weight-excited oscillations follows where both periodic and even chaotic motions may occur. Some concluding remarks summarize the contribution.

1 Introduction

Well understood is the vibrational behaviour of arches and similar structures under pulsating load /1-3/. Both small motions about the equilibrium configuration including conditions of resonance and dynamic snap-through into an inverted position and back during motion of the arch have been investigated. Also the nonlinear dynamics of a rotating straight shaft model axially restrained has recently been studied in detail /4-6/. Possible instabilities have been predicted and the post-critical behaviour in form of a synchronous whirl, non-synchronous whirl, and competing-mode types have been determined.

Much more complicated vibrational phenomena can be expected, if a naturally pre-curved slender structure rotates in a horizontal position about its longitudinal axis so that weight force and centrifugal one interact. The objective of the present work is to give some first results of the dynamic response of such systems for a restricted range of parameters.

2 Formulation

Consider a uniform, slender, flexible, symmetric shaft of non-circular cross-section shown in Fig. 1. Its neutral axis is assumed to be naturally pre-deformed in form of a planar curve. The ends of the shaft are simply supported at a distance ℓ apart and are constrained against translation. In a body-fixed coordinate frame (x, y, z), where x denotes the axis through the ends supported, the unloaded shape of the shaft is described by the function $v_0(x)$. The shaft is made of a Voigt visco-elastic material and external damping is modelled by assuming that the resulting dissipative force on a shaft element is proportional to its absolute velocity. The centroid of the oval cross-section is allowed to be different from the centre of mass, measured in the body-fixed reference frame by the eccentricity e and the location angle ε. The shaft has mass per unit length μ, extensional rigidity EA and bending stiffnesses $EI_{1,2}$. The damping parameters are d_i (representing material viscosity) and d_e (external dissipation). The shaft rotates about the horizontal X-axis of a space-fixed coordinate frame (X, Y, Z) in the gravitational field of the earth (constant of gravity g) at a fixed rate $\omega = const.$, where the longitudinal x-axis of the shaft and the inertial X-axis coincide. The shaft undergoes motion which can be represented by space- and time-dependent longitudinal and transverse displacements $u(x,t)$ and $v(x,t), w(x,t)$, respectively, measured in the rotating frame (x, y, z).

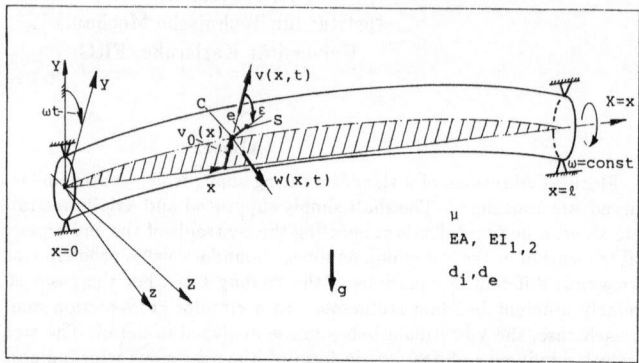

Fig. 1: Shaft model

The governing boundary value problem is generated by Hamilton's principle following /1,5,7/. The x component of the resulting equations of motion is simplified by neglecting the longitudinal inertial force and assuming that the shaft carries no end load; this leads to

$$\{u_x + \frac{v_x^2}{2} + \frac{w_x^2}{2} + v_x\,v_{0,x} + d_i\,[u_{tx} + (v_x + v_{0,x})\,v_{xt} + w_x\,w_{x,t}]\,\}_x = 0 \tag{1}$$

where subscripts x and t denote differentiation with respect to space and time, respectively. Integrating this with respect to x using the boundary conditions $u(0,t) = u(\ell,t) = 0$ and substituting the result into the other two equations of motion (for y and z) yields

$$\mu\,(v_{tt} - 2\,\omega\,w_t - \omega^2\,v) + d_e\,\mu\,(v_t - \omega\,w) + EI_1(v_{xxxx} + d_i\,v_{txxxx}) - \frac{EA}{\ell}(1 + d_i\frac{\partial}{\partial t}).$$
$$\int_0^\ell(\frac{v_x^2}{2} + \frac{w_x^2}{2} + v_{0,x}\,v_x)\,dx \cdot (v_{0,xx} + v_{xx}) \quad = \quad \mu\,(e\cos\varepsilon + v_0)\,\omega^2 - \mu\,g\,\cos\omega\,t,$$
$$\mu\,(w_{tt} + 2\omega v_t - \omega^2 w) + d_e\mu\,(w_t + \omega v) + EI_2(w_{xxxx} + d_i w_{txxxx}) - \frac{EA}{\ell}(1 + d_i\frac{\partial}{\partial t}). \tag{2}$$
$$\int_0^\ell(\frac{v_x^2}{2} + \frac{w_x^2}{2} + v_{0,x}\,v_x)\,dx \cdot w_{xx} \quad = \quad \mu\,e\,\omega^2\sin\varepsilon - d_e\,\mu\,\omega\,v_0 + \mu g\sin\omega t.$$

The corresponding boundary conditions are specified as

$$v = v_{xx} = w = w_{xx} = 0 \quad \text{at} \quad x = 0, \ell. \tag{3}$$

Assuming a sinusoidal unloaded shape of the shaft

$$v_0(x) = \sqrt{2}\,a_0\,\sin\pi\,\frac{x}{\ell}, \tag{4}$$

a modal truncation results in a pair of ordinary differential equations which govern the first mode vibrations. After some rescaling the equations are

$$\ddot{R}_1 - 2\,\Omega\,\dot{T}_1 - \Omega^2 R_1 + \beta_e\,(\dot{R} - \Omega\,T_1) + \Omega_1^2\,(R_1 + \beta_i\,\dot{R}_1) + \alpha\,\Omega_1^2\,R_1[\,\tfrac{1}{2}\,((R_1 - A_0)^2 + T_1^2) +$$
$$A_0(R_1 - A_0) + \beta_i\,\Big((R_1 - A_0)\dot{R}_1 + T_1\,\dot{T}_1 + A_0\,\dot{R}_1\Big)\,] = A_0\Omega_1^2 - \gamma_{1'}\cos\Omega\,\tau,$$
$$\ddot{T}_1 + 2\,\Omega\,\dot{R}_1 - \Omega^2\,T_1 + \beta_e\,(\dot{T} + \Omega\,R_1) + \varrho\,\Omega_1^2\,(\dot{T}_1 + \beta_i\,\dot{T}_1) + \alpha\,\Omega_1^2\,T_1. \tag{5}$$
$$[\,\tfrac{1}{2}\,((R_1 - A_0)^2 + T_1^2) + A_0(R_1 - A_0) + \beta_i\,\Big((R_1 - A_0)\dot{R}_1 + T_1\,\dot{T}_1 + A_0\,\dot{R}_1\Big)\,] = \gamma_1\,\sin\Omega\,\tau$$

where overdots correspond to derivatives with respect to nondimensional time τ. The variables $S_1 = R_1 - A_0, T_1$ represent the transverse vibrations, the quantity A_0 denotes the amplitude of the unloaded shape. The parameters β_i, β_e characterize the two damping mechanisms, $\varrho \geq 1$ measures the deviation from a circular cross section ($\varrho = 1$) and $\alpha \geq 0$ is a measure of slenderness of the shaft which models the structural nonlinearity. The nondimensional rotational speed is denoted Ω and the fundamental eigenvalue of a circular (straight) shaft with the smaller (EI_1) of the two bending stiffnesses is $\Omega_1 = \pi^2$. γ_1 represents the weight influence and E_1 the shaft unbalance.

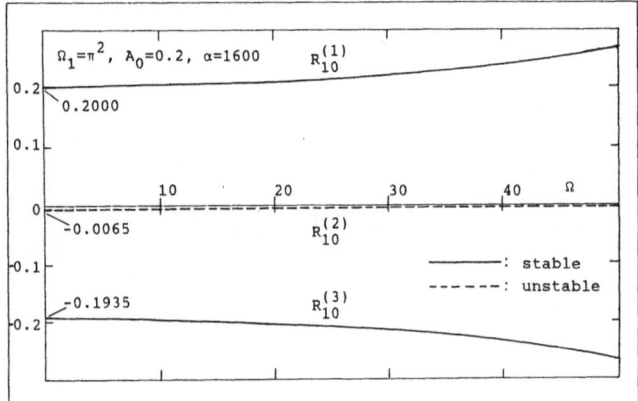

Fig. 2: Stationary response without influence of gravity

In all further calculations of this paper, only the initially perfectly balanced shaft ($E_1 = 0$) is considered. Since complicated interactions between resonance and instability are discussed in detail by Shaw and Shaw /6/, here the parameters α, A_0 are fixed to such values, $\alpha = 1600, A_0 = 0.2$, that the steady state response (for a neglected weight influence) is a circle of equilibrium and amplitude varying subsynchronous responses do not exist.

3   Non-circular Cross-section with Extremely Different Bending Stiffnesses ($\varrho \to \infty$)

The pair of gyroscopic differential equations (5) change in this case to a single differential equation

$$\ddot{R}_1 - \Omega^2 R_1 + \beta_e \dot{R}_1 + \Omega_1^2(R_1 + \beta_i \dot{R}_1) + \alpha \Omega_1^2 R_1[\tfrac{1}{2}(R_1 - A_0)^2 + A_0(R_1 - A_0)$$
$$+\beta_i \left((R_1 - A_0)\dot{R}_1 + A_0 \dot{R}_1\right)] = A_0\Omega_1^2 - \gamma_1 \cos \Omega \tau. \tag{6}$$

In order to determine the synchronous response ($\gamma_1 \equiv 0$), the fact is exploited that R_1 will attain a constant value R_{10} to be computed as solution of the algebraic relation

$$\Omega_1^2 \left[1 + \alpha R_{10}(A_0 + \frac{R_{10} - A_0}{2})\right](R_{10} - A_0) = R_{10} \Omega^2 \tag{7}$$

which is cubic in R_{10} and quadratic in Ω. For a positive Ω, three real solutions $R_{10}^{(i)}, i = 1, 2, 3$, can be determined (see Fig. 2). The feature of the response curves of importance is their stability. A straightforward linearization method, verified by computer simulations of the full nonlinear equation of motion (6), clarify the onset of instability. Obviously, two unique stable solution branches

$R_{10}^{(1,3)}$ (denoted by solid lines) corresponding to relative minima of the governing potential appear where $R_{10}^{(3)}$ is associated with a snapped-through configuration; the intermediate synchronous response branch $R_{10}^{(2)}$ is completely unstable and corresponds to a relative maximum of the potential function.

If the influence of gravity is included in the investigation, a harmonically pulsating excitation works and forced vibrations of a nonlinear system with a non-unique, non-symmetric stationary response curve have to be studied. Fig. 3 shows for a selected speed rate $\Omega = 15$ computer simulations of the equation of motion (6) for two different weight parameters γ_1.

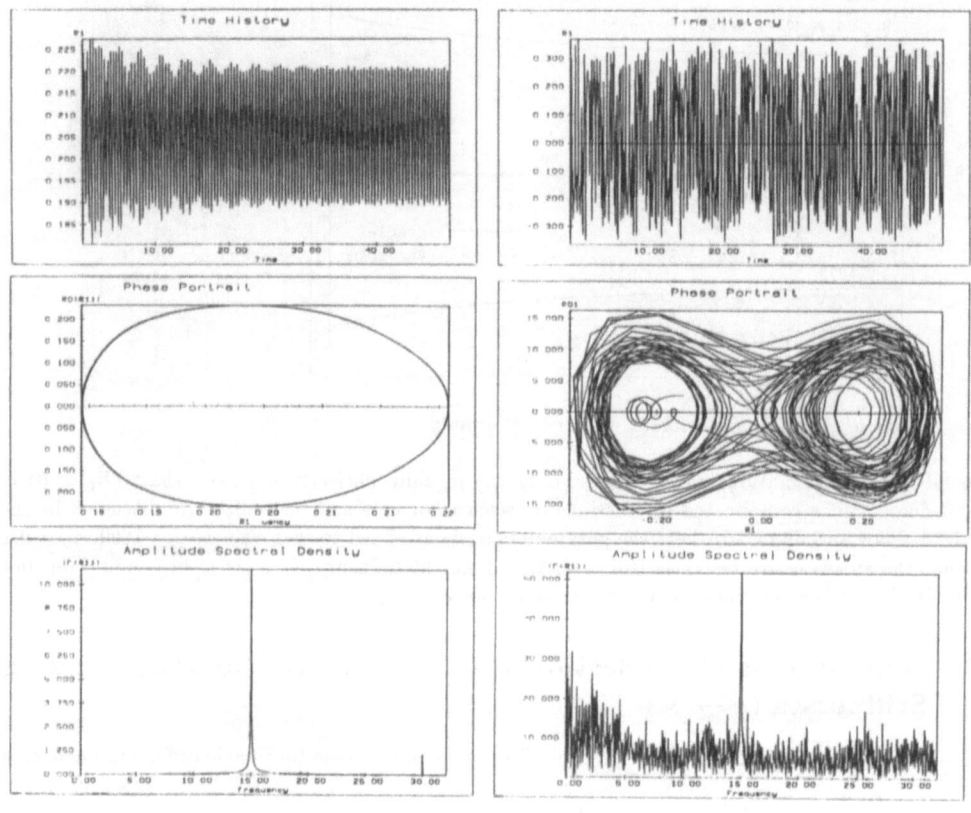

Fig. 3: Weight-excited vibrations   (a) $\gamma_1 = 100$ (b) $\gamma_1 = 300$

Besides the time history $R_1(\tau)$, a phase diagram $\dot{R}_1(R_1)$ and the amplitude spectral density $|\mathcal{F}(R_1(\tau))|$ is depicted. For a sufficiently low weight parameter γ_1, the centrifugal force dominates and small, nonlinear, but periodic vibrations about the initial equilibrium configuration (without snap-through phenomena) appear. On the other hand, for a heavy weight the gravity preponderates and chaotic snap-through into an inverted position and back during motion results. In an intermediate transition region not represented here, the dynamic behaviour can not so clearly be associated.

4 Circular Cross-section ($\varrho = 1$)

In this case, the equations of motion (5) where $\varrho = 1$ govern the dynamics of the shaft. But due to the gyroscopic coupling, much more complicated and manifold dynamic phenomena occur. The results will be presented in great detail in a future paper /8/; here only the most important ones are shortly reported.

For a constant amplitude response, R_1 and T_1 will attain constant values R_{10} and T_{10} to be obtained from a set of two algebraic equations setting all time derivatives in (5) to zero. It can be shown that now for a fixed rate Ω not three but seven solution combinations R_{10}, T_{10} (depending on $\beta_e > 0$) as the governing singular points exist; two of them represent stable sinks, the other five are unstable (one is an unstable source and the remaining four are different types of saddle points).

Similar as before, the study of the weight-excited vibrations exhibits two main categories of motion. For small values of γ_1, periodic oscillations about the natural stable fixpoint appear while for large γ_1 a pronounced snap-through behaviour can be observed. In contrast to the shaft with a cross-section of extremely different bending stiffnesses, here the border-line between these two main regions is shifted to significantly lower speed rates Ω and the patterns of motion are very complex in general. Remarkable, in particular, is the new fact that in the snap-through range a low- and a high-speed band appear. While in the first one the snap-through vibrations are chaotic, in the second one they are periodic.

5 Conclusions

The nonlinear oscillations of a naturally curved shaft rotating about its longitudinal axis in the gravity field of the earth have been studied. The results obtained give a first survey of the dynamics for such a parameter range that additional interactions between instabilities and resonance effects are avoided.

References

1. Mettler, E. and Weidenhammer, F., *Kinetisches Durchschlagen des schwach gekrümmten Stabes*, Ing.-Archiv 29, 1960, 302-314.

2. Huang, K.-Y. and Plaut, R.H., *Snap-through of a shallow arch under pulsating load*, in: Stability in the Mechanics of Continua (F.H. Schröder, ed.), 1982, Springer, Berlin, 215-223.

3. Clemens, H. and Wauer, J., *Free and forced vibrations of a snap-through oscillator*, in: Proc. 9th Int. Conf. on Nonlinear Oscillations, Vol. 3 (Y.A. Mitropolsky, ed.), 1984, Naukova Dumka, Kiev, 128-133.

4. Shaw, S.W., *Chaotic dynamics of a slender beam rotating about its longitudinal axis*, J. Sound & Vibration 124, 1988, 329-344.

5. Shaw, J. and Shaw, S.W., *Instabilities and bifurcations in a rotating shaft*, J. Sound & Vibration 132, 1989, 227-244.

6. Shaw, S.W. and Shaw, J., *Nonlinear interactions between resonance and instability in a symmetric rotor*, in: Proc. 3rd Int. Symp. on Transport Phenomena and Dynamics of Rotating Machinery, Vol. 2, 1990, Honolulu, 198-212.

7. Wauer, J., *Modelling and formulation of equations of motion for cracked rotating shafts*, Int. J. Solids Structures 26, 1990, 901-914.

8. Wauer, J. and Wei, H., *Dynamik eines Rotors mit krummer Welle*, in: Proc. Schwingungen in rotierenden Maschinen, 1991, Kassel, to be published.

International Series of Numerical Mathematics, Vol. 97, © 1991 Birkhäuser Verlag Basel

LYAPUNOV EXPONENTS AND INVARIANT MEASURES
OF DYNAMIC SYSTEMS

W.V. Wedig, University of Karlsruhe

Abstract:

According to the multiplicative ergodic theorem, Lyapunov exponents are numerically evaluated from time series of simulated dynamic systems. Introducing polar coordinates, one can split off the stationary parts of the system solution which determine the Lyapunov exponents by mean values performed in the time domain. This inifinte time integration can be reduced to a finite integral by means of associated invariant measures which are defined on the periodic ranges of the system angles.

1. Introduction to the problem

To investigate asymptotic stability and robustness of dynamic systems, one needs the top Lyapunov exponent which is calculable by means of the multiplicative ergodic theorem of Oseledec [1]. Accordingly, a nonstationary simulation of the system is performed in order to obtain a time series of a norm of the system vector. The natural logarithm of the norm divided by the time leads to the top Lyapunov exponent for infinitely increasing times.

This limiting procedure can considerably be simplified by means of the invariant measures of dynamic systems. For this purpose, we introduce polar or spherical coordinates. According to Khasminskii [2], they separate the angle equations from the nonstationary behaviour of amplitudes. Since the angles are Markovian and defined in a finite domain, the associated Liouville or Fokker-Planck equation leads to a periodic density of the invariant measure of the dynamic system.

The scalar amplitude equation can easily be integrated. Inserted into the multiplicative ergodic theorem, the top Lyapunov exponent is determined by a simpler mean value. This time average can finally be replaced by an ensemble average defined on the invariant measure of the system. In the present paper, we show relations between invariant measures and Lyapunov exponents for the stability investigations of equilibria or periodic limit cycles.

2. Mathieu equation

Basic models of oscillators with harmonic parameter excitations are described by the following Mathieu differential equation.

$$\ddot{x} + \omega_1^2(1 + \varepsilon \cos\varphi)x = 0, \qquad \dot{\varphi} = \omega. \tag{1}$$

Herein, ω_1 is the natural frequency of the oscillator and ε is the intensity of the cosine excitation. Its angle argument φ is defined by the equation $\dot{\varphi} = \omega$, where ω is the circle frequency of the parametric excitation. Dots denote derivatives with respect to the time t. Obviously, $x(t) = 0$ is a strong solution of (1). Its stability can be investigated by means of the Floquet theory [3] or equivalently, by means of the Markov theory [4].

2.1. Lyapunov exponents

As already mentioned, the top Lyapunov exponent λ of the trivial solution $x(t) = 0$ is calculable by the multiplicative ergodic theorem of Oseledec.

$$\lambda = \lim_{t \to \infty} \frac{1}{t} \log(\| \vec{x}(t) \| / \| \vec{x}(0) \|), \qquad \| \vec{x} \| = \sqrt{x_1^2 + x_2^2}. \qquad (2)$$

Herein, log denotes the natural logarithm and $\| \vec{x} \|$ is the Euclidean norm of the system vector $\vec{x} = (x_1, x_2)^T$. Introducing polar coordinates into the Mathieu equation leads to

$$x = x_1 = a \cos \psi, \qquad \dot{a} = -\tfrac{1}{2}\varepsilon \omega_1 a \cos \varphi \sin 2\psi, \qquad (3)$$

$$\dot{x} = \omega_1 x_2 = \omega_1 a \sin \psi, \qquad \dot{\psi} = -\omega_1(1 + \varepsilon \cos \varphi \cos^2 \psi). \qquad (4)$$

The amplitude a represents the Euclidean norm in (2) and ψ is the phase angle of the system.

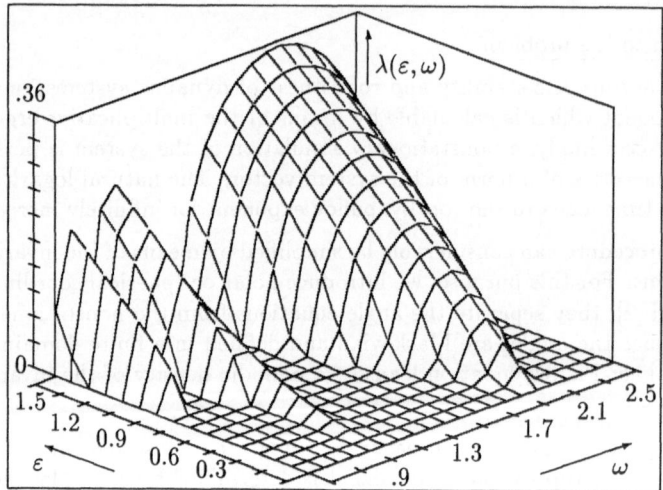

Figure 1: Distribution of the Lyapunov exponent

Obviously, the phase angle ψ in (4) is decoupled from the growth behaviour of the amplitude. Hence, the amplitude equation (3) can easily be integrated. We take the natural logarithm of the amplitude solution and insert it into the multiplicative ergodic theorem (2).

$$\lambda = \lim_{t \to \infty} \frac{1}{t} \log \frac{a(t)}{a(0)} = -\frac{1}{2}\varepsilon \omega_1 \lim_{t \to \infty} \frac{1}{t} \int_0^t \cos \varphi(\tau) \sin 2\psi(\tau) \, d\tau. \qquad (5)$$

In this form, the Lyapunov exponent λ is completely determined by the excitation angle φ and the phase angle ψ. Applying an Euler forward scheme to equation (4) and $\dot{\varphi} = \omega$, both angles are simulated in the time domain for the time step $\Delta t = 0.01$ and for $N_s = 100,000$ sample points. Subsequently, the time series are inserted into the mean value (5) in order to determine the top Lyapunov exponent $\lambda(\varepsilon, \omega)$. Figure 1 shows corresponding numerical results in the parameter range $0 \leq \varepsilon \leq 1.5$ and $0.5 \leq \omega \leq 2.5$ for the natural frequency $\omega_1 = 1$. We recognize the typical instability regions $\lambda(\varepsilon, \omega) > 0$ around the parameter resonances $\omega = 2\omega_1/k$ for $k = 1, 2, 3, ...$ Outside the instability regions, the Lyapunov exponents are uniformly distributed with the value $\lambda(\varepsilon, \omega) = 0$.

The sharp boundaries between these two $\lambda(\varepsilon, \omega)$- distributions indicate bifurcations of the stationary angle solutions $\varphi(t)$ and $\psi(t)$. To visualize them we take the cosine of both angles and plot the coordinates $y = \cos\varphi$ and $z = \cos 2\psi$ in the phase plane $-1 \leq y, z \leq 1$. Figure 2 shows simulation results for $\omega_1 = 1, \varepsilon = 1$ and for the decreasing excitation frequencies $\omega = 2.0, 1.3, 0.9$ and 0.6. The second excitation frequency $\omega = 1.3$ belongs to the Lyapunov exponent $\lambda(\varepsilon, \omega) = 0$. It leads to almost periodic solutions in the x, y-plane. The other three frequencies belong to $\lambda(\varepsilon, \omega) > 0$. They produce periodic solutions with period-doubling effects. For decreasing excitation frequencies, there is a cascade of such periodic solution ranges separated by almost periodic regions. The upper right picture in figure 2 shows typical realizations of almost periodic solutions. Their long time simulation will cover the entire y, z-plane. A corresponding Poincaré map indicates the almost periodic solution character for $\omega = 1.3$.

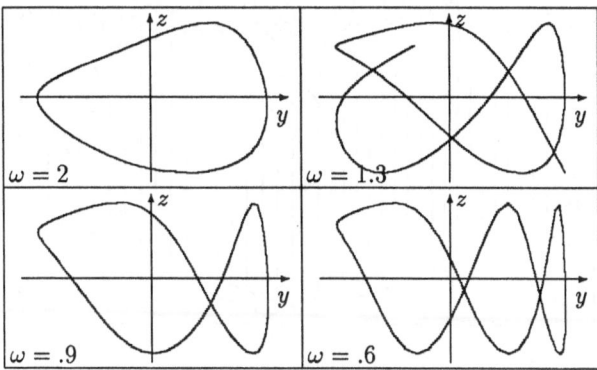

Figure 2: Periodic and almost periodic solutions

2.2. Invariant measures

As already mentioned, the phase angle ψ and the excitation angle φ are Markov processes and are stationary in the finite angle range $-\pi \leq \varphi, \psi \leq \pi$. Consequently, they possess a joint invariant measure, the density $p(\varphi, \psi)$ of which is periodic with respect to angle range, mentioned above. This density is determined by the Liouville equation which is a special form of the Fokker-Planck equation [5] associated with $\dot{\varphi} = \omega$ and equation (4).

$$\omega\frac{\partial}{\partial\varphi}p(\varphi, \psi) - \omega_1\frac{\partial}{\partial\psi}(1 + \varepsilon\cos\varphi\cos^2\psi)p(\varphi, \psi) = 0, \tag{6}$$

$$p(\varphi, \psi) = p_0(\varphi, \psi) + \varepsilon p_1(\varphi, \psi) + \varepsilon^2 p_2(\varphi, \psi) + ... \ . \tag{7}$$

Provided that $\lambda(\varepsilon, \omega) = 0$, the density $p(\varphi, \psi)$ is regular. It can be calculated via the perturbation expansion (7). This leads to a sequence of partial differential equations obtained by comparing all coefficients of ε^i for $i = 0, 1, 2, 3, \ldots$. The equation sequence can be solved step by step. The first three solution terms in (7) yield

$$p(\varphi, \psi) = \frac{1}{4\pi^2}\langle 1 + \frac{\varepsilon\omega_1}{\omega^2 - 4\omega_1^2}(2\omega_1 \cos\varphi \cos 2\psi - \omega \sin\varphi \sin 2\psi)$$

$$+ \frac{\varepsilon^2\omega_1^2/4}{(\omega^2 - 4\omega_1^2)(\omega^2 - \omega_1^2)}\{[(\omega^2 + 2\omega_1^2)\cos 2\psi + (\omega^2 + 4\omega_1^2)\cos 4\psi]\cos 2\varphi$$

$$- \omega\omega_1(2\sin 2\psi + 4\sin 4\psi)\sin 2\varphi\} - \frac{\varepsilon^2\omega_1^2/4}{\omega^2 - 4\omega_1^2}(2\cos 2\psi + \cos 4\psi) + \ldots). \qquad (8)$$

This solution holds provided that $p(\varphi, \psi) \geq 0$ in the entire angle range $-\pi \leq \varphi, \psi \leq \pi$. Figure 3 shows an evaluation of (8) for the parameters $\omega_1 = 1, \varepsilon = .5$ and $\omega = 1.5$.

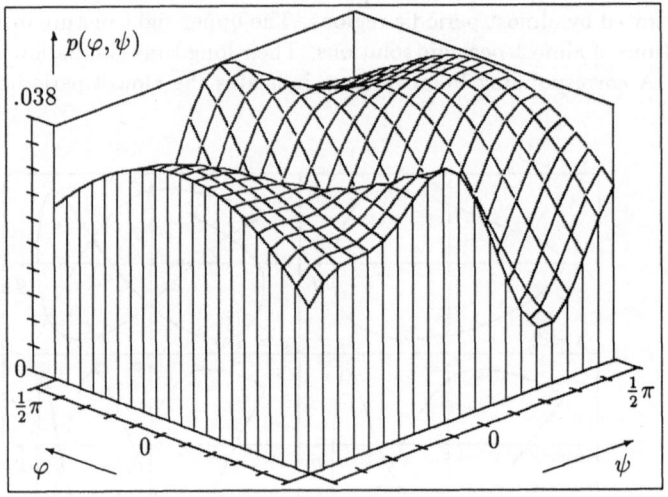

Figure 3: Density of the invariant angle measure

Applying the two-dimensional density $p(\varphi, \psi)$ to the top Lyapunov exponent $\lambda(\varepsilon, \omega)$, the time average in (5) can be replaced by the corresponding ensemble average.

$$\lambda = -\frac{1}{2}\varepsilon\omega_1 \int_{-\pi}^{\pi}\int_{-\pi}^{\pi} \cos\varphi \sin 2\psi \, p(\varphi, \psi) \, d\varphi d\psi = 0. \qquad (9)$$

Therewith, the infinite time integration in (5) is reduced to a finite double integral over the angle domain $-\pi \leq \varphi, \psi \leq \pi$. Obviously, this results in a vanishing Lyapunov exponent because there are no $\cos\varphi \sin 2\psi$-terms in the expanded $p(\varphi, \psi)$.

The calculated density (8) becomes negative for ε, ω- parameters near the resonance lines $\omega = 2\omega_1/k$ for $k = 1, 2, 3, \ldots$ where $\lambda(\varepsilon, \omega) > 0$. In this case, $p(\varphi, \psi)$ is singular and has to be recalculated by eliminating the singularities. The singular lines $\varphi(\psi)$ represent periodic solutions of $\dot\varphi = \omega$ and equation (4). Hence, $\varphi(\psi)$ is derived by dividing both equations.

$$\frac{d\varphi}{d\psi} = \frac{-\omega/\omega_1}{1 + \varepsilon \cos\varphi(\psi)\cos^2\psi}, \qquad \varphi(\psi) = \varphi(\psi + k\pi) + 2\pi, \qquad k = 1, 2, \ldots. \qquad (10)$$

Accordingly, the Liouville equation (6) and its solution (7) is reduced to

$$\frac{\partial}{\partial\psi}[1 + \varepsilon\cos\varphi(\psi)\cos^2\psi]p(\psi) = 0, \qquad \rightarrow p(\psi) = \frac{C}{1 + \varepsilon\cos\varphi(\psi)\cos^2\psi}, \qquad (11)$$

$$\lambda = -\frac{1}{2}\varepsilon\omega_1\int_{-k\pi/2}^{k\pi/2}\cos\varphi(\psi)\sin 2\psi\, p(\psi)d\psi, \qquad \int_{-k\pi/2}^{k\pi/2}p(\psi)d\psi = 1. \qquad (12)$$

In the density distribution (11), C is an integration constant determined by the normalization of $p(\psi)$ in one period $|\psi| \le k\pi/2$. The density $p(\psi)$ reduces the Lyapunov exponent (9) to the single integral, noted in (12).

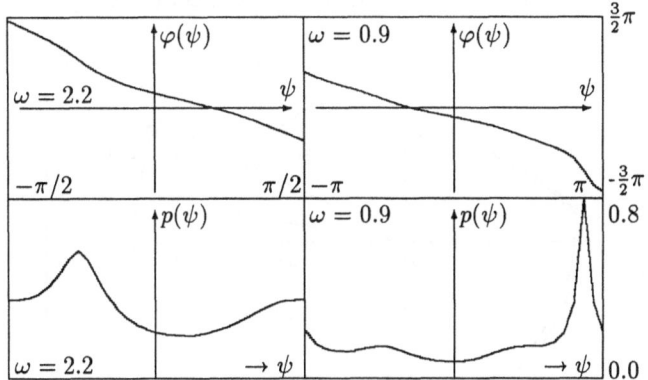

Figure 4: Periodic angle solutions and their densities

The evaluation of the top Lyapunov exponent, noted in (12), starts with the simulation of the equation (10). Applying a shooting procedure we determine those initial values $\varphi(-k\pi/2)$ which satisfy the periodicity condition $\varphi(-k\pi/2) = \varphi(k\pi/2)+2\pi$. Subsequently, we perform the normalization of $p(\psi)$ and the integration of the Lyapunov exponent $\lambda(\varepsilon,\omega)$, noted in (12). Figure 4 shows corresponding results for $\omega_1 = 1, \varepsilon = 1$ and for the excitation frequencies $\omega = 2.2$ and $\omega = 0.9$.

3. Van der Pol equation

To extend the analysis of Lyapunov exponents and of related invariant measures to stability investigations of non-trivial solutions, we consider the following Van der Pol equation.

$$\ddot{x} + x + (x^2 - \gamma)\dot{x} = 0, \qquad |\gamma| < \infty. \qquad (13)$$

For positive parameters $\gamma > 0$, its equilibrium $x = 0$ is unstable and bifurcates into a non-isotropic limit cycle. The simulation of this closed orbit can be performed in the rectangular coordinates $x_1 = x$ and $x_2 = \dot{x}$ or in polar coordinates. Introducing the amplitude a and the phase angle ψ into equation (13), we obtain a first order system, as follows.

$$x = x_1 = a\cos\psi, \qquad \dot{a} = a(\gamma - a^2\cos^2\psi)\sin^2\psi, \qquad (14)$$

$$\dot{x} = x_2 = a\sin\psi, \qquad \dot{\psi} = (\gamma - a^2\cos^2\psi)\sin\psi\cos\psi - 1. \qquad (15)$$

In figure 5, we show numerical results for the parameter $\gamma = .5$. The unperturbed limit cycle is denoted by $\bar{x}_1(t), \bar{x}_2(t)$ or by $\bar{a}(\psi)$, respectively. The perturbed solutions $x_1(t), x_2(t)$ or $a(t), \psi(t)$ are obtained by applying initial conditions outside the closed orbit $\bar{a}(\psi)$.

3.1. Orbital centre manifold

According to figure 5, there are two different stability concepts. The first one is the kinematic concept introduced by Lyapunov. It measures the distance between perturbed solutions $x_1(t), x_2(t)$ and unperturbed ones $\bar{x}_1(t), \bar{x}_2(t)$ at the same time t. This Eucledian distance of both points is marked by filled in circles in figure 5. Since the associated perturbation equations are coupled with the time-dependent limit cycle solutions $\bar{x}_1(t)$ and $\bar{x}_2(t)$, the kinematic concept of Lyapunov leads to a four dimensional dynamic problem.

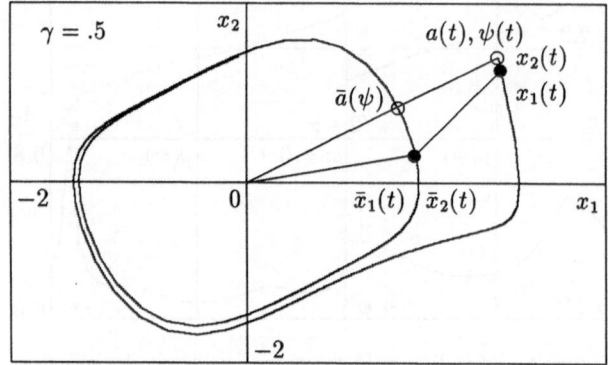

Figure 5: Simulation of the Van der Pol equation

The second stability concept goes back to Poincaré. It measures the distance between the perturbed time solution $a(t), \psi(t)$ and the unperturbed limit cycle $\bar{a}(\psi)$ at different times but at the same phase angle $\psi(t)$. In figure 5, this amplitude metric is represented by hollow circles. It possesses the advantage that there exists an invariant measure which allows one to derive associated Lyapunov exponents without any knowledge of the perturbed solutions $a(t)$ and $\psi(t)$. To calculate the amplitude distance between $a(t), \psi(t)$ and $\bar{a}(\psi)$, we need the geometrical configuration of the closed orbit $\bar{a}(\psi)$. For this purpose, we eliminate the time variable t by dividing the amplitude equation (14) and the phase equation (15).

$$\frac{d\bar{a}}{d\psi} = \frac{\bar{a}(\gamma - \bar{a}^2 \cos^2 \psi) \sin^2 \psi}{(\gamma - \bar{a}^2 \cos^2 \psi) \sin \psi \cos \psi - 1} = Q(\bar{a}, \psi). \tag{16}$$

Thus, we obtain a first order equation which gives the unperturbed limit cycle $\bar{a}(\psi)$ provided its integration is started in any known initial point $\bar{a}(\psi_0)$ of the closed orbit. Note that the right-hand side of this first order equation is abbreviated by $Q(\bar{a}, \psi)$.

The function $Q(\bar{a}, \psi)$ is related to an orbital centre manifold $\psi = h(a)$ in the following sense.

$$\begin{aligned} \dot{a} &= f^c(a, \psi), \\ \dot{\psi} &= f^s(a, \psi), \end{aligned} \qquad \rightarrow \qquad \frac{dh}{da} = \frac{f^s(a, h(a))}{f^c(a, h(a))} = \frac{1}{Q(a, h(a))} . \tag{17}$$

Herein, $f^s(a, \psi)$ is defined by (15) and belongs to the stable phase variable which takes only values in the finite angle range $-\pi \leq \psi \leq \pi$. If $\psi(t)$ exceeds this range during a time simulation, we can reset it by the periodicity condition $\psi(t + 2\pi) = \psi(t)$ without any influence on the amplitude behaviour. With this mod 2π restriction, the amplitude $a(t)$ is the only critical variable defined by $f^c(a, \psi)$ according to equation (14). Obviuosly, the orbital center manifold $\psi = h(a)$ is determined by the ratio of both functions f^s and f^c which is inverse to the Q- function, noted in (16). The orbital center manifold is calculable by Taylor expansions around $\bar{a}(\psi)$. According to [6], an explicit evaluation of $\psi = h(a)$ can be avoided. For this purpose, we replace the independent angle increment $d\psi$ in equation (16) by the equation (15) which gives the perturbed angle increment $d\psi(t)$ in dependence on the time t.

$$\dot{a} = Q(\bar{a}, \psi)[(\gamma - a^2 \cos^2 \psi) \sin \psi \cos \psi - 1]. \tag{18}$$

Note that the equation (18), now obtained, contains two different amplitudes: the unperturbed amplitude $\bar{a}(t)$ and the perturbed one $a(t)$. Taking the difference $\Delta = a - \bar{a}$ of both amplitudes and introducing it into (16), (15) and (14), we eliminate a and derive the following three equations for the variables $\bar{a}, \psi$ and Δ.

$$\dot{\bar{a}} = Q\{[\gamma - (\bar{a} + \Delta)^2 \cos^2 \psi] \sin \psi \cos \psi - 1\}, \tag{19}$$
$$\dot{\psi} = [\gamma - (\bar{a} + \Delta)^2 \cos^2 \psi] \sin \psi \cos \psi - 1, \tag{20}$$
$$\dot{\Delta} = \Delta[\gamma - (3\bar{a}^2 + 3\bar{a}\Delta + \Delta^2) \cos^2 \psi] \sin^2 \psi$$
$$+ \Delta Q(\bar{a}, \psi)(2\bar{a} + \Delta) \cos^3 \psi \sin \psi. \tag{21}$$

These equations can be simulated in the time domain with the initial conditions $\Delta(0) = \Delta_0, \psi(0) = \psi_0$ and $\bar{a}(0) = \bar{a}(\psi_0)$. Δ_0 and ψ_0 are arbitrary, meanwhile $\bar{a}(\psi_0)$ represents one limit cycle point to be known.

3.2. Orbital Lyapunov exponents

According to the multiplicative ergodic theorem, an orbital Lyapunov exponent λ is introduced by the amplitude metric defined by the deviation $\Delta(t)$ of both amplitudes, as follows.

$$\lambda = \lim_{t \to \infty} \frac{1}{t} \log \frac{\Delta(t)}{\Delta(0)}, \qquad \Delta(t) = a(t) - \bar{a}(\psi(t)), \tag{22}$$

$$\dot{\Delta} = \Delta[(\gamma - 3\bar{a}^2 \cos^2 \psi) \sin \psi + 2\bar{a}Q(\bar{a}, \psi) \cos^3 \psi] \sin \psi. \tag{23}$$

The amplitude difference $\Delta(t)$, given by equation (21), can be linearized to the form (23) for the asymptotic stability investigation of interest. This scalar equation (23) is integrated with respect to the initial condition $\Delta(0)$. Taking the natural logarithm of the solution $\Delta(t)$ and inserting it into the multiplicative ergodic theorem (22) yields

$$\lambda = \lim_{t \to \infty} \frac{1}{t} \int_0^t [(\gamma - 3\bar{a}^2 \cos^2 \psi) \sin \psi + 2\bar{a}Q(\bar{a}, \psi) \cos^3 \psi] \sin \psi \, d\tau. \tag{24}$$

Herein, $\psi(\tau)$ and $\bar{a}(\psi(\tau))$ are determined by the equations (19) and (20). They reduce to the original equations (14) and (15) for the asymptotic stability investigation $\Delta(t) \to 0$.

Provided that the unperturbed amplitude $\bar{a}(\psi)$ is periodic with respect to the phase angle

ψ, there exists an invariant measure of the stationary phase process which is determined by the Liouville equation associated with (15).

$$\frac{\partial}{\partial \psi}\{[\gamma - \bar{a}^2(\psi)\cos^2\psi]\sin\psi\cos\psi - 1\}p(\psi) = 0, \qquad |\psi| \leq \pi/2, \qquad (25)$$

$$p(\psi) = \frac{C}{1 - [\gamma - \bar{a}^2(\psi)\cos^2\psi]\sin\psi\cos\psi}, \qquad \int_{-\pi/2}^{\pi/2} p(\psi)d\psi = 1. \qquad (26)$$

The density equation (25) possesses the π-periodic solution (26). Calculating the integral in (26), the density $p(\psi)$ can be normalized in the angle range $|\psi| \leq \pi/2$ by means of the integration constant C.

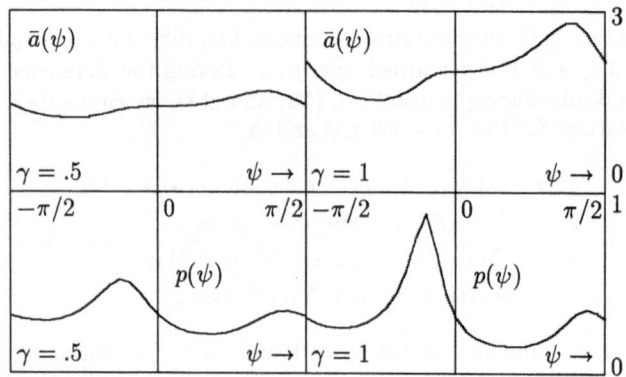

Figure 6: Determination of the Lyapunov exponent

The phase density $p(\psi)$ reduces the time averaging procedure in (24) to the corresponding ensemble average, as follows.

$$\lambda = \int_{-\pi/2}^{\pi/2}[(\gamma - 3\bar{a}^2\cos^2\psi)\sin\psi + 2\bar{a}Q(\bar{a},\psi)\cos^3\psi]\sin\psi p(\psi)d\psi, \qquad (27)$$

$$\lambda = \int_{-\pi/2}^{\pi/2}\{[\gamma\sin^2\psi - \bar{a}^2(\psi)\cos^4\psi]p(\psi) - \bar{a}^2(\psi)\sin\psi\cos^3\psi\frac{dp}{d\psi}\}d\psi. \qquad (28)$$

Finally, we can insert the $Q(\bar{a},\psi)$-function into (27) and perform a partial integration in arriving at (28). Note that both forms (27) and (28) determine the orbital Lyapunov exponent without any knowledge of the perturbed solutions $a(t)$ and $\psi(t)$. The evaluation of (27) starts with the simulation of $\bar{a}(\psi)$ or $\bar{a}^2(\psi)$ by means of a forward Euler scheme applied to equation (16). After some iteration steps we find closed orbits or that initial value $\bar{a}^2(-\pi/2)$ which satisfies the periodicity condition $\bar{a}^2(-\pi/2) = \bar{a}^2(\pi/2)$. The numerical solution $\bar{a}^2(\psi)$ determines the density (26) of the invariant measure. Subsequently, we normalize $p(\psi)$ according to (26) and evaluate the orbital Lyapunov exponent (27) by numerical integration. Figure 6 shows typical results of $\bar{a}(\psi)$ and $p(\psi)$ for the parameters $\gamma = .5$ and $\gamma = 1$. For negative γ-parameters, a periodic solution $\bar{a}^2(\psi)$ can be simulated by starting with negative initial values $\bar{a}^2(\psi_0) < 0$ and applying a negative angle increment in (16).

3.3. Stability of trivial solutions

To complete the stability investigation, we finally derive invariant measures and Lyapunov exponents for the trivial limit cycle $\bar{a}_0(\psi) = 0$. In this case, there is a critical damping $\gamma_c = 2$ which separates two different solutions of the Liouville equation (25). For overdamped systems $|\gamma| \geq 2$, the phase density $p(\psi)$ degenerates to a singular delta function around $\psi_0 = -\frac{1}{2}\arcsin 2/\gamma$ and leads to the following top Lyapunov exponent λ_0.

$$\lambda_0 = \int_{-\pi/2}^{\pi/2} \gamma \sin^2 \psi p(\psi) d\psi = \frac{1}{2}\gamma + \sqrt{\frac{1}{4}\gamma^2 - 1}, \qquad |\gamma| \geq 2, \qquad (29)$$

$$\lambda_0 = \int_{-\pi/2}^{\pi/2} \gamma \sin^2 \psi p(\psi) d\psi = \frac{1}{2}\gamma, \qquad |\gamma| \leq 2, \qquad (30)$$

For $|\gamma| \leq 2$, the density $p(\psi)$ is regular describing a stationary phase motion with the rotation number $\alpha = -[1 - \gamma^2/4]^{1/2}$. Naturally, the results (29) and (30) can also be calculated by means of the classical eigenvalue theory.

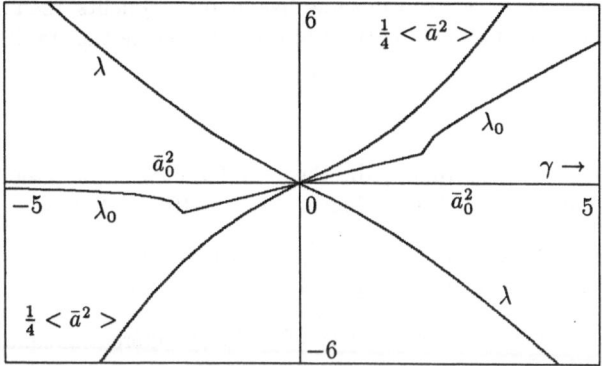

Figure 7: Lyapunov exponents of the limit cycles

Figure 7 shows the Lyapunov exponents λ_0 of the trivial limit cycles, the Lyapunov exponents λ and the averaged squared amplitude $< \bar{a}^2(\psi) >$ of the non-trivial limit cycles in dependence on the γ-parameter of the Van der Pol equation. For $\gamma < 0$, λ_0 is negative and λ is positive. Consequently, the trivial solution $\bar{a}_0 = 0$ is asymptotically stable. The limit cycle is unstable. Vice versa, for positive γ-parameters, the mean square of the closed orbit becomes positive and asymptotically stable.

Both Lyapunov exponents λ and λ_0 are vanishing at the bifurcation point $\gamma = 0$ which represents the critical case where the linearized perturbation equation (23) fails. To investigate this critical case, we go back to the nonlinear perturbation equation (21) and solve it in the time domain for $\gamma = 0, \bar{a}_0 = 0$ and $\psi = -t$.

$$\dot{\Delta} = -\Delta^3 \cos^2 \psi \sin^2 \psi, \qquad \Delta(t) = \Delta_0/\sqrt{1 + \frac{1}{2}(t - \frac{1}{4}\sin 4t)\Delta_0^2}. \qquad (31)$$

The nonlinear perturbation solution (31) vanishes for infinitely increasing times. Hence, the trivial solution $\bar{a}_0 = 0$ is asymptotically stable in the critical case $\gamma = 0$.

4. Conclusions

Classical eigenvalues of linear time-invariant systems are generalized by Lyapunov exponents and rotation numbers which substitute the real parts and the imaginary parts, respectively. In case of stability investigations of trivial solutions, linear time-variant systems are reduced to an amplitude centre manifold by means of Khasminskii's projection on circles, spheres or hyperspheres[7]. The paper explains this technique by means of the Mathieu equation introducing polar coordinates for the projection. Associated to the angle equation, there is an invariant measure which determines the top Lyapunov exponent by a finite integral over the angle range.

To extend this technique to non-trivial solutions, the Van der Pol equation is considered. Applying Poincaré's stability concept, the periodic limit cycle leads to an invariant measure which allows one to calculate the orbital Lyapunov exponent of the closed orbit without any knowledge of the perturbed solution of the Van der Pol equation. Relations to the orbital centre manifold are discussed. In particular, the orbital centre manifold is derivable by means of polar coordinates without any knowledge of the eigenvalues of the linearized state equations. In the critical case of vanishing Lyapunov exponents, the analysis has to be extended to nonlinear perturbation equations in order to decide the stability of stationary solutions.

References

1. Oseldec, V.I., A multiplicative ergodic theorem, Lyapunov characteristic numbers for dynamical systems, Trans. Moscow Math. Soc. 19 (1968) 197-231.

2. Khasminskii, R.Z., Necessary and sufficient conditions for asymptotic stability of linear stochastic systems. Theory Prob. Appl., 12 (1967) 144-147.

3. Coddington, E.A., Levinson, N., Theory of ordinary differential equations. McGraw-Hill, New York, 1955.

4. Wedig, W., Vom Chaos zur Ordnung, Mitteilungen GAMM (R. Mennicken ed.), ISSN 0936-7195, 2 (1989) 3-31.

5. Arnold L., Stochastic Differential Equations. Wiley, New York, 1974.

6. Wedig, W., Invariant measures and Lyapunov exponents of equilibria and limit cycles. Oberwolfach-Tagung 1990 on Lyapunov Exponents (L.Arnold, J.P. Eckmann).

7. Wedig, W., Dynamic stability of beams under axial forces - Lyapunov exponents for general fluctuating loads. Eurodyn '90, Conference on Structural Dynamics, Preprint of Proceedings, Ruhr-University Bochum, SFB 151, Vol. 1 (1990) 57-64.

Address: Prof. Dr.-Ing. W. Wedig, Institute for Technical Mechanics
University of Karlsruhe, D-7500 Karlsruhe 1, BRD

International Series of Numerical Mathematics, Vol. 97, © 1991 Birkhäuser Verlag Basel

COMPUTATION OF HOPF BRANCHES BIFURCATING FROM
TAKENS-BOGDANOV POINTS FOR PROBLEMS WITH SYMMETRIES

B. Werner and V. Janovsky

AMS (MOS) subject classification: 65H10, 20C30, 58F14.

Key words: *Hopf bifurcation, real and imaginary Hopf points, pitchforks, turning points, Takens Bogdanov bifurcation, TB-points, subspace breaking bifurcation, computation of bifurcation points, $\mathbb{Z}_2$-symmetry, Brusselator.*

1 Introduction

We consider dynamical systems of O.D.E.'s

$$\dot{x} = g(x, \lambda, \alpha), \tag{1.1}$$

where $x \in X = \mathbb{R}^N$ is the state variable, $\lambda \in \mathbb{R}$ is a distinguished bifurcation parameter, α is an additional imperfection or control parameter, and $g : X \times \mathbb{R} \times \mathbb{R} \to X$ is sufficiently smooth.

Takens-Bogdanov points $(x_0, \lambda_0, \alpha_0)$ of (1.1) (*TB-points*) are essentially defined by the equilibrium condition $g(x_0, \lambda_0, \alpha_0) = 0$ and by the degeneracy that the Jacobian $g_x^0 := g_x(x_0, \lambda_0, \alpha_0)$ has an algebraically double and geometrically simple eigenvalue $\mu = 0$ ([17] and [1]).

For fixed $\alpha = \alpha_0$, TB-points correspond to simple λ-singular points of $g = 0$. Since generically the only λ-singular points are (simple) quadratic turning points, mainly *TB-turning points* are considered in the literature ([11], [5], [14]).

But in systems with symmetries (see [6]), certain symmetry breaking bifurcation points will occur generically. The simplest, but most important one, is the *pitchfork* in case of $\mathbb{Z}_2$-symmetry (see [16]). Using the concept of *bifurcation subgroups* in [4] in connection with the equivariant branching lemma, multiple bifurcation points can be reduced to simple ones - either of pitchfork type for symmetric subgroups or of a more general, possibly transcritical type (*subspace breaking bifurcation point*) for asymmetric subgroups.

Hence we focus on *TB-pitchforks, subspace breaking TB-points* and also on TB-turning points (sec. 4).

TB-points are of interest because of the occurrence of Hopf points and of global bifurcations of (1.1) for fixed α close to the TB-parameter α_0, see [7], p.364f, and [3] for the $O(2)$-symmetry case.

We will contribute to the analysis and to the computation of Hopf points near a TB-point. The main analytical and numerical tool will be the following extended system $G_H = 0$ for Hopf points, where α will be used as a bifurcation parameter.

$$G_H : \begin{cases} X \times X \times \mathbb{R} \times \mathbb{R} \times \mathbb{R} \times \mathbb{R} \to X \times X \times \mathbb{R} \\ (x, \xi, \lambda, \nu, \alpha) \mapsto (g(x, \lambda, \alpha), F_H(g_x(x, \lambda, \alpha), \xi, \nu)) \end{cases}, \tag{1.2}$$

$$F_H : \mathbb{R}^{N,N} \times X \times \mathbb{R} \to X \times \mathbb{R}^2, \quad (A, \xi, \nu) \mapsto \left((A^2 + \nu I)\xi, l^T \xi, l^T A \xi - 1\right), \quad l^T \in X'. \tag{1.3}$$

This system (with ω^2 instead of ν and other normalizing conditions for ξ) has been investigated in [12]. But we have replaced the term ω^2 (with the Hopf frequency ω) by the term ν which we now allow to vanish or to become negative. It turns out that the extended system $G_H = 0$ is very appropriate to treat *real* Hopf points ($\nu = \omega^2 > 0$), TB-points ($\nu = 0$, ξ is the first principal vector of g_x^0) and *imaginary* Hopf points ($\nu = -\omega^2 < 0$, g_x^0 has algebraically simple eigenvalues $\pm\omega$) in a unifying way.

It might be puzzling that *imaginary (real)* Hopf points are associated with *real (imaginary)* eigenvalues. Our notation is motivated by the fact that a *Hopf frequency f* defined by $\nu = f^2$, is *imaginary (real)* in the case of *imaginary (real)* Hopf points. (In [9] the notation *neutral saddle point* is used instead of imaginary Hopf points.)

The reasons for introducing imaginary Hopf points are the following. As already mentioned in [9], the branch of real Hopf points (if α is varied) can be smoothly continued through a TB-point into a branch of imaginary Hopf points (Th. 4.3 makes this statement rigorous). The branch of imaginary Hopf points may contain interesting global codimension 2-bifurcation points (see [2]).

Moreover, numerical computations (see sec.6) show that a branch of imaginary Hopf points bifurcating for instance in a TB-pitchfork, can connect with a branch of real Hopf points.

Our main theoretical result is given by Th.5.5. It says that - under some non-degeneracy conditions (TB) - TB-pitchforks (based on a $\mathbb{Z}_2$-symmetry), resp. subspace breaking TB-points (based on the existence of a nonlinearly invariant subspace) are pitchforks, resp. subspace breaking bifurcation points of $G_H = 0$ w.r.t. α. This allows a bifurcation analysis - useful for numerical computations - for Hopf points on the primary (symmetric) and on the secondary (asymmetric) branch as well.

To make our approach more transparent we characterize nondegenerate steady state and Hopf bifurcation by the fact that the one-dimensional manifold of Jacobians $A(s)$ along a primary branch $C = \{(x(s), \lambda(s)) : s \in I\}$ of equilibria intersects certain *manifolds of codimension one*, namely $\mathcal{M}_S$ and $\mathcal{M}_H$, in the space of all real $N \times N$-matrices *transversally* for $s = s_0$. Here $\mathcal{M}_S$ is given by all matrices of rank $N - 1$, while $\mathcal{M}_H$ contains certain *generalized Hopf matrices* A satisfying $F_H(A, \xi, \nu) = 0$ with F_H defined in (1.3).

We formulate transversality conditions (T1) and (T2) for the transversal crossing of $\mathcal{M}_S$ and of $\mathcal{M}_H$ by any smooth path of matrices $A(s)$, which (non-surprisingly) turn out to be equivalent to *regularity* conditions for various extended systems for turning points, pitchforks, subspace breaking bifurcation points and Hopf points (Th. 3.1).

Non-degenerate Takens-Bogdanov bifurcation (for $\alpha = \alpha_0$) will be defined by the simultaneous transversal crossing of $\mathcal{M}_H$ *and* of $\mathcal{M}_S$ by the path $A_{\alpha_0}(s)$ (sec. 4).

Our approach is related to that in [14] for TB-turning points and to that in [20] for TB-pitchforks. But we think that we have used a more appropriate extended system $G_H = 0$ for the characterization of Hopf points.

We illustrate our results and their numerical consequences for the two- and three-box-Brusselator system where we compute (symmetric and asymmetric) Hopf branches intersecting in TB-points.

2 Singular and Hopf matrices, transversality conditions.

Let $\mathcal{M}_S$ contain all real $N \times N$-matrices of rank $N - 1$. $\mathcal{M}_S$ is a manifold of codimension 1. We remark that $A \in \mathcal{M}_S$ iff there is a normalizing linear functional $l^T \in X'$ such that

$$F_S : \mathbb{R}^{N,N} \times X \to X \times \mathbb{R}, \quad (A, \varphi) \mapsto (A\varphi, l^T\varphi - 1) \tag{2.1}$$

has a solution (A, φ), satisfying $\text{rank} \frac{\partial F_S}{\partial \varphi}(A, \varphi) = N$. To each $A_0 \in \mathcal{M}_S$ we associate vectors $\varphi_0, \psi_0 \neq 0$ spanning the kernel, resp. the cokernel of A_0:

$$A_0 \varphi_0 = 0, \ \psi_0^T A_0 = 0. \tag{2.2}$$

The set of *generalized Hopf matrices* $\mathcal{M}_H$ consists of all matrices A such that there is an associated *Hopf number* $\nu \in \mathbb{R}$ with the property that for

- $\nu := \omega^2 > 0$: A has algebraically simple purely imaginary eigenvalues $\pm i\omega$.

- $\nu = 0$: $\mu = 0$ is an algebraically double and geometrically simple eigenvalue of A.

- $\nu := -\omega^2 < 0$: A has algebraically simple real eigenvalues $\pm \omega$.

Obviously $\mathcal{M}_H = \mathcal{M}_H^+ \cup \mathcal{M}_H^0 \cup \mathcal{M}_H^-$, where the superscript is related to the sign of the Hopf number ν. We call $A \in \mathcal{M}_H$ a *real Hopf matrix* if $A \in \mathcal{M}_H^+$, an *imaginary Hopf matrix* if $A \in \mathcal{M}_H^-$ and a *TB-matrix*, if $A \in \mathcal{M}_H^0$.

We remark without proof that if $A \in \mathcal{M}_H$ then there is a normalizing linear functional $l^T \in X'$ such that F_H (see (1.3)) has a root (A, ξ, ν) satisfying $\text{rank} \frac{\partial F_H}{\partial(\xi, \nu)}(A, \xi, \nu) = N + 1$.

$A \in \mathcal{M}_H$ can have more than one Hopf number. But if we restrict to those Hopf matrices with *unique* Hopf number, it follows from the last remark that $\mathcal{M}_H$ is a manifold of codimension 1. (Think for $N = 2$ of $\text{trace}(A) = 0$ as defining equation).

To each $A_0 \in \mathcal{M}_H$ we can associate not only the Hopf number ν_0 but also vectors φ_0, ξ_0 and ψ_0, χ_0 spanning the null space of $A_0^2 + \nu_0 I$ and of $(A_0^T)^2 + \nu_0 I$ respectively and satisfying

$$(A_0^2 + \nu_0 I)\xi_0 = 0, \ \chi_0^T(A_0^2 + \nu_0 I) = 0,$$
$$A_0 \xi_0 = \varphi_0, \ \chi_0^T A_0 = \psi_0^T, \ \psi_0^T \varphi_0 = 0, \ \psi_0^T \xi_0 = 1, \ \chi_0^T \varphi_0 = 1, \ \chi_0^T \xi_0 = 0. \tag{2.3}$$

Now consider a smooth *path* (one-dimensional manifold) of matrices $A(s)$ where s is the path parameter, intersecting $\mathcal{M}_S$ or $\mathcal{M}_H$ for $s = s_0$. The transversality conditions are given by scalar inequalities and are rather easily obtained by requiring that (s_0, φ_0) $((s_0, \xi_0, \nu_0))$ are regular solutions of $f_S(s, \varphi) := F_S(A(s), \varphi) = 0$, or of $f_H(s, \xi, \nu) := F_H(A(s), \xi, \nu) = 0$ respectively.

$$\mathcal{M}_S: \qquad \qquad \psi_0^T A'(s_0)\varphi_0 \neq 0, \tag{T1}$$

$$\mathcal{M}_H: \qquad \psi_0^T A'(s_0)\xi_0 + \chi_0^T A'(s_0)\varphi_0 \neq 0. \tag{T2}$$

To show (T2) assume that $f_H(s_0, \xi_0, \nu_0) = 0$. Set $A_0 := A(s_0)$, $A_0' := A'(s_0)$. If (t, u, β) lies in the kernel of Df_H^0, then

$$t(A_0 A_0' + A_0' A_0)\xi_0 + (A_0^2 + \nu_0 I)u + \beta \xi_0 = 0.$$

Multiplying this equation by ψ_0^T and χ_0^T and using (2.3) it follows that (T2) implies $t = 0$ and $\beta = 0$. $u = 0$ follows then easily.

On the other hand, if (T2) is not true, the construction of a nontrivial kernel vector of Df_H^0 is possible.

If $A(s_0)$ is a real Hopf matrix $(\nu > 0)$, then (T2) can be shown to be equivalent to the well known eigenvalue crossing condition: there is pair of conjugate complex eigenvalues $\alpha(s) \pm i\beta(s)$ of $A(s)$ with $\alpha(s_0) = 0$, $\alpha'(s_0) \neq 0$.

3   Turning Points, Pitchforks, Subspace breaking bifurcation points and Hopf points

The following bifurcation points (x_0, λ_0) of one-parameter problems $\dot{x} = g(x, \lambda)$ are all defined by two conditions. The first one ensures that there is a smooth primary branch $\mathcal{C} = \{(x(s), \lambda(s)), s \in I\}$ through (x_0, λ_0) defining a smooth path of matrices $A(s) := g_x(x(s), \lambda(s))$. The second one is the *transversality condition* (T1) or (T2).

We will consider (simple) λ-*singular* points (x_0, λ_0) - relevant for steady state bifurcation - and (generalized simple) Hopf points - relevant for Hopf bifurcation - defined by $g(x_0, \lambda_0) = 0$ and by $g_x^0 \in \mathcal{M}_S$, resp. $g_x^0 \in \mathcal{M}_H$.

Then in the first case φ_0, ψ_0 (see (2.2)) and in the second case additionally ξ_0, χ_0 are defined (see (2.3)).

For future applications we include the possibility that g depends also on a second parameter α. Subscripts of g denote partial derivatives, while the superscript 0 indicates evaluation in the bifurcation point. We will use the following notations.

$$A_0 := g_x^0, \; A_1 := g_{xx}^0 \varphi_0, \; A_\lambda := g_{xx}^0 v_\lambda + g_{x\lambda}^0, \; A_\alpha := g_{xx}^0 v_\alpha + g_{x\alpha}^0 \tag{3.1}$$

where A_λ and A_α have a meaning only if v_λ and v_α can be defined by

$$g_x^0 v_\lambda + g_\lambda^0 = 0, \; g_x^0 v_\alpha + g_\alpha^0 = 0. \tag{3.2}$$

Our analysis will involve certain numbers

$$b_\lambda := \psi_0^T g_\lambda^0, \quad c_1 := \psi_0^T A_1 \varphi_0, \; c_\lambda := \psi_0^T A_\lambda \varphi_0, \; c_\alpha := \psi_0^T A_\alpha \varphi_0, \tag{3.3}$$

$$d_1 := \psi_0^T A_1 \xi_0 + \chi_0^T A_1 \varphi_0, \; d_\lambda := \psi_0^T A_\lambda \xi_0 + \chi_0^T A_\lambda \varphi_0, \; d_\alpha := \psi_0^T A_\alpha \xi_0 + \chi_0^T A_\alpha \varphi_0. \tag{3.4}$$

3.1   Steady State Bifurcation points

Quadratic turning points: The conditions are

$$b_\lambda \, (= \psi_0^T g_\lambda^0) \neq 0, \quad c_1 \, (= \psi_0^T g_{xx}^0 \varphi_0 \varphi_0) \neq 0. \tag{3.5}$$

The first condition ensures the locally unique existence of a smooth branch $\mathcal{C}$ where the path parameter s can be chosen in such a way that $x'(s_0) = \varphi_0$. The second condition in (3.5) is the transversality condition (T1) since $A'(s_0) = A_1$.

Symmetry breaking bifurcation points (pitchforks):   These bifurcation points arise generically in problems, where $g(x, \lambda)$ satisfies the $\mathbb{Z}_2$-symmetry

$$g(Sx, \lambda) = Sg(x, \lambda) \quad \text{for all } x, \lambda \tag{3.6}$$

with $S \neq I, S^2 = I$. The symmetry transformation S decomposes X into

$$X := X_+ \oplus X_-, \quad X_+ := \{x \in X : Sx = x\}, \quad X_- := \{x \in X : Sx = -x\}.$$

The conditions for a pitchfork (x_0, λ_0) (c. [16]) are given by

$$x_0 \in X_+, \varphi_0 \in X_-, \quad c_\lambda \, (= \psi_0^T A_\lambda \varphi_0) \neq 0. \tag{3.7}$$

The first two conditions ensure the unique existence of a smooth *symmetric* branch $\mathcal{C} \subset X_+ \times \mathbb{R}$ through (x_0, λ_0) which can be parametrized by $s = \lambda$. The last condition in (3.7) is the transversality condition (T1), since $A'(s_0) = A_\lambda$ (with $v_\lambda \in X_+$, see (3.2)). Observe that $b_\lambda = 0$, $c_1 = 0$ (and $d_1 = 0$).

A pitchfork is a symmetry breaking bifurcation point in the sense that there is a unique *asymmetric* branch C^a bifurcating from C in (x_0, λ_0).

Observe that our definition of a pitchfork does not include that the bifurcation is strictly sub- or supercritical. For this information one needs $c_2 := -(1/3)\psi_0^T(g_{xxx}^0\varphi_0^3 + 3g_{xx}^0 w_0\varphi_0) \neq 0$, where $w_0 \in X_+$ is defined by $g_x^0 w_0 + g_{xx}^0\varphi_0^2 = 0$.

Subspace breaking bifurcation points: These bifurcation points occur if there is a *nonlinearly* invariant linear subspace X_0 of X:

$$g(x, \lambda) \in X_0 \quad \text{for all } x \in X_0, \lambda \in \mathbb{R}$$

(see [18]). The conditions are in analogy to (3.7)

$$x_0 \in X_0, \quad \varphi_0 \notin X_0, \quad c_\lambda \ (= \psi_0^T A_\lambda \varphi_0) \neq 0. \tag{3.8}$$

Again the first two conditions ensure the unique existence of a primary branch C in $X_0 \times \mathbb{R}$ which can be parametrized by $s = \lambda$. The transversality condition (T1) can be stated exactly as in (3.7), where now $v_\lambda \in X_0$. (A pitchfork is a subspace breaking bifurcation point setting $X_0 := X_+$).

A unique branch C^a lying *not* in $X_0 \times \mathbb{R}$ (hence *subspace breaking*) bifurcates from C in (x_0, λ_0). The bifurcation is *transcritical* iff $c_1 \ (= \psi_0^T g_{xx}^0 \varphi_0^2) \neq 0$. As in the pitchfork case, we will call C the *symmetric* and C^a the *asymmetric* branch.

3.2 REAL AND IMAGINARY HOPF POINTS.

Assume that $g(x_0, \lambda_0) = 0$ and that g_x^0 is a Hopf matrix with Hopf number $\nu_0 \neq 0$. The conditions for a Hopf point are

$$g_x^0 \text{ is regular }, \quad d_\lambda \ (= \psi_0^T A_\lambda \xi_0 + \chi_0^T A_\lambda \varphi_0) \neq 0. \tag{3.9}$$

The first condition trivially guarantees the unique existence of a branch C which can be parametrized by $s = \lambda$. The last condition in (3.9) is the transversality condition (T2), since $A'(s_0) = A_\lambda \ (= g_{xx}^0 v_\lambda + g_{x\lambda}^0)$. We call (x_0, λ_0) a *real* (an *imaginary*) Hopf point of g if the Hopf number ν_0 is positive (negative).

In the case of $\mathbb{Z}_2$-symmetry one can distinguish between *symmetry preserving* Hopf points, where $x_0, \varphi_0, \xi_0 \in X_+$, and *symmetry breaking* Hopf points, where $x_0 \in X_+, \varphi_0, \xi_0 \in X_-$. Similarly *subspace preserving* and *subspace breaking* Hopf points can be defined in the case of a subspace invariance.

3.3 REGULARITY OF EXTENDED SYSTEMS

From geometrical point of view it is not surprising that the transversality conditions formulated in (3.5),(3.7),(3.8) and (3.9) are precisely the conditions which ensure the regularity of the solutions of certain *extended systems* $G_S = 0$ (3.10) for steady state bifurcation points and of $G_H = 0$ defined by (1.2) in the introduction for Hopf points (in this section we suppress the second parameter α).

With (2.1) we define

$$G_S : X \times X \times \mathbb{R} \to X \times X \times \mathbb{R}, \quad (x, \varphi, \lambda) \mapsto (g(x, \lambda), F_S(g_x(x, \lambda), \varphi)). \tag{3.10}$$

In case of $\mathbb{Z}_2$-symmetry or X_0-invariance the regularity of zeros of G_S is enforced by considering certain restrictions of G_S, namely

$$G_{S,\mathbb{Z}_2} : X_+ \times X_- \times \mathbb{R} \to X_+ \times X_- \times \mathbb{R}, \quad G_{S,X_0} : X_0 \times X \times \mathbb{R} \to X_0 \times X \times \mathbb{R}.$$

The following theorem recalls regularity results which are essentially known ([10], [15], [16], [18], [12]), but not yet formulated for imaginary Hopf points.

Theorem 3.1 *The conditions above for a quadratic turning point (a pitchfork or a subspace break-ing bifurcation point) hold if and only if there is a normalizing linear functional $l^T \in X'$ (satisfying $l^T S = -l^T$ or $l^T X_0 = 0$) such that $(x_0, \varphi_0, \lambda_0)$ is a regular root of $G_S = 0$ (of $G_{S,\mathbb{Z}_2} = 0$ or of $G_{S,X_0} = 0$).*

For fixed α, (x_0, λ_0) is a Hopf point with Hopf number $\nu_0 \neq 0$ iff there is a normalizing functional $l^T \in X'$ such that $(x_0, \xi_0, \lambda_0, \nu_0)$ is a regular root of $G_H = 0$.

Th.3.1 is important for numerical computations (quadratic convergence of Newton's method) and for path following of the bifurcation points w.r.t. the second parameter α.

Though in analogy to the restrictions $G_{S,\mathbb{Z}_2}$ and G_{S,X_0} of G_S also $G_{H,\mathbb{Z}_2}$ and G_{H,X_0} of G_H can be defined (which we will use later), these restrictions are not necessary to enforce regularity in case of symmetry (subspace) breaking Hopf points (but their use reduces the numerical effort).

Regular roots $(x_0, \xi_0, \lambda_0, \nu_0)$ of $G_H = 0$ can also exist if $\nu_0 = 0$ (see Th. 4.3).

4   TB-turning points, TB-pitchforks and subspace breaking TB-points

For simple λ-singular points, $\mu = 0$ is not necessarily an algebraically simple eigenvalue of g_x^0. Hence the additional property that g_x^0 is a TB-matrix, though not generic in one-parameter problems, is not excluded in the cases of turning points, pitchforks and subspace breaking bifurcation points. Since the geometry of branches of equilibria through (x_0, λ_0) is not influenced by this degeneracy, it is natural to assume - beside of (T1) - the additional transversality condition (T2) for the path $A(s) := g_x(x(s), \lambda(s))$.

Definition 4.2 *A quadratic turning point (x_0, λ_0) defined by (3.5) is called a (quadratic) TB-turning point if $g_x^0 \in \mathcal{M}_H^0$ and $d_1 (= \psi_0^T A_1 \xi_0 + \chi_0^T A_1 \varphi_0) \neq 0$.*

A pitchfork (a subspace breaking bifurcation point) defined by (3.7) (by (3.8)) is called a TB-pitchfork (a subspace breaking TB-point) if $g_x^0 \in \mathcal{M}_H^0$ and $d_\lambda \neq 0$. Note that $\xi_0 \in X_-$ ($\xi_0 \notin X_0$).

Now we study the regularity of the root $(x_0, \xi_0, \lambda_0, \nu_0)$ of $G_H = 0$ ((1.2)) if $\nu_0 = 0$, again suppressing the last entry α. To enforce regularity in the case of a TB-pitchfork or of a subspace breaking TB-point, we use restrictions $G_{H,\mathbb{Z}_2}$ and G_{H,X_0} in analogy to $G_{S,\mathbb{Z}_2}$ and G_{S,X_0}.

Theorem 4.3 *If the conditions for a (quadratic) TB-turning point (TB-pitchfork, subspace break-ing TB-point) hold, then there is a normalizing linear functional $l^T \in X'$ (satisfying $l^T S = -l^T$ or $l^T X_0 = 0$) such that $(x_0, \xi_0, \lambda_0, 0)$ is a regular root of $G_H = 0$ ($G_{H,\mathbb{Z}_2} = 0$, $G_{H,X_0} = 0$).*

This theorem will immediately prove the character of TB-points as an origin of real (and of imaginary) Hopf points on the primary (symmetric) branch (see next section).

Now consider TB-pitchforks or subspace breaking TB-points. The question whether - by variation of α - there rise up also Hopf points on the bifurcating secondary (asymmetric) branch, can be attacked by means of the following proposition.

Proposition 4.4 *Let* (x_0, λ_0) *be a TB-pitchfork or a subspace breaking TB-point. Without the restriction of* G_H *above,* $(x_0, \xi_0, \lambda_0, 0)$ *is a simple singular root of* G_H.

In case of a pitchfork, the kernel of DG_H^0 *is spanned by* $\Phi_0 := (\varphi_0, w_1, 0, 0)$, *where* $w_1 \in X_+$ *is defined by* $g_x^0 w_1 + g_{xx}^0 \xi_0 \varphi_0 = w_0$, $g_x^0 w_0 + g_{xx}^0 \varphi_0^2 = 0$.

PROOF: : Using $g_x^0 \xi_0 = \varphi_0$, $(\varphi, w, \sigma, \tau) \in Kernel(DG_H^0)$ iff

$$g_x^0 \varphi + \sigma g_\lambda^0 = 0, \tag{4.1}$$

$$g_x^0 g_{xx}^0 \xi_0 \varphi + g_{xx}^0 \varphi_0 \varphi + (g_x^0)^2 w + \sigma(g_x^0 g_{x\lambda}^0 \xi_0 + g_{x\lambda}^0 \varphi_0) + \tau \xi_0 = 0, \tag{4.2}$$

$$l^T w = 0, \tag{4.3}$$

$$l^T g_{xx}^0 \varphi \xi_0 + l^T g_x^0 w + \sigma l^T g_{x\lambda}^0 \xi_0 = 0. \tag{4.4}$$

From (4.1) and the definition of v_λ one gets $\varphi = \beta \varphi_0 + \sigma v_\lambda$, with some $\beta \in \mathbb{R}$. Now from (4.2) we obtain

$$\beta(A_0 A_1 \xi_0 + A_1 \varphi_0) + A_0^2 w + \sigma(A_0 A_\lambda \xi_0 + A_\lambda \varphi_0) + \tau \xi_0 = 0. \tag{4.5}$$

Multiplying (4.5) from the left with ψ_0^T, χ_0^T and using (2.3) it follows that $\beta c_1 + \sigma c_\lambda + \tau = 0$, $\beta d_1 + \sigma d_\lambda = 0$ and $\beta(l^T A_1 \xi_0) + \sigma(l^T A_\lambda \xi_0) + l^T A_0 w = 0$.

Now in the pitchfork case, $c_1 = 0, d_1 = 0$. Since $d_\lambda \neq 0$, it follows that $\sigma = 0$ and also $\tau = 0$. The only nontrivial solution is essentially given by $\beta = 1$ and $w = w_1$.

In the case of a subspace breaking bifurcation point, $\beta = 0$ would imply $\sigma = 0, \tau = 0, w = 0$. Hence without loss of generality $\beta = 1$. Then $\sigma = -d_1/d_\lambda$ and $\tau = -c_1 - \sigma c_\lambda$ while w can be uniquely determined by these quantities. ∎

5   TB-points and Hopf branches

Here we re-introduce the second parameter α, adding α in the last position of G_S in (3.10) and of G_H. We will assume that for $\alpha = \alpha_0$, (x_0, λ_0) is a TB-turning point, a TB-pitchfork or a subspace breaking TB-point. Moreover, we can assume that for fixed α there exist primary (symmetric) branches $C(\alpha)$ close to α_0.

The Implicit Function Theorem in combination with the regularity results of Th.3.1 and Th.4.3 implies the existence of two smooth branches of bifurcation points. Namely,

$$C_S = \{(x_S(\alpha), \lambda_S(\alpha)) : |\alpha - \alpha_0| < \delta\},$$

a branch of steady state bifurcation points (quadratic turning points, pitchforks or subspace breaking bifurcation points) and

$$C_H = \{(x_H(\alpha), \lambda_H(\alpha)) : |\alpha - \alpha_0| < \delta\}\},$$

a Hopf branch of real or imaginary Hopf points satisfying $x_S(\alpha_0) = x_0 = x_H(\alpha_0)$, $\lambda_S(\alpha_0) = \lambda_0 = \lambda_H(\alpha_0)$ ($\delta > 0$ sufficiently small). Moreover there exist Hopf numbers $\nu_H(\alpha)$ depending smoothly on α with $\nu_H(\alpha_0) = 0$.

For each α, both the steady state bifurcation points $(x_S(\alpha), \lambda_S(\alpha))$ and the Hopf points $(x_H(\alpha), \lambda_H(\alpha))$ are lying on the primary branches $C(\alpha)$.

If the non-degeneracy condition

$$e_0 := \frac{d}{d\alpha} \nu_H'(\alpha)|_{\alpha=\alpha_0} \neq 0 \tag{5.1}$$

is fulfilled, then the TB-point (x_0, λ_0) is an isolated point on $\mathcal{C}_H$ and just seperates real and imaginary Hopf points on the Hopf branch $\mathcal{C}_H$ - a strict change from real to imaginary Hopf points or vice versa.

(5.1) can be equivalently formulated in various ways, (see sec.5.1). It is geometrically instructive to look at the pathes of matrices

$$A_S(\alpha) := g_x(x_S(\alpha), \lambda_S(\alpha), \alpha) \in \mathcal{M}_S, \; A_H(\alpha) := g_x(x_H(\alpha), \lambda_H(\alpha), \alpha) \in \mathcal{M}_H,$$

crossing $\mathcal{M}_H$, respectively $\mathcal{M}_S$ at the TB-matrix A_0. The transversality conditions (T1) and (T2),

$$\psi_0^T A_H'(\alpha_0)\varphi_0 \neq 0, \quad \psi_0^T A_S'(\alpha_0)\xi_0 + \chi_0^T A_S'(\alpha_0)\varphi_0 \neq 0, \tag{5.2}$$

can both be shown to be equivalent with condition (5.1).

5.1 Bifurcation of Hopf branches in TB-pitchforks and subspace breaking TB-points

The transversality conditions (T1) and (T2) for TB-pitchforks and subspace breaking TB-points are $c_\lambda \neq 0$ and $d_\lambda \neq 0$. After some lengthy calculations we obtain $e_0 = \nu'(\alpha_0) = c_\lambda d_\alpha / d_\lambda - c_\alpha$.

Hence the essential non-degeneracy conditions for a TB-pitchfork and a subspace breaking TB-point are

$$c_\lambda \neq 0, \; d_\lambda \neq 0, \; D := \mathrm{Det} \begin{pmatrix} c_\lambda & c_\alpha \\ d_\lambda & d_\alpha \end{pmatrix} \neq 0. \tag{TB}$$

$\mathcal{C}_S$ is a branch of symmetry breaking bifurcation points $(x_S(\alpha), \lambda_S(\alpha))$, in which the asymmetric branches $\mathcal{C}^a(\alpha)$ bifurcate from the symmetric branches $C(\alpha)$. Analogously $\mathcal{C}_H$ is a branch of symmetry (subspace) breaking Hopf points. We call $\mathcal{C}_H$ the *symmetric Hopf branch*. Next we will investigate conditions under which Hopf points also exist on the asymmetric branches $C^a(\alpha)$. This leads to the *asymmetric Hopf branch C_H^a*.

Assume first that g satisfies a $\mathbb{Z}_2$-symmetry. Let $\tilde{S}$ operate on $Y := (x, \xi, \lambda, \nu)$ by $\tilde{S}Y = (Sx, -S\xi, \lambda, \nu)$. Then G_H satisfies a $\mathbb{Z}_2$-equivariance $G_H(\tilde{S}Y, \alpha) = \tilde{S}G_H(Y, \alpha)$. Now Prop. 4.4 claims that the kernel vector Φ_0 is anti-symmetric with respect to $\tilde{S}$.

Similarly the X_0-invariance of g implies a $\tilde{X}_0$-invariance of G_H setting $\tilde{X}_0 := X_0 \times X \times \mathbb{R} \times \mathbb{R}$. The kernel vector Φ_0 is not in $\tilde{X}_0$.

Now it can be shown that (5.1) is just the unfolding condition under which $(Y_0, \alpha_0) := (x_0, \xi_0, \lambda_0, 0, \alpha_0)$ is a pitchfork (subspace breaking bifurcation point) of the extended system $G_H(Y, \alpha) = 0$.

Theorem 5.5 *Let for $\alpha = \alpha_0$, (x_0, λ_0) be a TB-pitchfork (or a subspace breaking TB-point) satisfying (TB). Then $(x_0, \xi_0, \lambda_0, 0, \alpha_0)$ is a pitchfork (subspace breaking bifurcation point) of the extended system $G_H = 0$ w.r.t. α.*

In case of a TB-pitchfork we sketch the Proof: It is easy to show that the vector Ψ_0 spanning the cokernel of the Y-derivative $G_{H,Y}^0$ is given by $\Psi_0 = (\psi_0, 0, 0, 0)$. The pitchfork non-degeneracy is given by

$$\frac{d}{d\alpha}\Psi_0^T G_{H,Y}(x_H(\alpha), \lambda_H(\alpha), \nu(\alpha), \xi(\alpha), \alpha)\Phi_0 \mid_{\alpha=\alpha_0} \neq 0.$$

Then

$$\Psi_0^T G_{H,Y}(.)\Phi_0 = \psi_0^T g_x(.)\varphi_0,$$

and the condition for a pitchfork is given by $\frac{d}{d\alpha}\psi_0^T A_H(\alpha)\varphi_0|_{\alpha=\alpha_0} \neq 0$ which we remarked in (5.2) to be equivalent with $e_0 \neq 0$. ∎

From the bifurcation analysis of a pitchfork or a subspace breaking bifurcation point it follows that there exists a branch

$$\{(x_H^a(\tau), \xi_H^a(\tau), \lambda_H^a(\tau), \nu_H^a(\tau), \alpha_H^a(\tau)) : \tau \in I\}$$

of asymmetric solutions of $G_H = 0$, where τ_0 corresponds to the TB-point. Under the assumptions $(\alpha_H^a)''(\tau_0) \neq 0$ and $(\nu_H^a)''(\tau_0) \neq 0$, resp. $(\alpha_H^a)'(\tau_0) \neq 0$ and $(\nu_H^a)'(\tau_0) \neq 0$,

$$C_H^a := \{(x_H^a(\tau), \lambda_H^a(\tau)) : \tau \in I\}$$

is an *asymmetric* Hopf branch of either real or imaginary Hopf points with Hopf numbers $\nu_H^a(\tau)$.

In the case of a TB-pitchfork, one should note that - due to the pitchfork character - the (asymmetric) Hopf points exist either for $\alpha \geq \alpha_0$ or for $\alpha \leq \alpha_0$ depending on the sign of $(\alpha_H^a)''(\tau_0)$. Whether the asymmetric Hopf points are real or imaginary, depends on the sign of the second derivative $(\nu_H^a)''(\tau_0)$. These terms can be expressed by certain third derivatives of g and can be computed numerically ([8]).

In the case of a subspace breaking TB-point, the intersection of the asymmetric Hopf branch with the symmetric one can be expected to be transcritical if $(\alpha_H^a)'(\tau_0) \neq 0$. If additionally $(\nu_H^a)'(\tau_0) \neq 0$, the TB-point will separate real and imaginary Hopf points on the asymmetric Hopf branch.

The detection of TB-points (compare [11] for TB-turning points) can be easily performed during path following of the steady state bifurcation points $(x_S(\alpha), \lambda_S(\alpha))$. The sign of $\psi_0^T(\alpha)\varphi_0(\alpha)$ strictly changes for $\alpha = \alpha_0$, since we can show that

$$\frac{d}{d\alpha}(\psi_0^T(\alpha)\varphi_0(\alpha)) \mid_{\alpha=\alpha_0} = 2(\frac{d_\lambda c_\alpha}{c_\lambda} - d_\alpha) \neq 0, \tag{5.3}$$

being equivalent with (5.1) or with $D \neq 0$, see (TB).

6   Numerical Computations

We are mainly interested in the computation of the Hopf branches C_H and C_H^a bifurcating in a TB-pitchfork or subspace breaking TB-point using the extended system $G_H = 0$. The efficient techniques for solving $G_H = 0$, presented in [12], apply also for our system.

Concerning the direct computation of TB-points, we remark that the determining system in [11] for the numerical computation of TB-turning points can also been used for TB-pitchforks and subspace breaking TB-points. We suggest a system which is almost identical with $G_H = 0$, but with $\nu = 0$ fixed and α released. Restrictions $G_{H,\mathbb{Z}_2}$ and G_{H,X_0} decrease the effort. The transversality conditions (TB) are just the regularity conditions.

For numerical demonstrations we consider the n-box Brusselator system for $n = 2$ and $n = 3$, c. [13],[4],[19]. The state z and the Brusselator reaction law $f(z)$ for each box is given by

$$z := (x, y) \in \mathbb{R}^2, \; f(z) := \begin{pmatrix} A - (B+1)x + x^2 y \\ Bx - x^2 y \end{pmatrix}.$$

If $z_j \in \mathbb{R}^2$ is the state vector in box j, j=1,...,n, then the dynamic of the whole reaction is modelled by the equations

$$\dot{z}_j = f(z_j) + \lambda^{-2}[D(z_{j+1} + z_{j-1} - 2z_j)], \; j = 1, \ldots, n, \; (z_0 := z_n, z_{n+1} := z_1), \tag{6.1}$$

where $D \in \mathbb{R}^{2 \times 2}$ is a diagonal matrix with diagonal entries d_1, d_2 describing diffusion. The bifurcation parameter λ controls the size of diffusion of the reactants between the boxes. For our numerical computations we fix $A = 2, d_1 = 1, d_2 = 10$, while $\alpha := B$ is varied.

Obviously this problem has the symmetry of the dihedral group D_n, acting on $X = \mathbb{R}^{2n}$. There is a symmetric trivial branch parametrized by λ on which $z_j \equiv (A, B/A), \quad j = 1, \ldots n$.

6.1 2-BOX BRUSSELATOR

Due to the $\mathbb{Z}_2$-symmetry in this problem, there are pitchforks on the trivial branch (which vary with $\alpha = B$). The coordinates of a TB-pitchfork are

$$x_1 = x_3 = 2, \quad x_2 = x_4 = 2.98989, \quad \lambda = 4.7385255, \quad B = 5.979797.$$

Using a pathfollowing algorithm, the symmetric Hopf branch $\mathcal{C}_H$ and the asymmetric Hopf branch $\mathcal{C}_H^a$ emanating in the TB-point can be computed using the system $G_H = 0$. Here $G_H = 0$ simplifies for the computation of symmetric Hopf points using the restriction $G_{H,\mathbb{Z}_2}$ with $l^T = (1, 0, -1, 0)$ and leaving us with the five variables $x_1(= x_3), x_2(= x_4), \lambda, B$ and $\xi_2(= -\xi_4)$ (note that $\xi_1 = -\xi_3 = 0$).

Fig.1a shows the symmetric Hopf branch $\mathcal{C}_H$ (dashed line), the asymmetric Hopf branch $\mathcal{C}_H^a$ (solid line) and the branch of pitchforks $\mathcal{C}_S$ (dashdotted line), intersecting in the TB-pitchfork, in the parameter plane. Real (imaginary) Hopf points are marked by the symbol +, resp. −.

Note that the asymmetric Hopf branch $\mathcal{C}_H^a$ emanates (according to Th.5.5) pitchfork like from the TB-pitchfork, turns back and changes from imaginary to real Hopf points after having passed a TB-turning point (TB-points are labed as ●).

Fig.1b shows for different fixed $\alpha = B$, asymmetric branches of equilibria $\mathcal{C}^a(\alpha)$ (dashdotted line) bifurcating in pitchforks (labeled as ○), together with the asymmetric Hopf branch (solid line), projected into the λ-x_2 - plane.

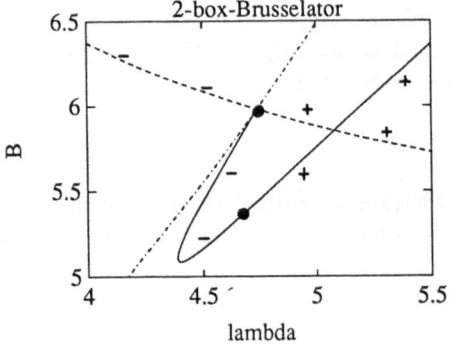
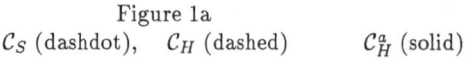
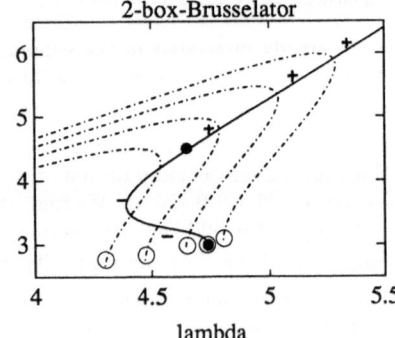

Figure 1a   Figure 1b

$\mathcal{C}_S$ (dashdot), $\mathcal{C}_H$ (dashed) $\mathcal{C}_H^a$ (solid) $\mathcal{C}^a(\alpha)$ (dashdot), $\alpha = 5.2, 5.5, 5.8, 6.1$

6.2 3-BOX BRUSSELATOR

The dihedral group D_3 has one two-dimensional irreducible representation, hence one expects eigenvalues of multiplicity two of the Jacobians along the trivial solution branch and corresponding bifurcation points. An associated bifurcation subgroup Σ containing the identity and one reflection exists and is asymmetric ([4]). Hence, if we restrict our problem to the 4-dimensional fixed point space X defined by $x_1 = x_5, x_2 = x_6$ and by arbitrary x_3, x_4, we expect and find transcritical subspace breaking bifurcation points, where the invariant subspace X_0 is given by the (fully) symmetric states $x_1 = x_3$ and $x_2 = x_4$. Asymmetric states are characterized by $x_1 \neq x_3$ or $x_2 \neq x_4$ (though still $x_1 = x_5, x_2 = x_6$.)

The coordinates of a subspace breaking TB-point are

$$x_1 = x_3 = 2, \ x_2 = x_4 = 2.9899, \ \lambda = 5.8034848, \ B = 5.979797.$$

Again $G_H = 0$ simplifies for the computation of symmetric Hopf points using the restriction G_{H,X_0} with $l^T = (1, 0, -1, 0)$ and leaving us (utilizing an additional symmetry information about ξ, see [19]) with the five variables $x_1(= x_3), x_2(= x_4), \lambda, B$ and $\xi_2(= -\xi_4/2)$ (note that $\xi_1 = -\xi_3/2 = 0$).

Fig.2 is completely analogous to Fig.1. Instead of pitchforks, transcritical subspace breaking bifurcation points occur. Note that, according to Th.5.5, the symmetric and the asymmetric Hopf branches intersect transcritically in the subspace breaking TB-point. There is a strict change from real to imaginary Hopf points on both branches. Again the asymmetric Hopf branch changes from imaginary to real Hopf points after having passed the TB-turning point

$$x_1 = 2.6174, x_2 = 2.2629, x_3 = 0.7652, x_4 = 5.1580, \ \lambda = 6.6251119, \ B = 5.6709060.$$

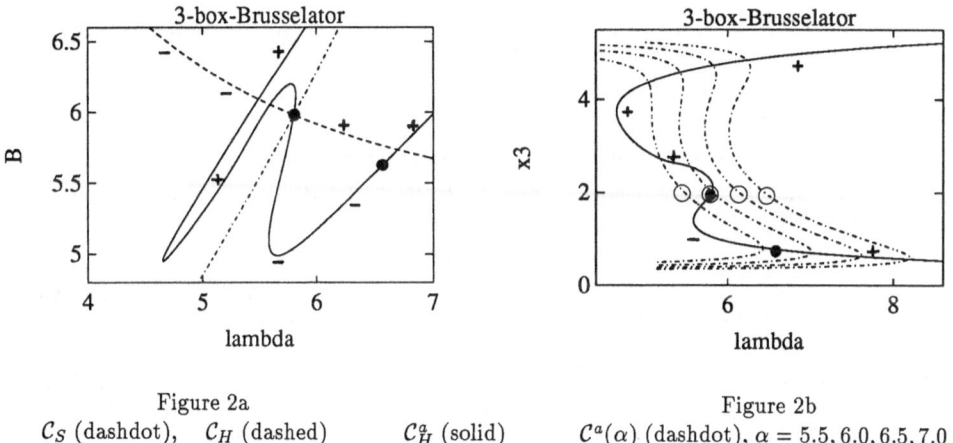

Figure 2a
C_S (dashdot), C_H (dashed) C_H^a (solid)

Figure 2b
$C^a(\alpha)$ (dashdot), $\alpha = 5.5, 6.0, 6.5, 7.0$

References

[1] R.I. Bogdanov. *Versal Deformations of a singular point on the plane in the case of zero eigenvalues.* Functional Anal. and Applications **9** (2), 144-145, 1975.

[2] S.N. Chow, B. Deng, B. Fiedler. *Homoclinic Bifurcation at Resonant Eigenvalues.* Preprint SC 88-10, Konrad-Zuse-Zentrum für Informationstechnik Berlin, 1988.

[3] G. Dangelmayr, E. Knobloch. *The Takens-Bogdanov Bifurcation with O(2)-Symmetry.* Phil. Trans. R. Soc. Lond. **A 322**, 243-279, 1987.

[4] M. Dellnitz, B. Werner. *Computational methods for bifurcation problems with symmetries - with special attention to steady state and Hopf bifurcation points.* J. of Comp. and Appl. Math. **26**, 97-123, 1989.

[5] B. Fiedler, P. Kunkel. *A quick multiparameter test for periodical solutions.* In: **Bifurcation, Analysis, Algorithms, Applications**, T. Küpper, R. Seydel, H. Troger (eds.), ISNM **79**, 61-70, Birkhäuser, Basel, 1987.

[6] M. Golubitsky, I. Stewart, D. Schaeffer. **Singularities and Groups in Bifurcation Theory**, Vol. 2, Springer 1988.

[7] J. Guckenheimer, P. Holmes. **Nonlinear Oscillations, Dynamical Systems and Bifurcations of Vector Fields.** Springer, New-York, 1983.

[8] V. Janovsky, B. Werner. *Constructive analysis of Takens-Bogdanov points with $\mathbb{Z}_2$-symmetry.* In preparation.

[9] A. Khibnik. *LINLBF: A program for continuation and bifurcation analysis of equilibria up to codimension three.* In: **Continuation and Bifurcations: Numerical Techniques and Applications**, D. Roose, B. de Dier, A. Spence (eds.), NATO ASI Series C, Vol. **313**, 283-296, Kluwer Academic Publishers, Dordrecht, 1990.

[10] G. Moore, A. Spence. *The calculation of turning points of nonlinear equations.* SIAM J. Numer. Math. **17**, 567-576, 1980.

[11] D. Roose. *Numerical computations of origins for Hopf bifurcation in a two parameter problem.* In: **Bifurcation, Analysis, Algorithms, Applications**, T. Küpper, R. Seydel, H. Troger (eds.), ISNM **79**, 268-276, Birkhäuser, Basel, 1987.

[12] Roose, D., Hlavacek, V. *A direct method for the computation of Hopf bifurcation points.* SIAM J. Appl. Math. **45**, 879-894, 1985.

[13] I. Schreiber, M. Holodniok, M. Kubicek, M. Marek. *Periodic and aperiodic regimes in coupled dissipative chemical oscillators.* J. of Statist. Phys. **43**, 489-518, 1986.

[14] A. Spence, K.A. Cliffe, A.D. Jepson. *A Note on the Calculation of Paths of Hopf Bifurcations.* J. of Comp. and Appl. Math. **26**, 125-131, 1989.

[15] A. Spence, B. Werner. *Nonsimple turning points and cusps.* IMA J. Numer. Anal. **2**, 413-427, 1982.

[16] B. Werner, A. Spence. *The computation of symmetry-breaking bifurcation points.* SIAM J. Numer. Anal. **21**, 388-399, 1984.

[17] F. Takens. *Singularities of vector fields.* Publ. Math. I.H.E.S. **43**, 47-100, 1974.

[18] B. Werner. *Regular systems for bifurcation points with underlying symmetries.* In: **Numerical Methods for Bifurcation Problems**, T. Küpper, H.D. Mittelmann, H. Weber (eds.), ISNM **70**, 562-584, Birkhäuser, Basel, 1984.

[19] B. Werner. *Eigenvalue problems with the symmetry of a group and bifurcations.* In: **Continuation and Bifurcations: Numerical Techniques and Applications**, D. Roose, B. de Dier, A. Spence (eds.), NATO ASI Series C, Vol. **313**, 71-88, Kluwer Academic Publishers, Dordrecht, 1990.

[20] W. Wu, A. Spence, A.K. Cliffe. *Steady-State/Hopf mode interaction at a symmetry breaking Takens-Bogdanov point.* Preprint University of Bath, 1990.

Bodo Werner
Institut für Angewandte Mathematik
Universität Hamburg
Bundesstr. 55
D 2000 Hamburg 13

Vladimir Janovsky
Department of Numerical Analysis
Charles University of Prague
Malostranska nam. 2/25
118 00 Praha 1, CSFR